M. Chofflon · L. Steinman (Eds.) Immunoneurology

Springer

Berlin
Heidelberg
New York
Barcelona
Budapest
Hong Kong
London
Milan
Paris
Santa Clara
Singapore
Tokyo

Michel Chofflon Lawrence Steinman (Eds.)

Immunoneurology

With 14 Figures and 24 Tables

Springer

Dr. MICHEL CHOFFLON
Clinique de Neurologie
Hôpital Cantonal Universitaire
CH-1211 Genève 14
Switzerland

Dr. LAWRENCE STEINMAN
Department of Neurology
Stanford University
Stanford, California 94305-5235
USA

ISBN-13: 978-3-642-64741-3 e-ISBN-13: 978-3-642-61191-9
DOI: 10.1007/978-3-642-61191-9

Librairy of Congress Cataloging-in-Publication Data applied for

Die Deutsche Bibliothek – CIP-Einheitsaufnahme
Immunoneurology: with 24 tables / Michel Chofflon; Lawrence Steinman (ed.).
- Berlin; Heidelberg; New York: Springer, 1996
ISBN-13: 978-3-642-64741-3
NE: Chofflon, Michel [Hrsg.]

Softcover reprint of the hardcover 1st edition 1996

Production: PRO EDIT GmbH, D-69126 Heidelberg
Typesetting: Satztechnik K. Steingräber, D-69126 Heidelberg
Cover Design: design & production, D-69121 Heidelberg
SPIN: 10543563 27/3136-543210 - Printed on acid-free paper

Contents

Introduction

Michel Chofflon[1], **Lawrence Steinman**[2]

[1] Clinique de Neurologie, Hôpital Cantonal Universitaire, CH-1211 Genève 14, Suisse
[2] Department of Neurology, Stanford University, Stanford, CA 94305-5235, USA

Most papers on the neuroimmune diseases, especially those discussing multiple sclerosis, begin with sentences such as: "the unknown etiology of disease," "the unpredictable natural history," or "the uncertain response to therapies."

If we still agree with this kind of statement, it is in order not to be prematurely enthusiastic about the considerable amount of information gathered over recent years that has enabled us to understand some of the pathogenic events. It is now possible to think about modifications of these pathways and to develop better therapies. We still do not know why the immune response against central myelin, as in multiple sclerosis, or against peripheral myelin, as in the Guillain-Barré syndrome, starts without any apparent reason. However, we are starting to understand some of the various obligatory milestones in these diseases. Specific antigenic T cells or antibodies, the interplay of cytokines, the role of monocytes as antigen-presenting cells or as partners in a final common pathway to execute the orders of the immune response are all being investigated.

All this information has been gathered for two purposes: (1) to try new therapies that make sense even if we are not sure of the triggering event and (2) to guess at, find proof of, and finally confirm the nature of this triggering event.

Immunopathology of the Lambert-Eaton myasthenic syndrome

Bethan Lang, John Newsom-Davis

Neurosciences Group, Institute of Molecular Medicine, University of Oxford, John Radcliffe Hospital, Oxford OX3 9DU, UK

Introduction

Autoimmune diseases account for an increasing number of neurological disorders, both acute and chronic, many of which have only recently been identified. Myasthenia gravis (MG) was one of the first diseases to be confirmed as autoimmune [13, 61]. Here, antibodies are directed against the acetylcholine receptor (AChR) on the postsynaptic membrane of the neuromuscular junction causing a decrease in the number of functional receptors. Other antibody-mediated examples are stiff man syndrome [63], acquired neuromyotonia [62] and the Miller Fisher syndrome, a variant of the Guillain-Barré syndrome [57]. The immune system has also been implicated in a number of disorders that associate with particular cancers, such as subacute cerebellar degeneration [18], subacute sensory neuropathy [17], opsoclonus/myoclonus [8] and the Lambert-Eaton myasthenic syndrome (LEMS). This often associates with small cell carcinoma of the lung (SCLC) and is one of the best understood of these paraneoplastic disorders, with the early identification of the antigen helping to elucidate the aetiological role of the cancer.

Clinical features of LEMS

The syndrome was first reported by Anderson and co-workers [1], who described a myasthenic syndrome associated with a bronchial neoplasm. Its electromyographic distinction from MG was shown by Lambert and colleagues [24]. In patients with LEMS, the amplitude of the compound muscle action potential (CMAP) following supramaximal nerve stimulation is typically greatly reduced. However, immediately after a brief period of maximum voluntary contraction of the muscle, or of high-frequency nerve stimulation, the CMAP amplitude increases by greater than 100%.

Correspondence to: B. Lang

This presynaptic facilitation is characteristic of the disease and, together with the reduced resting amplitude, distinguishes LEMS from MG.

Most patients present with proximal muscle weakness and fatiguability, especially affecting the lower limbs and gait [47]. In some patients a mild degree of ptosis and bulbar weakness may be present, but diplopia or substantial involvement of respiratory muscles is rare. Depressed tendon reflexes are common, and may increase after maximum voluntary contraction (post-tetanic potentiation). Dysfunction of the autonomic nervous system is also common, which may be manifest as dry mouth, sexual impotence or constipation. There is an increased frequency of other autoimmune diseases particularly vitiligo, and thyroid disease [19]. This autoimmune connection can be accounted for, at least in part, by an immunogenetic association with HLA-B8 and the immunoglobulin heavy chain marker Glm 2 [69].

The knowledge that LEMS is an autoimmune disorder has been exploited in the treatment of the disorder. Patients respond well to an intensive course of plasma exchange, which acts by temporarily depleting the circulating autoantibodies [45]. Patients without a carcinoma have been successfully treated with immunosuppressive drugs such as prednisolone and azathioprine [45]. Intravenous immunoglobulin infusion has also recently been used successfully [2].

Tumour association

Sixty percent of LEMS patients have an SCLC. The neurological symptoms are often the earliest clue to an underlying tumour. Typically they may precede the radiological appearance of the cancer by 2 years and sometimes by 5 years or more. Muscle weakness often improves if the tumour is treated or excised [3].

The SCLC is a tumour of neuroectodermal origin that accounts for approximately 25% of all lung carcinomas. It is fast-growing, metastasises rapidly, and is very strongly associated with smoking. It is thought to be derived from the Kulchitsky cells which are endocrine cells of the bronchial epithelium. SCLC cells have a large nucleus, with a high nuclear-to-cytoplasm ratio. The tumour contains neurosecretory granules, expresses neuron-specific enolase and may secrete biologically active peptides, such as bombesin, adrenocorticotrophic hormone (ACTH) and anti-diuretic hormone (ADH). The prevalence of LEMS amongst SCLC patients appears to be around 3% [12]. Other tumours such as lymphoma, breast and ovarian cancer have been reported with LEMS, but the association is far weaker and not necessarily significant.

Electrophysiological characteristics

Depolarisation of the presynaptic nerve terminal by a nerve impulse causes the opening of voltage-gated calcium channels (VGCCs), which allows the local influx of calcium ions into the terminal. This rise in calcium ion concentration results in the release of discrete packages (quanta) of acetylcholine (ACh), that in man number approximately 30. This is referred to as the quantal content of the end-plate potential (EPP) and differs between species. The ACh crosses the synaptic gap, binds to the AChR on the postsynaptic membrane and elicits an EPP which triggers an action potential when the critical firing threshold for the muscle cell is reached. Propagation of the action potential along the muscle fibre membranes leads to muscle contraction.

The physiological defect underlying the disorder was discovered by Elmqvist and Lambert [11, 23], who demonstrated a substantial decrease in the number of ACh quanta released in response to a single nerve impulse, i.e. a reduction in the quantal content of the EPP. During a high-frequency train of stimuli, the quantal content increased (accounting for the observed post-tetanic potentiation) but was still substantially lower than that in healthy muscle. Direct chemical measurement of ACh release confirmed this reduction in LEMS patients, and also showed that the levels of ACh in the nerve terminal and choline acetyltransferase activity were normal [40].

Evidence for antibody mediation

The autoimmune nature of the disease has been clearly demonstrated by the passive transfer to mice of all the electrophysiological features of LEMS following injection of immunoglobulins (IgG) prepared from the plasma of patients undergoing therapeutic plasma exchange. A highly significant decrease in the quantal content of the EPP was observed on comparing these mice to mice injected with an equivalent amount of control IgG [21, 25, 26]. The mice injected with LEMS IgG also showed presynaptic facilitation following high-frequency (40 Hz) nerve stimulation [46]. The time course of the effect was shown to follow closely the rise in the level of human IgG in the mouse sera, reaching a maximum after 10 days of injections [53]. The effect was also shown to be independent of the later components of the complement cascade as mice genetically deficient in the fifth component (C_5) were as susceptible as normal mice [53].

Injection of mice with LEMS IgG reduces the amount of ACh released in response to a nerve stimulus. Failure of the nerve impulse to reach the nerve terminal is excluded as a cause since this reduction is also observed with K^+-stimulated release [27]. However, if the external Ca^{2+} ion concentration is raised, whether in muscle biopsies taken from patients, or in LEMS IgG-injected mice, the defect can be overcome [27]. These results are consistent with a reduction in the number of functional VGCCs.

Morphological characteristics

Freeze-fracture electron microscopic studies of the nerve terminal reveals a distinctive double parallel array of membrane particles 10–12 nm in diameter that comprises an "active zone." These active zone particles (AZPs) are thought to represent VGCCs [54]. Similar studies on LEMS patients demonstrated a depletion of the active zones and disorganisation of AZPs [14]. Similar changes were observed in mice receiving daily injections of LEMS IgG [15]. The passive transfer model of LEMS has allowed early events in the process of disorganisation of AZPs to be studied. The particles initially cluster, and then decline in total number. These changes can be explained by the cross-linking of adjacent AZPs by the divalent IgG molecule, resulting in clustering and an increased rate of internalisation [16]. Passive transfer of divalent $F(ab')_2$ fragments produced similar morphological and electrophysiological changes as whole IgG. However, monovalent Fab fragments were without significant effect [44, 52]. Thus, LEMS IgG induces down-regulation of VGCCs at the nerve terminal by cross-linking of adjacent channels. These changes would be consistent with the clinical defect seen in LEMS.

Calcium channels as the antigenic target

Calcium channel subtypes

The VGCC is a multimeric transmembrane protein consisting of alpha 1, alpha 2, beta, and delta subunits (Fig. 1). Depending on the tissue of origin, a fifth subunit such as the skeletal muscle gamma or the neuronal p95 may also be present. The alpha 1 subunit is thought to possess both channel function and drug binding sites. However, the disulphide-linked alpha 2-delta subunit and intracellular beta are both important in modifying the characteristics of the channels encoded by the alpha 1 subunit [41].

VGCCs can be divided into a number of subtypes (T, L, N, P and Q) based on their localisation, activation range, conductance and pharmacological profile. The T-type VGCC is a transient channel that is widespread in its cellular distribution and appears to have a role in pacemaking in cardiac and neuronal cells. The L-type VGCC, characterised by large, long-lasting conductance, is found both in excitable cells and in the cardiovascular system and is particularly sensitive to the dihydropyridines. In marked contrast, the N-, P- and Q-type channels appear to be confined to neurones and are important for neurotransmitter release. N-type channels are involved in neurotransmitter release in a number of systems such as the amphibian neuromuscular junction [4], mammalian brain [55] and hippocampal CA1 neurones [9]. They also appear to play an important role in the autonomic nervous system, where noradrenergic, cholinergic and non-adrenergic, non-cholinergic (NANC) transmission is controlled by N-type VGCCs [7].

P-type VGCCs were first identified in the Purkinje cells of the cerebellum [37], while the electrophysiologically distinct Q channels have been localised in the hippocampus. Recent studies have implicated P- (and/or Q-) type VGCC in the control of ACh release at the mammalian neuromuscular junction [68].

Toxins that specifically block these three channels have been identified. ω-Conotoxin (ω-CgTx) GVIA derived from the fish-eating snail *Conus geographus* shows high affinity for N-type channels, whilst ω-Aga IVA, from the spider *Agelenopsis aperta* blocks P-type VGCC at nanomolar concentrations. Uncertainty still exists over the nature of the Q-type channel, which appears to be electrophysiologically distinct from the P type. It is blocked by ω-Aga IVA at micromolar concentrations, a concentration 10–100 times that needed to block P-type channels. Q-type channels are completely blocked by ω-CmTx MVIIC derived from another cone snail, *Conus magus*.

Similar heterogeneity in the SCLC channel subtypes has been found by other methods. Partial cDNA sequences that encode for P- and L-type VGCCs have been identified using the polymerase chain reaction (PCR) in a prototypic SCLC line established from a patient without evidence of neurological disease [48]. L- and N-type VGCCs have been demonstrated electrophysiologically in SCLCs using whole cell and patch clamp techniques [49], and low levels of ^{125}I-labelled ω-CgTx GVIA binding have also been found in various SCLC cell lines and tumours, indicating the presence of N-type channels [31, 36, 60].

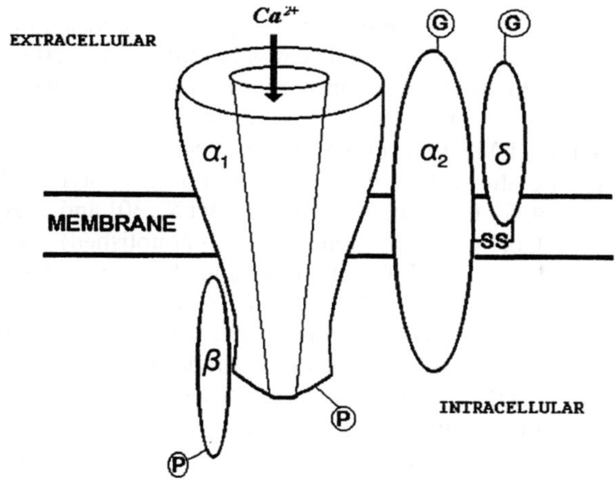

Fig. 1. Schematic representation of the subunits of the voltage-gated calcium channel (VGCC). VGCCs are multimeric transmembrane proteins. The channel-forming alpha 1 subunit (α_1) is thought to possess many of the drug binding sites. The other subunits, beta (β) and the disulphide-linked alpha 2-delta ($\alpha_2\delta$), are important in modifying the characteristics of the channel. A fifth subunit such as the gamma subunit of the dihydropyridine-sensitive skeletal muscle VGCC or the neuronal p95 may also be involved. Sites of cAMP-dependent phosphorylation *P*, and glycosylation *G*, are indicated

Antibodies to VGCCs

The antibodies found in LEMS are heterogeneous in their ability to block or bind to different calcium channel subtypes. LEMS IgG blocks to various degrees L-type VGCCs in bovine adrenal chromaffin cells [22], rat pituitary glands [38] and the rodent neuroblastoma cell line NG108 [51]. However, LEMS IgG is without effect on T-type channels [51]. Similarly, LEMS IgG has no action on L- and T-type VGCC in mouse cardiac muscle [28] and insect skeletal muscle [50].

The observation that many LEMS patients have an underlying SCLC raised the possibility that expression of the specific antigen by the SCLC cells may act as a trigger for the autoantibody response in this cancer-associated form of LEMS. Calcium channels have been shown to be present in SCLC using several different techniques. SCLC has been shown to generate short-duration action potentials [66] that can be blocked by the inorganic calcium channel antagonists cobalt and manganese [39]. Functional VGCCs in SCLC were confirmed by Roberts et al. [56], who demonstrated that depolarisation caused by a high external concentration of K^+ ions [96 mM] caused an influx of Ca^{2+} into the cells (measured as $^{45}Ca^{2+}$), which could be inhibited by VGCC antagonists. This provided a means of investigating the physiological actions of LEMS IgG in vitro.

When SCLC cells were cultured for 2–4 days in media containing IgG prepared from LEMS patients, the K^+-stimulated Ca^{2+} flux was significantly decreased compared to cells grown in control IgG or media alone [6, 56]. The inhibition of Ca^{2+} flux caused by a patient's IgG specifically correlated with an index of disease severity (the amplitude of the compound muscle action potential) [29]. Using a SCLC cell line that we derived from the lung tumour of a LEMS patient, we found that the influx of $^{45}Ca^{2+}$ correlated with the increase in K^+ concentration. A maximum concentration of

144 mM K^+ ions elicited a flux of 2.73 ± 0.14 nmol Ca^{2+}/mg protein, representing a sevenfold increase over basal flux (2.7 mM K^+). This flux was inhibited up to 80% by pre-incubation in IgG from LEMS patients (Fig. 2), but was only partially blocked by dihydropyridines (22%) or by ω-CgTx GVIA (2%). These results imply that in this cell line the channel affected by the LEMS IgG represents up to 80% of the VGCCs present. The pharmacological profile shows that it is a non-N, non-L subtype [20]. These channels were also blocked (80%) by P-channel blockers FTX [30] and ω-Aga IVA and Q-type channel blocker ω-CmTx MVIIC (Lang, et al., unpublished). These functional studies strongly suggest that LEMS IgG acts on P- and/or Q-type VGCCs.

Serum antibodies to N-type calcium channels have been found in many LEMS patients. The peptide toxin ω-CgTx GVIA, which binds specifically and essentially irreversibly to the N-type VGCCs, has been used as a basis for a radioimmunoassay by several groups for the diagnosis of LEMS. The assay was first described by Sher et al. [59], who used VGCCs extracted from a human neuroblastoma cell line. Antibodies able to precipitate ^{125}I-ω-CgTx GVIA-labelled receptors were detected in 90% of LEMS patients. However, in many the titres were extremely low, and positive titres were also detected in 42% of SCLC patients without LEMS and 10% of other controls. Two other groups have reported similar assays. Lennon and Lambert [31] found positive titres in 52% of patients, while Leys et al. report 44% [35, 36]. Both groups found low but positive titres in a significant number of control patients. Antibodies directed against L-type VGCCs labelled with the dihydropyridine [^{3}H]PN200-110 were also detected in a subpopulation of LEMS patients with high anti-N-type VGCC antibody titres [10]. In view of these findings, it seems unlikely that N- or L-type VGCCs are the main immunogenic target in LEMS.

The finding that ω-CmTx MVIIC blocked K^+-stimulated Ca^{2+} flux in SCLC led us to use this toxin in a radioimmunoassay in which calcium channel extracts from human cerebellum were labelled with ^{125}I-ω-CmTx. A similar assay has been described independently by Lennon et al. [32]. We detected antibodies in 85% of LEMS patients; no positive titres have been detected in the controls (healthy, neurological and SCLC) [43] (Fig. 3). The higher percentage of antibody-positive patients and the lack of positivity in the controls indicates its strong specificity for LEMS and implies that antibodies directed against the ^{125}I-ω-CmTx MVIIC-labelled VGCCs are important in the pathogenesis of the disease.

Antibodies have also been found directed against the beta subunit of the VGCC. Rosenfeld et al. [58] used the serum from a patient with a high anti-ω-CgTx GVIA antibody titre to screen a protein expression cDNA library derived from human fetal brain. They isolated from this library a clone that coded for a neuronal beta subunit [58]. El Far et al. [10] also demonstrated antibodies that react with the beta subunit of the skeletal muscle calcium channel by Western blotting. However, the beta subunit is believed to be intracellular (see Fig. 1), so that its role in the primary pathogenesis of LEMS is presently uncertain. It is possible that the production of anti-beta-subunit antibodies may be a secondary response to cell damage resulting from the initial antibody-mediated attack on extracellular portions of the VGCCs. Ultrastructural studies of the presynaptic region of the motor end-plate of LEMS patients showed no tissue damage, thus making this hypothesis unlikely [67]. This does not completely rule out the hypothesis, however, since damage might be occurring to VGCC-expressing cells elsewhere in the body.

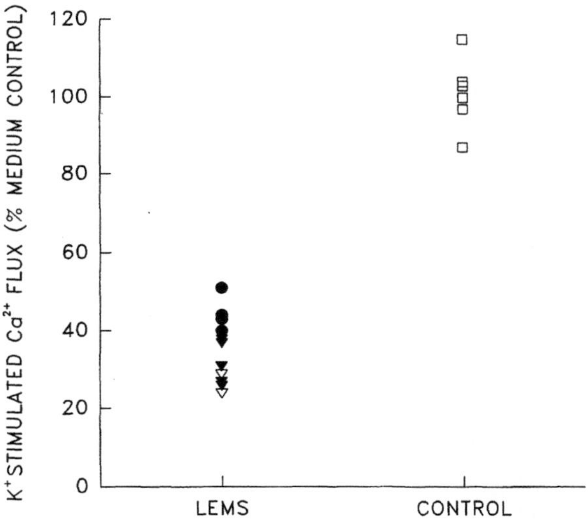

Fig. 2. The effect of Lambert-Eaton myasthenic syndrome (LEMS) IgG on K^+-stimulated Ca^{2+} flux. K^+-stimulated Ca^{2+} flux was measured in small cell lung carcinoma (SCLC) cells (MB), grown for 2–3 days in medium containing IgG (2 mg/ml culture medium), prepared from LEMS patients (SCLC-LEMS •, NCD-LEMS , SU-LEMS ▽), or patients with other neurological diseases . LEMS IgG significantly inhibited K^+-stimulated Ca^{2+} flux, while the control IgG was without effect. No significant difference was found in the ability of the subgroups of LEMS patients to block the Ca^{2+} flux. (Data adapted from [20]). (NCD, no cancer detected. SU, status uncertain)

Other synaptic proteins

As reviewed above, current evidence strongly implicates the voltage-gated calcium channel as the main target for autoantibodies in LEMS. However, antibodies directed against calcium channel-associated proteins may also play a part in the disease aetiology.

Neurotransmission requires the docking of synaptic vesicles to the presynaptic plasma membrane, their signal-dependent fusion and subsequent exocytosis. This involves the formation of a multimeric protein complex comprising synaptobrevin (also known as VAMP or synaptic vesicle-associated membrane protein) located in the synaptic vesicle, syntaxin and SNAP25 (synaptasome-associated protein 25) located in the presynaptic membrane. These proteins, together with cytosolic proteins *N*-ethylmaleimide-sensitive fusion protein (NSF) and SNAP (soluble NSF-attachment proteins) in the presence of non-hydrolysable ATP form a core particle which is responsible for stable docking (for a review see [64]). Another synaptic vesicle protein, synaptotagmin, is thought to be involved in the negative regulation of this process. Synaptotagmin is a Ca^{2+}-binding transmembrane protein present in synaptic vesicles. Binding of synaptotagmin in the multi-protein complex appears to prevent vesicular fusion. This negative effect ("fusion clamp") is relieved by Ca^{2+} influx through VGCCs following nerve stimulation.

Serum antibodies to synaptotagmin have been reported by Leveque et al. [33] in a minority of LEMS patients, all of whom had high anti-ω-CgTx GVIA receptor antibody titres (anti-N-type antibodies). This association may not be simply fortuitous as

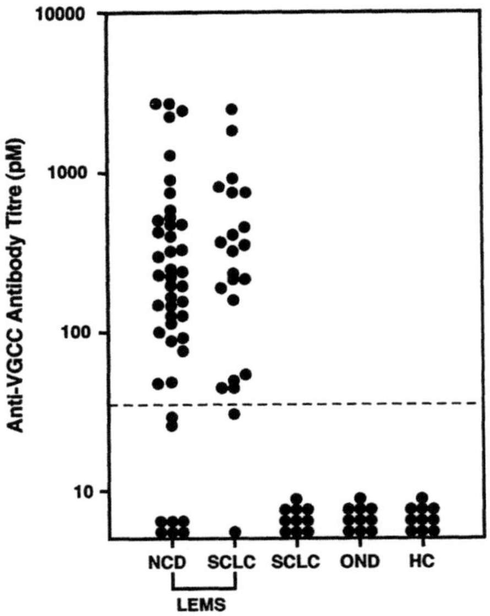

Fig. 3. Serum antibodies in LEMS and controls measured by radioimmunoprecipitation. Anti-ω-CmTx-MVIIC-VGCC antibody titres of serum taken from LEMS or control patients were measured by immunoprecipitation of VGCCs extracted from human cerebellum. In each case non-specific binding of each serum measured in the presence of excess cold ω-CmTx MVIIC has been subtracted. Antibody titres were considered positive if greater than 40 pM (mean+ 3SDs for the healthy controls; *dashed line*). SCLC, small cell lung carcinoma; NCD, no cancer detected; HC, healthy control; OND, other neurological diseases

syntaxin – an integral part of the docking mechanism – is also found complexed with ω-CgTX GVIA receptor, supporting the hypothesis of a close relationship between these proteins and the presynaptic VGCCs [34].

A model for LEMS has also been raised by injecting rats with peptides derived from the N-terminal sequence of synaptotagmin [65]. Rats were immunised with one of two peptides representing either residues 1–30 or 20–53. Both elicited strong anti-peptide antibodies, but only the second peptide produced electrophysiological changes. Some of the rats immunised with this peptide showed reduced quantal content of the EPP and post-tetanic facilitation, similar to that observed in LEMS patients and in mice injected with LEMS IgG. Although synaptotagmin is a protein associated with the membrane of the synaptic vesicle, the N terminus (residues 1–53) is intravesicular. Therefore, this part of the molecule would be exposed at the external surface of the nerve terminal during exocytosis and could then be accessible to circulating antibodies.

Synaptotagmin has also been identified in eight of ten SCLC cell lines tested by Western blotting, as has syntaxin [5]. These findings raise the possibility that antibodies may have initially been raised against synaptotagmin inappropriately expressed by the tumour, their cross-reaction with synaptotagmin at the nerve terminal leading to interference with ACh release.

Antibodies directed against other synaptic proteins may also exist. Synaptobrevin which like synaptotagmin has an intravesicular portion, may be a prime candidate for investigation. The antibodies directed against synaptic vesicle proteins have not yet been fully characterised, nor is it known what role, if any, they play in the aetiology of the disease.

Initiation of the autoimmune response

The evidence is now strong that the antigenic target in LEMS is the VGCC, whether or not the patient has an associated SCLC. Antibodies against other components of the neurotransmitter release mechanism may also be implicated. Two important questions need to be considered: where is autoantibody production initiated against these proteins, and why?

In LEMS patients with associated cancer, the answer appears clear. The strong association with SCLC (> 60%) indicates its importance. If the tumour is excised or treated, the neurological symptoms often improve [3]. It is a tumour of neuroectodermal origin (like the nervous system) and has been shown electrophysiologically and through K^+-stimulated $^{45}Ca^{2+}$ flux assays to have functional VGCCs. Molecular biological techniques have confirmed expression of L- and P-type VGCCs. Binding studies using specific toxins have established the presence of N- and Q-type channels. Expression of the synaptic vesicle protein synaptotagmin has been demonstrated by Western blotting techniques. Therefore, the hypothesis [56] that antibodies are initially raised against tumour determinants, which can then cross-react with the equivalent proteins at the neuromuscular junction (and at the autonomic ganglia) looks increasingly convincing. The tumour appears to initiate the response early in its existence, since the neurological syndrome may precede X-ray evidence of the tumour by up to 5 years. At this time the SCLC is presumably immunogenic and patients, who are genetically high responders, are capable of mounting a strong immune response. This is shown by the association of SCLC-LEMS patients with certain immunoglobulin heavy chain markers and HLA-B8 and DR3 [69].

The tumours themselves may be subtly different in the patients who develop neurological symptoms. In a comparison of the immunocytochemical characteristics of SCLC in patients with or without evidence of LEMS, marked infiltration of the tumour by activated macrophages was observed in the LEMS-associated tumours, indicating an ongoing aggressive autoimmune response [42]. However, in these tumours there was a decrease in the expression of 200-kDa neurofilament and of MHC class I antigens compared with tumours from patients without LEMS. These findings result from the production of antibodies to tumour VGCCs, and perhaps also from cytotoxic T cells which, by killing tumour cells strongly expressing class I antigens, would leave behind the less-differentiated cells.

LEMS patients who do not have an SCLC (approximately 40% of all cases) appear to have anti-VGCC antibodies of the same specificity and at a similar incidence to those with LEMS and SCLC. They are, however, an epidemiologically distinct group, often being young, female and non-smokers. Some have been followed up to 20 years after the onset of neurological symptoms without evidence of a tumour. The antigenic stimulus in this group is as yet unknown. The Kulchitsky cells of the bronchial epithelium, which are thought to be the progenitor cell from which the SCLC is derived, may be involved but as in many other autoimmune diseases the reason for loss of self tolerance is unknown. It may, therefore, be first necessary to identify the exact epitopes to elucidate the autoimmune mechanisms involved. To do this it will be important to identify the different subtypes of VGCCs and the individual subunits involved. Identification of the 'epitopes' on the primary sequence could then lead to specific immunotherapies in which elimination of sensitised T and B cells may be possible.

Summary and conclusions

LEMS is an antibody-mediated autoimmune disease that can occur in isolation, or as a paraneoplastic disorder in association with SCLC (60% of patients). The underlying defect is a reduction in the quantal release of the neurotransmitter ACh from the presynaptic nerve terminal at the neuromuscular junction. Experimental evidence indicates the autoantibodies are directed against nerve terminal VGCCs causing down-regulation in the number of functional channels by cross-linkage. Functional VGCCs have been detected in SCLC cell lines. In cancer-associated LEMS it appears likely that antibodies initially provoked by tumour VGCCs cross-react with VGCCs at the nerve terminal, causing the clinical disorder.

Antibodies against L-, N- and P-/Q- subtypes of the calcium channels have been identified and radioimmunoassays have been developed to help diagnose the disease. Using peptide toxin ^{125}I-ω-CmTx MVIIC to label P-/Q-type VGCC solubilised from human cerebellum, positive antibody titres can be detected in 85% of patients. However, autoantibodies in LEMS are heterogenous; the antigenic targets include different VGCC subtypes, the intracellular beta subunit and the synaptic vesicle protein synaptotagmin. The disease phenotype may reflect the diversity and titre of these different antibodies.

Acknowledgements. This work was supported by the Medical Research Council and the Sir Jules Thorn Charitable Trust.

References

1. Anderson HJ, Churchill-Davidson HC, Richardson AT (1983) Bronchial neoplasm with myasthenia. Lancet II:1291
2. Bain P, Elrington G, Goodger E, Misbah S, Panegyres P, MacPherson K, Chapel H, Newsom-Davis J (1994) A randomised double-blind controlled study of IVIG in the Lambert-Eaton myasthenic syndrome. J Neurol Neurosurg Psychiatry (in press)
3. Chalk CH, Murray NM, Newsom-Davis J, O'Neill JH, Spiro SG (1990) Response of the Lambert-Eaton myasthenic syndrome to treatment of associated small-cell lung carcinoma. Neurology 40:1552
4. Cohen MW, Jones OT, Angelides KJ (1991) Distribution of Ca^{2+} channels on frog motor nerve terminals revealed by fluorescent ω-conotoxin. J Neurosci 11:1032
5. David P, El Far O, Martin-Moutot N, Poupon MF, Takahashi M, Seagar MJ (1993) Expression of synaptotagmin and syntaxin associated with N-type calcium channels in small cell lung cancer. FEBS Lett 326:135
6. De Aizpurua HJ, Lambert EH, Griesmann GE, Olivera BO, Lennon VA (1988) Antagonism of voltage-gated calcium channels in small cell carcinomas of patients with and without Lambert-Eaton myasthenic syndrome by autoantibodies, ω-conotoxin and adenosine. Cancer Res 48:4719
7. De Luca A, Li CG, Rand MJ, Reid JJ, Thaina P, Wong-Dusting HK (1990) Effects of ω-conotoxin GVIA on autonomic neuroeffector transmission in various tissues. Br J Pharmacol 101:437
8. Dropcho E, Payne R (1986) Paraneoplastic opsoclonus-myoclonus. Association with medullary thyroid carcinoma and review of the literature. Arch Neurol 43:410
9. Dutar P, Rascol O, Lamour Y (1989) ω-Conotoxin GVIA blocks synaptic transmission in the CA1 field of the hippocampus. Eur J Pharmacol 174:261
10. El Far O, Marqueze B, Leveque C, Martin-Moutot N, Lang B, Newsom-Davis J, Yoshida A, Takahashi M, Seagar MJ (1995) Antigens associated with N and L type calcium channels in Lambert-Eaton myasthenic syndrome. J Neurochem (in press)
11. Elmqvist D, Lambert EH (1968) Detailed analysis of neuromuscular transmission in a patient with the myasthenic syndrome sometimes associated with bronchogenic carcinoma. Mayo Clin Proc 43:689

12. Elrington GM, Murray NMF, Spiro SG, Newsom-Davis J (1991) Neurological paraneoplastic syndromes in patients with small cell lung cancer: a prospective survey of 150 patients. J Neurol Neurosurg Psychiatry 54:764
13. Fambrough DM, Drachman DB, Satyamurti S (1973) Neuromuscular junction in myasthenia gravis: decreased acetylcholine receptors. Science 182:293
14. Fukunaga H, Engel AG, Osame M, Lambert EH (1982) Paucity and disorganisation of presynaptic membrane active zones in the Lambert-Eaton myasthenic syndrome. Muscle Nerve 5:686
15. Fukunaga H, Engel AG, Lang B, Newsom-Davis J, Vincent A (1983) Passive transfer of Lambert-Eaton myasthenic syndrome with IgG from man to mouse depletes the presynaptic membrane active zones. Proc Natl Acad Sci USA 80:7636
16. Fukuoka T, Engel AG, Lang B, Newsom-Davis J, Prior C, W.-Wray D (1987) Lambert-Eaton myasthenic syndrome: early morphological effects of IgG on the presynaptic membrane active zones. Ann Neurol 22:193
17. Graus F, Elkon KB, Cordon-Cardo C, Posner JB (1986) Sensory neuronopathy and small cell lung cancer. Antineuronal antibody that also reacts with the tumor. Am J Med 80:45
18. Greenlee JE, Brashear HR (1983) Antibodies to cerebellar Purkinje cells in patients with paraneoplastic cerebellar degeneration and ovarian carcinoma. Ann Neurol 14:609
19. Guttmann L, Crosby TW, Takamori M, Martin JD (1972) The Eaton-Lambert syndrome and autoimmune disorders. Am J Med 53:354
20. Johnston I, Lang B, Leys K, Newsom-Davis J (1994) Heterogeneity of calcium channel autoantibodies detected using a small-cell lung cancer line derived from a Lambert-Eaton myasthenic syndrome patient. Neurology 44:334
21. Kim YI (1985) Passive transfer of the Lambert-Eaton myasthenic syndrome: neuromuscular transmission in mice injected with plasma. Muscle Nerve 8:162
22. Kim YI, Neher E (1988) IgG from patients with Lambert-Eaton syndrome blocks voltage dependent calcium channels. Science 239:405
23. Lambert EH, Elmqvist D (1971) Quantal components of end-plate potentials in the myasthenic syndrome. Ann NY Acad Sci 183:183
24. Lambert EH, Eaton LM, Rooke ED (1956) Defect of neuromuscular conduction associated with malignant neoplasma. Am J Physiol 187:612
25. Lang B, Newsom-Davis J, Wray D, Vincent A, Murray NMF (1981) Autoimmune aetiology for myasthenic (Eaton-Lambert) syndrome. Lancet II:224
26. Lang B, Newsom-Davis J, Prior C, Wray D (1983) Antibodies to motor nerve terminals: an electrophysiological study of a human myasthenic syndrome transferred to mouse. J Physiol (Lond) 344:335
27. Lang B, Newsom-Davis J, Peers C, Prior C, Wray DW (1987) The effect of myasthenic syndrome antibody on presynaptic calcium channels in the mouse. J Physiol (Lond) 390:257
28. Lang B, Newsom-Davis J, Wray DW (1988) The effect of Lambert-Eaton myasthenic syndrome antibody on slow action potentials in mouse cardiac ventricle. Proc R Soc Lond [Biol] 235:103
29. Lang B, Vincent A, Murray NM, Newsom-Davis J (1989) Lambert-Eaton myasthenic syndrome: immunoglobulin G inhibition of Ca^{2+} flux in tumor cells correlates with disease severity. Ann Neurol 25:265
30. Lang B, Johnston I, Leys K, Elrington G, Marqueze B, Leveque C, Martin-Moutot N, Seagar M, Hoshino T, Takahashi M, Sugimori M, Cherksey BD, Llinas R, Newsom-Davis J (1993) Autoantibody specificities in Lambert-Eaton myasthenic syndrome. Ann NY Acad Sci 681:382
31. Lennon VA, Lambert EH (1989) Autoantibodies bind solubilised calcium channel-omega-conotoxin complexes from small cell lung carcinoma: a diagnostic aid for Lambert-Eaton myasthenic syndrome. Mayo Clin Proc 64:1498
32. Lennon VA, Krvzer T, Griesmann GE, Pinsky NA, Millianich G, Ramachandran J, Lambert EH (1993) P-like Ca^2+ channel in human cerebellum and small cell carcinoma bind *omega*-conotoxin MVIIC and are precipitable by Lambert-Eaton myasthenic syndrome (LES) IgGs that react highly with N-type Ca^{2+} channels. Soc Neurosci Abstr 19:1753
33. Leveque C, Hoshino T, David P, Shoji-Kasai Y, Leys K, Omori A, Lang B, El Far O, Sato K, Martin-Moutot N, Newsom-Davis J, Takahashi M, Seagar MJ (1992) The synaptic vesicle protein synaptotagmin associates with calcium channels and is a putative Lambert-Eaton myasthenic syndrome antigen. Proc Natl Acad Sci USA 89:3625

34. Leveque C, El Far O, Martin-Moutot N, Sato K, Kato R, Takahashi M, Seagar MJ (1994) Purification of the N-type calcium channel associated with syntaxin and synaptotagmin. A complex implicated in synaptic vesicle exocytosis. J Biol Chem 269:6306
35. Leys K, Lang B, Vincent A, Newsom-Davis J (1989) Calcium channel antibodies in Lambert-Eaton myasthenic syndrome. Lancet II:1107
36. Leys K, Lang B, Johnston I, Newsom-Davis J (1991) Calcium channel autoantibodies in the Lambert-Eaton myasthenic syndrome. Ann Neurol 29:307
37. Llinas R, Sugimori M, Lin J-W, Cherskey B (1989) Blocking and isolation of a calcium channel from neurons in mammals and cephalopods utilising a toxin fraction (FTX) from funnel-web spider poison. Proc Natl Acad Sci USA 86:1689
38. Login IS, Kim YI, Judd AM, Spangelo BL, MacLeod RM (1987) Immunoglobulins of Lambert-Eaton myasthenic syndrome inhibit rat pituitary hormone release. Ann Neurol 22:610
39. McCann FV, Pettengill OS, Cole JJ, Russell JAG, Sorenson GD (1981) Calcium spike electrogenesis and other electrical activity in continuously cultured small cell carcinoma of the lung. Science 212:1155
40. Molenaar PC, Newsom-Davis J, Polak RL, Vincent A (1982) Eaton-Lambert syndrome: acetylcholine and choline acetyltransferase in skeletal muscle. Neurology 32:1062
41. Mori Y, Friedrich T, Kim M-S, Mikami A, Nakai J, Ruth P, Bosse E, Hofmann F, Flockerzi V, Furuichi T, Mikoshiba K, Imoto K, Tanabe T, Numa S (1991) Primary structure and functional expression from complementary DNA of a brain calcium channel. Nature 350:398
42. Morris CS, Esiri MM, Marx A, Newsom-Davis J (1992) Immunocytochemical characteristics of small cell lung carcinoma associated with the Lambert-Eaton myasthenic syndrome. Am J Pathol 140:839
43. Motomura M, Johnston I, Lang B, Vincent A, Newsom-Davis J (1994) An improved diagnostic assay for Lambert-Eaton myasthenic syndrome. J Neurol Neurosurg Psychiatry (in press)
44. Nagel A, Engel AG, Lang B, Newsom-Davis J, Fukuoka T (1988) Lambert-Eaton myasthenic syndrome IgG depletes presynaptic membrane active zone particles by antigenic modulation. Ann Neurol 24:552
45. Newsom-Davis J, Murray NM (1984) Plasma exchange and immunosuppressive drug treatment in the Lambert-Eaton myasthenic syndrome. Neurology 34:480
46. Newsom-Davis J, Murray N, Wray D, Lang B, Prior C, Gwilt M, Vincent A (1982) Lambert-Eaton myasthenic syndrome: electrophysiological evidence for a humoral factor. Muscle Nerve 5:517
47. O'Neill JH, Murray NM, Newsom-Davis J (1988) The Lambert-Eaton myasthenic syndrome. A review of 50 cases. Brain 111:577
48. Oguro-Okano M, Griesmann GE, Wieben ED, Slaymaker SJ, Snutch TP, Lennon VA (1992) Molecular diversity of neuronal-type calcium channels identified in small cell lung carcinoma. Mayo Clin Proc 67:1150
49. Pancrazio JJ, Oie HK, Kim YI (1992) Voltage-sensitive calcium channels in human small-cell lung cancer cell line. Acta Physiol Scand 144:463
50. Pearson HA, Newsom-Davis J, Lees G, Wray D (1989) Lack of action of Lambert-Eaton myasthenic syndrome antibody on calcium channels in insect muscle. Ann NY Acad Sci 560:291
51. Peers C, Lang B, Newsom-Davis J, Wray D (1990) Selective action of myasthenic syndrome antibodies on calcium channels in a rodent neuroblastoma x glioma cell line. J Physiol (Lond) 421:293
52. Peers C, Johnston I, Lang B, Wray D (1993) Cross-linking of presynaptic calcium channels: a mechanism of action for Lambert-Eaton myasthenic syndrome antibodies at the mouse neuromuscular junction. Neurosci Lett 153:45
53. Prior C, Lang B, Wray D, Newsom-Davis J (1985) Action of Lambert-Eaton myasthenic syndrome IgG at mouse motor nerve terminals. Ann Neurol 17:587
54. Pumplin DW, Reese TS, Llinas R (1981) Are the presynaptic membrane particles the calcium channels? Proc Natl Acad Sci USA 78:7210
55. Reynolds IJ, Wagner JA, Synder SH, Thayer SA, Olivera BM, Miller RJ (1986) Brain voltage-sensitive Ca Channels subtypes differentiated by ω-conotoxin fraction GVIA. Proc Natl Acad Sci USA 83:8804
56. Roberts A, Perera S, Lang B, Vincent A, Newsom-Davis J (1985) Paraneoplastic myasthenic syndrome IgG inhibits $^{45}Ca^{2+}$ flux in a human small cell carcinoma line. Nature 317:737
57. Roberts M, Willison H, Vincent A, Newsom-Davis J (1994) Serum factor in Miller Fisher variant of Guillain-Barré syndrome and neurotransmitter release. Lancet 343:454
58. Rosenfeld MR, Wong E, Dalmau J, Manley G, Posner JB, Sher E, Furneaux HM (1993) Cloning and characterization of a Lambert-Eaton myasthenic syndrome antigen. Ann Neurol 33:113
59. Sher E, Canal N, Piccolo G, Gotti C, Scoppetta C, Evoli A, Clementi F (1989) Specificity of calcium channel autoantibodies in Lambert-Eaton myasthenic syndrome. Lancet II:640

60. Sher E, Pandiella A, Clementi F (1990) Voltage-operated calcium channels in small cell lung carcinoma cell lines: pharmacological, functional, and immunological properties. Cancer Res 50:3892
61. Simpson JA (1960) Myasthenia gravis: a new hypothesis. Scott Med J 5:419
62. Sinha S, Newsom-Davis J, Mills K, Byrne N, Lang B, Vincent A (1991) Autoimmune aetiology for acquired neuromyotonia (Isaacs' syndrome). Lancet 338:75
63. Solimena M, Folli F, Aparisi R, Pozza G, De Camilli P (1990) Autoantibodies to gaba-ergic neurons and pancreatic beta cells in stiff-man syndrome. N Engl J Med 322:1555
64. Söllner T, Rothman JE (1994) Neurotransmission: harnessing fusion machinery at the synapse. Trends Neurosci 17:344
65. Takamori M, Hamada T, Komai K, Takahashi M, Yoshida A (1994) Synaptotagmin can cause an immune-mediated model of Lambert-Eaton myasthenic syndrome in rats. Ann Neurol 35:74
66. Tischler AS, Dichter MA, Biales B (1977) Electrical excitability of oat cell carcinoma. J Pathol 122:153
67. Tsujihata M, Kinoshita I, Mori M, Mori K, Shirabe S, Satoh A, Nagataki S (1987) Ultrastructural study of the motor end-plate in botulism and Lambert-Eaton myasthenic syndrome. J Neurol Sci 81:197
68. Uchitel OD, Protti DA, Sanchez V, Cherskey BD, Sugimori M, Llinas R (1992) P-type voltage-dependent calcium channel mediates presynaptic calcium influx and transmitter release in mammalian synapses. Proc Natl Acad Sci USA 89:3330
69. Willcox N, Demaine AG, Newsom-Davis J, Welsh KI, Robb SA, Spiro SG (1985) Increased frequency of IgG heavy chain marker Glm(2) and of HLA-B8 in Lambert-Eaton myasthenic syndrome with and without associated lung carcinoma. Hum Immunol 14:29

Immunological findings in amyotrophic lateral sclerosis

Jack P. Antel, Neil R. Cashman

Montreal Neurological Institute, 3801 University Street, Montreal, Quebec H3A 2B4, Canada

Amyotrophic lateral sclerosis (ALS) remains a syndrome defined by clinical and pathological rather than etiological criteria. The definition includes loss of both upper motor neurons (UMNs), i.e., neurons in motor cortex with secondary degeneration of the corticospinal tracts, and lower motor neurons (LMNs), i.e., neurons in brain stem motor nuclei and anterior horns of the spinal cord. Other neuronal populations such as those of the spinocerebellar tract, dorsal column, and intermediolateral cell column may also be involved in some cases, as assessed by pathological more so than by clinical criteria [95]. The mean duration of survival is 3–4 years with extremes in both directions being encountered. The usual sporadic form of the disease occurs in a male:female ratio of 2–3:1 and has a rather even distribution throughout the world except for several discrete foci in the Western Pacific where incidence rates have been increased up to 1000-fold. In these foci, some of which seem to be disappearing, unique putative neuronotoxic environmental factors are postulated to be involved [58]. The peak incidence rate has usually been regarded as being in the 6th and 7th decade of life, although data from Olmstead County, Minnesota, suggest that incidence rates do not decline after this age [40]. A continued increase in incidence rate would be more consistent with a degenerative disorder than a primary autoimmune or infectious etiology.

Progressive motor neuron disease (MND) phenotypes are also well described in which the disease distribution exclusively or predominately involves only the UMNs or only the LMNs. The former is referred to as primary lateral sclerosis; the latter as spinal or bulbar muscular atrophy. The heterogeneity of clinical phenotypes associated with the autosomal dominant form of ALS with point mutations in the superoxide dysmutase (SOD) gene indicates that a common etiology can account for multiple MND clinical presentations [65]. The association of sporadic MND cases, particu-

Correspondence to: J.P. Antel

larly those with a LMN phenotype, with a diverse range of metabolic abnormalities indicates that clinical phenotype alone does not designate etiology [58]. In this report, we consider the contribution of the immune system to development of the range of disorders considered as MNDs or of clinical phenotypes mimicking the MNDs, particularly the motor neuropathies. Such considerations include: (1) are immune mediators the primary cause of particular MNDs and (2) can immune mediators be secondary contributors to neuronal injury.

Immune mediators as primary causes of MNDs

To establish that a disease is mediated by an immune mechanism, criteria akin to those put forth by Koch [47] for viral diseases should be applied. Such criteria would include presence of specific immune responses in patients and not controls, passage of the disease from patients into animals via the putative immune mediators or induction of a model disease by active immunization with a disease-related antigen, and response of the disease to therapies which inhibit the immune response.

Amyotrophic lateral sclerosis

Traditional pathological analysis of tissues from ALS cases has led to the disease being regarded as a degenerative disorder without a major inflammatory component. More recent analysis of ALS tissue using increasingly sensitive immunocytochemical techniques have identified the presence of immune mediators, raising the question of their potential contributions to the overall level of tissue injury [43, 49]. Similar considerations apply to other "neurodegenerative diseases", including Alzheimer's disease [56]. Both in the spinal cord (ventral horn and corticospinal tracts) and motor cortex of ALS patients, one can detect the presence of collections of T cells. These T cells are reported to be of restricted heterogeneity, perhaps implicating an antigen-restricted response. Troost et al. [83], in contrast to other studies [26], could only detect T cells of the CD8 or cytotoxic phenotype and not of the CD4 or helper phenotype in the spinal cord. In the study of Troost et al. [83], B cells were not detected. However, immunoglobulin (Ig) deposits in the tissues, in astrocytes, and possibly in neurons (see discussion below) are described, as is the presence of complement components [21]. The pathological findings in ALS are further characterized by the presence of macrophages, microglia, and reactive astrocytes [59]. The macrophage/microglia express major histocompatibility complex (MHC) class I and II antigens and Fc receptors, suggesting a state of activation [87]. Neuronophagia, which is observed in ALS, has been ascribed to perivascular macrophages rather than microglia [85]. T cells and macrophages are also found in denervated muscle in cases of ALS [84].

As regards ALS patient susceptibility to autoimmune diseases, some epidemiological surveys suggest an increased frequency of autoimmune phenomenon in this patient group [8]. There is, however, no apparent increased risk of disease in families outside of those which have a single gene autosomal or sex-linked inheritance pattern. Consensus from a series of studies is that any immunogenetic predisposition related to MHC or T cell receptor germ-line repertoire is at best weak (reviewed in [2, 3]). Development of ALS in immunodeficient individuals is reported but seems to be uncommon.

Tissues and body fluids from patients with ALS have been examined for presence of specific immune responses and products. Initial studies of post-mortem tissue failed to demonstrate specific accumulation of Ig in or on MNs in ALS cases [92]. Such findings can also be inferred from the array of studies which have used ALS tissue as "controls" to assess presence or binding of specific sera or derived antibodies to MNs in situ in cases of dysimmune states-associated LMN syndromes. Recently, Ig deposition was described within MNs of the spinal cord and motor cortex of ALS patients in a manner parallel to that found in the experimental autoimmune MND (EAMND) model induced in guinea pigs, as described below [24]. The means whereby the Ig gains access to the MNs remains speculative, although one notes the ability of MNs to transport Ig retrogradely from their nerve terminals to the cell soma [27]. High levels of Ig accumulation within neurons is a hallmark of selected paraneoplastic syndromes, such as subacute cerebellar degeneration, in which the involved neurons do not have peripheral connections [64].

Sera and CSF from ALS patients have been examined for presence of neuronotoxic activity. Kiernan and Hudson [45] found that sera from about 50% of ALS patients would bind to MNs in sections of human spinal cord; however, sera from a similar proportion of controls would also bind. ALS sera are reported to be cytotoxic to neurons maintained in dissociated and organotypic tissue culture (reviewed in [13]). The cytotoxic fraction has variably been attributed to an Ig and non-Ig fraction. Successful passive transfer of disease using human sera has not been accomplished. ALS sera are also reported to contain a neuroelectric blocking factor and to inhibit terminal axonal sprouting induced by botulinum toxin [20, 35, 39, 70]. In none of these assays can one distinguish sporadic and autosomal dominant familial forms of the disease. The CSF of ALS patients does not show consistent abnormalities of cell counts, Ig levels or heterogeneity of the Ig (oligoclonal bands) [7].

Most recently, Smith and colleagues [10,55,79] have identified an antibody directed at the L-type skeletal muscle voltage-gated calcium channel (VGCC) that is detectible in a high proportion of ALS patients' sera and only in a minority of patients with other neurological diseases, such as Guillain-Barré syndrome and Eaton-Lambert syndrome (ELS). The ALS sera differ from the ELS sera in that the former bind to the $\alpha 1$ subunit of the channel, while the former bind to both the $\alpha 1$and $\beta 1$ subunits [46]. Titers of this antibody correlate with rate of disease progression. This antibody is also found in patients with familial ALS and in animals with EAMND. The antibody or Fab fragments thereof were shown to reduce peak Ca^2+ current through dihydropyridine-sensitive Ca^2+ channels in mammalian skeletal muscle fibers or reconstituted bilayers [18, 19]. ALS serum-derived IgG was further shown to be cytotoxic to a MN-neuroblastoma hybrid cell line [79]. The effect was dependent on extracellular calcium and blocked by peptide antagonists of N or P-type VGCCs or by pre-incubating the ALS IgG with L-type VGCCs. The resultant conclusion of these studies was that ALS IgG can perturb physiological activity at the neuromuscular junction and can directly lead to MN cell death via interaction with neuronal-type calcium channels. Passive transfer of this antibody has not yet been shown to induce a MND in animals.

Scattered reports contain descriptions of other specific antibodies associated with ALS. Most of these are based on ALS sera or Ig fractions binding to specific gangliosides or to neuronal cell lines in vitro [59, 67, 80, 102]. Studies using monoclonal antibodies generated in vitro using a range of immunogens, including isolated MNs, spinal cord, and synaptosomes from neuromuscular junction preparations, indicate that discrete epitopes capable of being recognized by antibody exist on or in MNs [5, 57, 75, 101]. ALS sera have also been examined for presence of antibodies directed at exogenous agents, particularly viruses; to date, no distinct disease-associated response has been identified [94].

As regards immune-mediated animal models of MND, the term EAMND has been applied to a disease induced in guinea pigs by repeated immunization with purified bovine spinal MNs [79]. This disease is characterized clinically by the development of "slow and progressive" limb weakness and electrophysiological and morphological evidence of muscle denervation [32]. There is loss of motor nerve cells of the spinal cord and brain stem with neuronophagia but not inflammation. Deposits of Ig can be detected in MN cytoplasm and at neuromuscular junctions. The same anti-Ca^2+ channel antibody is found in this disease as is found in ALS patients. A second disease, termed EA gray matter disease (EAGMD), is induced by immunization with bovine gray matter homogenate [25]. In this disorder, there is rather abrupt onset of weakness and some animals develop spasticity. Pathologically, there are inflammatory infiltrates in the ventral horn, spinal cord white matter, and cerebral cortex. The distribution of inflammation would seem to differ from that observed in the prototype autoallergic disorder, experimental allergic encephalomyelitis (EAE). The EAGMD animals show high titers of antibody to MNs, akin to those found in EAMND, and also high titers of anti-myelin basic protein (MBP) antibody.

The above findings in ALS patients and in animals have formed the basis for clinical trials using immunosuppressive and immunomodulatory agents. To date no conclusive positive results have been achieved. The array of agents used include: total lymphoid irradiation, cyclosporine, cyclophosphamide, interferon (IFN), anti-retroviral therapy, plasmapheresis, isoprinosine and corticosteroids [9, 12, 22, 44, 93, 94].

LMN syndromes

These disorders would include those in which the primary site of pathology would be the MN cell body per se (neuronopathy) and those in which the motor nerve axon or its myelin sheath is the primary site. Extensive clinical laboratory studies have established the association between LMN syndromes and dysimmune states, particularly IgM paraproteinemias [51]. In some but not all of these cases there is evidence of multifocal conduction block (MCB), as determined by clinical electrophysiological testing [51]. Only rare cases of IgM paraprotein-associated MND have been examined pathologically. In such cases, the extent of spinal MN loss is less than in typical ALS or spinal muscle atrophy cases, suggesting that the primary insult is to the motor nerve axon rather than to the cell body or proximal axon segment, as is thought to occur in ALS. Thus, this entity is increasingly regarded as a motor neuropathy rather than a primary neuronopathy.

In many cases of motor neuropathy associated with IgM paraprotein, with or without MCB, a more specific antibody can be identified in the sera. Most commonly, this antibody reacts with carbohydrate moieties [Gal (β1-3) GalNAc epitope] of GM1 and GD1b gangliosides, which are present in both peripheral and central myelin [48]. GM is more abundant in the myelin of motor peripheral nerve myelin than in sensory nerve myelin (reviewed in [82]). The gal (β1-3) GalNAc epitope is also available at the pre-synaptic side of the neuromuscular junction. The titers of such antibody can be extraordinarily high, in some cases being detectible at serum dilutions of 10^6 or

10^7. High titer anti-GM1 antibody can be found without a monoclonal IgM being detected by the usual clinical laboratory techniques [66]. Conversely, Parry estimated that up to 50% of cases of LMN syndrome cases with MCB do not have the anti-GM1 antibody [63]. Peripheral blood mononuclear cells from patients with high titer anti-GM1 antibody in the serum produce high amounts of the antibody in response to the T cell-dependent polyclonal activator, pokeweed mitogen [37]. Limiting dilution analysis shows a high frequency of GM1-specific B cells [99].

The questions persist as to whether and how the anti-GM1 antibody produces the clinical syndrome [1, 50]. Kaji et al. [41, 42] reported two patients with MCB syndrome and moderately high levels of anti-GM1 antibody but no monoclonal Ig, in which nerve biopsy showed demyelination at the site of the conduction block. In only a minority of cases can one demonstrate that there is selective binding of Ig to the node of Ranvier region of the axons, despite the presence of MCB [69]. The prototype case in which such binding was demonstrated might now be termed ALS-probable (P)UMNs, as discussed below. In our experience, sera from cases of paraprotein-associated motor neuropathy or MCB neuropathy bind to myelin of peripheral nerve, whereas more diffuse staining is found with other forms of MND and neuropathies [31]. Corbo et al. [16] showed that the anti-GM1 antibodies from all their cases with LMN syndromes bound to the surface of bovine spinal MNs and immunostained the gray matter of spinal cord. The staining could be blocked by cholera toxin, indicating the specificity of the binding for GM1. Heimann-Patterson et al. [38] found that these M proteins bound to fetal human spinal cord-derived neurons maintained in co-culture with muscle and that survival of such neurons was reduced. Sera from some but not all patients with MCB and anti-GM1 antibodies, when injected into rat tibial nerve, result in reduced amplitude and dispersion of the compound action potential and demyelination of some fibers [90]. Interruption of neuromuscular transmission is also now reported (see [82]).

Although the mean titers of anti-GM1 antibody in the sera of the overall ALS population are increased over those of controls, the titers are markedly less than those in the paraprotein-associated LMN cases [48, 91]. Li and Pestronk [53] reported that the anti-GM1 antibodies of ALS patients differ in their binding affinities to membranes compared to patients with LMN syndromes.

The entity of motor neuropathy associated with autoantibody and/or conduction block has received considerable attention because of reports indicating the response of the syndrome to immunotherapy [86]. Feldman et al. [29] reported that prednisone therapy alone or plasmapheresis were rarely effective in treatment of the syndrome; however, cyclophosphamide was effective in cases which failed to respond to the previous therapies. Therapeutic response was usually associated with a fall in serum anti-IgM antibody titers. Salazar-Grueso et al. [68], however, reported a case of a LMN syndrome without MCB but with a high-titer polyclonal anti-GM1 antibody that responded to corticosteroid therapy in the face of rising antibody titers. Nobile-Orazio et al. [61] found that high-dose intravenous Ig therapy was also effective as a therapy for the multifocal motor neuropathy (MMN) syndrome. Azulay et al. [11] found that Ig treatment was effective for cases of LMN syndromes with high-titer anti-GM1 antibody, with or without MCB. The results of the therapeutic trials reported by Tan et al. [81] emphasize that not all cases of LMN syndromes associated with anti-GM1 antibody or MCB respond to immunotherapy. Furthermore, response to therapy did not correlate with anti-GM1 antibodies. Of note, however, is that, in this series as well, only patients with LMN syndromes or motor neuropathy with MCB responded to treatment. How the cases which respond to therapy differ from those which fail to respond remains to be clarified. Similar experience exists with other forms of paraprotein-associated neuropathies, such as those with anti-ganglioside antibody.

The relation between MNDs and non-IgM paraproteinemia remains less clearly established. The incidence of IgG paraproteinemia increases with age, being found in up to 10% of those over the age of 70 (reviewed in [4]). The concurrence of IgG paraproteinemia with a distinct clinical phenotype of MND or motor neuropathy is less clear than for the IgM cases[66, 77]. Kornberg and Pestronk [48] concluded from their analysis of over 5,000 serum samples that cases with high-titer IgG anti-GM1 antibody were strongly associated with acute motor neuropathies and rarely associated with a chronic MCB motor neuropathy syndrome. An individual case of IgA paraproteinemia associated with ALS has been described in which there is deposition of the paraprotein in spinal MNs and in vitro evidence of neuronal cytotoxicity mediated by the paraprotein [36].

Although the association of paraproteinemias/autoantibodies are most clear in LMN syndrome cases, such associations are also suggested to occur in a patient group with definite LMN dysfunction and probable (P) UMN dysfunction defined as retained or brisk reflexes in the presence of significant muscle atrophy [66, 100, 103]. Such cases have been labelled as ALS-PUMNS. Cases of ALS-PUMNS responding significantly to immunotherapy are difficult to identify. The frequency of ALS in patients with lymphoma seems to be increased and may or may not be accompanied by paraproteinemia [103].

Immune mediators as secondary contributors to ALS/MND pathology

The basis whereby exogenous cells (lymphocytes, macrophages) infiltrating the sites of ALS pathology or endogenous glial cells (microglia, astrocytes) could induce CNS neuronal injury remains to be defined [28]. A number of immune effector mechanisms involving either direct cell-cell interaction or soluble factors can be implicated. MNs throughout the neuraxis are not known to express either class I or II MHC antigens, making a traditional antigen-restricted cytotoxic T cell response unlikely. We have found, however, that human cortical neurons, in a similar manner to previous observations with oligodendrocytes, are susceptible to lysis by activated CD4 and CD8 T cells bearing alpha/beta (α/β) T cell receptors (TcRs), acting in a promiscuous (i.e., non-MHC-restricted) manner [57]. This in vitro lysis requires a time duration (18 h) greater than for classic cytotoxicity (5 h) and is not reproduced by supernatant from the activated effector cells. Gamma/delta (γ/δ) T cells bearing γ/δ TcRs, which can even be derived from the CSF of individuals without apparent neurological disease, exert potent cytotoxic effects on neural cell targets [30]. The recognition molecules for such cells are not established; however, members of the heat-shock protein (HSP) family of molecules remain possible candidates. HSP expression is described in MNs in ALS [33]. Macrophage/microglia ingestion of MNs (neuronophagia) is a well-described phenomenon in MND. Macrophages and microglia, particularly when activated, express Fc receptors in situ. Thus, an antibody-dependent cell cytotoxicity mechanism needs to be considered given the presence of both Ig and macrophages/microglia in the vicinity of MNs in the disease state [88, 89].

Multiple non-Ig soluble factors capable of effecting tissue injury within the CNS may be expected to be present in lesions containing lymphocytes, macrophages, and activated astrocytes and microglia. As regards the intrinsic glial cells of the CNS, the astrocytes and microglia, each has been shown largely in studies in vitro either to secrete an array of cytokines or express mRNA for such molecules [52, 71, 73]. In our studies of microglial cells isolated from surgically resected human adult temporal lobe tissue and maintained in dissociated cell culture, we can detect secretion of tumor necrosis factor (TNF), interleukin (IL)-1, IL-6, and IL-10; levels of secretion increase following stimulation with IFN-γ and lipopolysaccharide, activation via Fc receptors using immune complexes, or activation via ingestion of myelin [96, 98]. This array of cytokines closely parallels those of macrophages. Phenotypic differences between microglia and blood-derived monocytes do exist [89]. Data regarding cytokines secreted by astrocytes is largely derived from studies of fetal tissue-derived cells or from malignant astrocytes since techniques to prepare purified cultures of the slowly dividing adult astrocytes is still limited. The array of cytokines attributed to astrocytes based on the above studies include TNF, IL-3, and IL-6. Our in situ hybridization data confirm expression of mRNA for TNF and IL-6 in human adult tissue-derived astrocytes [57]. The glial cells produce additional arrays of soluble molecules which could contribute to neuronal injury (free radicals, nitric oxide) or alternately promote survival (neuronotrophic growth factors, including the cytokine-like molecule CNTF) [54].

In vitro studies have been conducted to determine whether soluble factors released by immune cells that infiltrate the CNS or endogenous glia can induce injury of neural cells. Most of the available data in this field are derived using mixed populations of fetal or neonatal tissue-derived cortical neurons or cell lines with neuronal features [14]. Lack of methods to isolate and maintain MNs in tissue culture limit studies of these cells in isolation. Most studies to date related to neuronal cell injury have used cell death as the index. The mechanisms of cell injury leading to cell death need clearer definition. Our studies, using both primary cerebral cortical neurons and the MN hybrid cell line NSC 34, indicate that such cells can be induced to undergo nuclear fragmentation or apoptosis under conditions of serum deprivation [57, 74]. Our initial studies indicate that activated T cell supernatants can induce nuclear fragmentation in human fetal cortical neurons [57]. This effect is more apparent than that found using purified TNF alone. TNF did induce nuclear fragmentation of human adult CNS-derived oligodendrocytes in the same assay system, in accord with previous reports using bovine and rodent cells [23, 72]. Guillian et al. [34] found that microglia-derived supernatants could lyse mouse hippocampal neurons but not oligodendrocytes; this effect could be blocked by N-methyl-D-aspartate receptor antagonists [34]. The microglia effect could not be attributed to TNF. The putative factor was considered to belong to a novel class of low molecular weight molecules.

Although the emphasis above has been on mechanisms whereby reactive glia and infiltrating hematogenous cells can contribute to neuronal injury, the possibility also exists that specific molecules released by such cells may protect from injury or promote regeneration [76]. We have previously shown that astrocytes can protect oligodendrocytes from free radical-induced injury mediated by addition of catecholamines to the cultures [62]. Macrophages and their supernatants are reported to promote axonal growth over sites of optic nerve lesions in vitro [17]. Thus, the particular state of glial calls may be critical for determining whether they exert a positive or negative influence on neuronal survival or recovery. In this regard, regulatory cytokines produced by infiltrating immune cells may play an important role [97]. The signal whereby injury or disease of neurons initiates the glial cell response remains unclear.

Neurons are reported to be capable of cytokine production, including IFN-γ -like molecules.

Conclusions

The sporadic form of ALS is the most common form of MND and remains a syndrome of unproven etiology or etiologies. The application of increasingly sensitive immunocyto-chemical techniques to pathological specimens indicates the presence of a humoral and cellular immune response at the site of MN destruction. The magnitude of the response would seem to be quantitatively much less than that found in traditional inflammatory disorders such as multiple sclerosis. The similarity of the response in the autosomal dominant form of the disease, which can be reproduced in mice expressing the mutant SOD transgene, suggests that this response is at most a secondary event in the disease pathogenesis. This pattern of immune response seems rather parallel to that seen in other "neurodegenerative" disorders. Whether the disease-specific calcium channel antibody which correlates with disease progression also arises in response to tissue injury is not defined, although it is not reported in patients with MN destruction as a result of polio. The failure of even intense forms of immunotherapy to alter the disease course in ALS does further question the primary or secondary role of immune effector mechanisms to the disease process. One cannot exclude that individual cases, such as those arising on the background of lymphoma, may have a more significant immunopathogenesis and be akin to other paraneoplastic syndromes which affect different neuronal populations.

The LMN syndromes associated with IgM paraproteins and anti-GM1 antibodies would seem to be examples of motor neuropathies rather than primary neuronopathies, although the neuronotoxic effects of some of these antibodies are of note. Whether the markedly high titers of antibody found represent an intrinsic B cell derangement, as those cases associated with lymphoma or myeloma suggest, or a defect in T cell regulation of the B cell response together with persistent antigen stimulation remains to be resolved. GM1 is not usually considered as a potent antigen, although high-titer responses can be induced with appropriate immunization protocols [15]. The clinical response of some of these cases to immunotherapy strengthens the hypothesis regarding the primary immune basis of the disorder. The inconsistency of the response remains, however, to be defined.

References

1. Adams D, Kuntzer T, Steck AJ, Lobrinus A, Janzer RC, Regli F (1993) Motor conduction block and high titres of anti-GM1 ganglioside antibodies: pathologicalal evidence of a motor neuropathy in a patient with lower motor neuron syndrome. J Neurol Neurosurg Psychiatry 56:982
2. Antel JP, Noronha ABC, Oger J, Arnason BGW (1982) Immunology of amyotrophic lateral sclerosis. Adv Neurol 36:395
3. Antel JP, Mikuk J (1991) Immunogenetics and motor neuron disease. Adv Neurol 56:357
4. Antel JP, Minuk J (1994) Neuroimmunology of aging. In: Albert M, Knoefel J (eds) Clinical neurology of aging. Oxford University Press, New York, second edition, 121
5. Antel JP, Kuchibhota J, Sefansson K (1985) Generation of monoclonal antibodies (mAbs) recognizing neuronal elements in formalin-fixed paraffin-embedded human tissue. J Neuropathol Exp Neurol 44:533
6. Antel JP, Williams K, Blain M, McCrea E, McLaurin J (1994) Oligodendrocyte lysis by CD4+ T cells independent of tumor necrosis factor. Ann Neurol 35:341

7. Apostolski S, Nikolic J, Bugarski-Prokepljevic C, Miletic V, Pavlovic S, Filipovic S (1991) Serum and CSF immunological findings in ALS. Acta Neurol Scand 83:96
8. Appel SH, Stockton-Appel V, Steward SS, et al (1986) Amyotrophic lateral sclerosis: associated clinical disorders and immunological evaluations. Arch Neurol 43:234
9. Appel SH, Stewart SS, Appel V, Harati Y, Mietlowski W, Weiss W, Belendiuk GW (1988) A double-blind study of the effectiveness of cyclosporine in amyotrophic lateral sclerosis. Arch Neurol 45:381
10. Appel SH, Engelhardt JL, Garcia J, Stefani E (1991) Immunoglobulins from animal models of motor neuron disease and from human amyotrophic lateral sclerosis patients passively transfer physiological abnormalities to the neuromuscular junction. Proc Natl Acad Sci USA 88:647
11. Azulay JP, Blin O, Pouget J, Boucrault J, Bille-Turc F, Carles G, Serratrice G (1994) Intravenous immunoglobulin treatment in patients with motor neuron syndromes associated with anti-GM1 antibodies: a doublee-blind, placebo-controlled study. Neurology 44:429
12. Brown RH, Hauser SL, Harrington H, Weiner HL (1986) Failure of immuno-suppression with a 10- to 14-day course of high-dose intravenous cyclophosphamide to alter the progression of amyotrophic lateral sclerosis. Arch Neurol 43:383
13. Cashman NR, Antel JP (1988) Amyotrophic lateral sclerosis: an immunological perspective. In: Goldstein R (ed) Immunology and allergy clinics of North America. Saunders, Philadelphia, pp 331–342
14. Cashman NR, Durham HM, Blustajn JK, Oda K, Tabira T, Shaw IT, Dahrouge S, Antel JP (1992) Neuroblastoma X spinal cord (NSC) hybrid cell lines resemble developing motor neurons. Dev Dyn 194:209
15. Chaudury V, Pestronk A (1992) Different patterns of glycolipid antibody reactivity: lower motor neuron syndromes vs. immunization. J Neuroimmunol 36:127
16. Corbo M, Quattrini A, Lugaresi A, Santoro M, Latov N, Hays AP (1992) Patterns of reactivity of human anti-GM1 antibodies with spinal cord and motor neurons. Ann Neurol 32:487
17. David S, Bouchard C, Tsatas O, Giftochristos N (1990) Macrophages can modify the nonpermissive nature of the adult mammalian central nervous system. Neuron 5:463
18. Delbono O, Garcia J, Appel SH, Stefani E (1991) Calcium current and charge movement of mammalian muscle: action of amyotrophic lateral sclerosis immunoglobulins. J Physiol (Lond) 444:723
19. Delbono O, Magnelli V, Sawada T, Smith RG, Appel SH, Stefani E (1993) Fab fragments from amyotrophic lateral sclerosis IgG affect calcium channels of skeletal muscle. Am J Physiol 264:C537
20. Donaghy M, Duchen LW (1986) Sera from patients with motor neuron disease and associated paraproteinemia fail to inhibit experimentally induced sprouting of motor nerve terminals. J Neurol Neurosurg Psychiatry 49:817
21. Donnenfeld H, Kascak R, Bartfeld H (1984) Deposits of IgG and C_3 in the spinal cord and motor cortex of ALS patients. J Neuroimmunol 6:51
22. Drachman DB, Chaudury V, Cornblath D, Kuncl RW, Pestronk A, Clawson L, Melllits ED, Quaskey S, Quinn R, Calkins A, Order S (1994) Trial of immunosuppression in amyotrophic lateral sclerosis using total lymphoid irradiation. Ann Neurol 35:142
23. D'Souza S, Wang E, Beaudet A, Antel JP (1994) Mechanisms of immune-mediated injury to oligodendrocytes in multiple sclerosis. Soc Neurosci Abstr 20:886
24. Engelhardt JI, Appel SH (1990) IgG reactivity on the spinal cord and motor cortex in amyotrophic lateral sclerosis. Arch Neurol 47:1210–1216
25. Engelhardt JI, Appel SH, Killian JM (1991) Motor neuron destruction in guinea pigs immunized with bovine spinal cord ventral horn homogenate: experimental autoimmune gray matter disease. J Neuroimmunol 27:21
26. Engelhardt JI, Tajti J, Appel SH (1993) Lymphocytic infiltrates in the spinal cord in amyotrophic lateral sclerosis. Arch Neurol 50:30
27. Fabian RH, Petroff G (1987) Intraneuronal IgG in the central nervous system: uptake by retrograde axonal transport. Neurology 37:1780
28. Fabry Z, Raine CS, and Hart MN (1994) Nervous tissue as an immune compartment: the dialect of the immune response in the CNS. Immunol Today 15:218
29. Feldman EL, Bromberg MB, Albers JW, Pestronk A (1991) Immunosuppressive treatment in multifocal motor neuropathy. Ann Neurol 30:493
30. Freedman MS, Ruijs TCG, Selin LK, Antel JP (1991) Peripheral blood $\gamma\delta$ cells lyse fresh human brain-derived oligodendrocytes. Ann Neurol 30:794

31. Freedman MS, Zielensky S, Grant I, Antel JP (1993) Correlation of anti-GM1 ganglioside antibody reactivity with peripheral nerves and clinical phenotype. Neurology 43[Suppl 2]:A343
32. Garcia J, Engelhardt JI, Appel SH, Stefani E (1990) Increased MEPP frequency as an early sign of experimental immune-mediated motoneuron disease. Ann Neurol 28:329
33. Garofalo O, Kennedy PG, Swash M, Martin JE, Luthert P, Anderton BH, Leigh PN (1991) Ubiquitin and heat shock protein expression in amyotrophic lateral sclerosis. Neuropathol Appl Neurobiol 17:39
34. Giulian D, Vaca K, Corpuz M (1993) Brain glia release factors with opposing actions upon neuronal survival. J Neurosci 13:29
35. Gurney ME, Belton AC, Cashman N, Antel JP (1984) Inhibition of terminal axonal sprouting by serun from patients with ALS. N Engl J Med 311:933
36. Hays AP, Roxas A, Sadiq SA, et al (1990) A monoclonal IgA in a patient with amyotrophic lateral sclerosis reacts with neurofilaments and surface antigen on neuroblastoma cells. J Neuropathol Exp Neurol 49:383
37. Heidenreich F, Leifeld L, Jovin T (1994) T cell-dependent activity of ganglioside GM1-specific B cells in Guillain-Barré syndrome and multifocal motor neuropathy in vivo. J Neuroimmunol 49:97
38. Heiman-Patterson T, Krupa T, Thompson P, Nobile-Orazio E, Tahmoush AJ, Shy ME (1993) Anti-GM1/GD1b M-proteion damage human spinal cord neurons co-cultured with muscle. J Neurol Sci 120:38
39. Henderson CE, Hauser SL, Hutchet M, Dessi F, Hentati F, Taguchi T, Changeux JP, Fardeau M (1987) Extracts of muscle biopsies from patients with spinal muscular atrophies inhibit neurite outgrowth from spinal neurons. Neurology 37:1361
40. Juergens SM, Kurland LT, Okazaki H, Mulder DW (1980) ALS in Rochester, Minnesota, 1927–1977. Neurology 30:463
41. Kaji R, Oka N, Tsuji T, Mezaki T, Nishio T, Akiguchi I, Kimura J (1993) Pathological findings at the site of conduction block in multifocal motor neuropathy. Ann Neurol 33:152
42. Kaji R, Hirota N, Oka N, Kohara N, Watnabe T, Nishio T, Kimura J (1994) Anti-GM1 antibodies and impaired blood-nerve barrier may interfere with remyelination in multifocal motor neuropathy. Muscle Nerve 17:108
43. Kawamata T, Akiyama H, Yamada T, McGeer PL (1992) Immunologic reactions in amyotrophic lateral sclerosis bain and spinal cord tissue. Am J Pathol 140:691
44. Kelemen J, Hedlund W, Orlin JB, Berkman EM (1983) Plasmapheresis with immunosuppression in amyotrophic lateral sclerosis. Arch Neurol 40:752
45. Kiernan JA, Hudson AJ (1993) Anti-neurone antibodies are not characteristic of amyotrophic lateral sclerosis. Neuroreport 4:427
46. Kimura F, Smith RG, Delbono O, Nyormoi O, Schneider T, Nastainczyk W, Hofmann F, Setfani E, Appel SH (1994) Amyotrophic lateral sclerosis patient antibodies label Ca^2+ channel α_1 subunit. Ann Neurol 35:164
47. Koch HHR (1878) Untersuchungen über die Ätiologie der Wundinfectionskrankheiten
48. Kornberg AJ, Prestronk A (1994) The clinical and diagnostic role of anti-GM1 antibody testing. Muscle Nerve 17:100
49. Lampson LA, Kushner PD, Sobel RA (1990) Major histocompatibility complex antigen expression in the affected tissues in amyotrophic lateral sclerosis. Ann Neurol 28:365
50. Lange DJ, Trojaborg W, Latov N, et al (1992) Multifocal motor neuropathy with conduction block: is it a distinct clinical entity? Neurology 42:497
51. Lange DJ, Trojaborg W (1994) Do GM1 antibodies induce demyelination? Muscle Nerve 17:105
52. Lee SC, Liu W, Dickson DW, Brosnan CF, Berman JW (1993) Cytokine production by human fetal microglia and astrocytes. Differential induction by lipopolysaccharide and Il-1 beta. J Immunol 150:2659
53. Li F, Pestronk A (1991) Autoantibodies to GM1 ganglioside: different reactivity to GM1-liposomes in amyotrophic lateral sclerosis and lower motor neuron disorders. J Neurol Sci 104:209
54. Louis JC, Magal E, Takayama S, Varon S (1993) CNTF protection of oligodendrocytes against natural and tumour necrosis factor-induced death. Science 259:689
55. Magnelli V, Sawada T, Delbono O, Smith RG, Appel SH, Stefani E (1993) Amyotrophic lateral sclerosis immunoglobulins action on mammalian single skeletal muscle Ca^2+ channels. J Physiol (Lond) 461:103
56. McGeer PL, Akiyama H, Itagaki S, McGeer EG (1989) Immune system response in Alzheimer's Disease. Can J Neurol Sci 16:516

57. McLaurin J, D'Souza S, Stewart J, Blain M, Antel JP (1995) Cytokine effect on human CNS-derived neural cells. Int J Dev Neurosci (in press)
58. Mitsumoto H, Hanson MR, Chad DA (1988) Amyotrophic lateral sclerosis: recent advances in pathogenesis and therapeutic trials. Arch Neurol 45:189
59. Murayama S, Inoue K, Kawakami H, Bouldin TW, Suzuki K (1991) A unique pattern of astrocytosis in the primary motor areas in amyotrophic lateral sclerosis. Acta Neuropathol 82:456
60. Nakao T, Kon K, Ando S, Miyatake T, Yuki N, Li YT, Fuyura S, Hirabayashi Y (1993) Novel lacto-ganglio type gangliosides with GM2-epitope in bovine brain which react with IgM from a patient of the amyotrophic lateral sclerosis-like disosrder. J Biol Chem 268:21028
61. Nobile-Orazio E, Meucci N, Barbieri S, Carpo M, Scarlato G (1993) High-dose intravenous immunoglobulin therapy in multifocal motor neuropathy. Neurology 43:152
62. Noble PG, Antel JP, Yong VW (1994) Astrocytes and catalase prevent the toxicity of catecholamines to oligodendrocytes. Brain Res 633:83
63. Parry GJ (1994) Antiganglioside antibodies do not necessarily play a role in multifocal motor neuropathy. Muscle Nerve 17:97
64. Posner JB (1989) Paraneoplastic syndromes. Curr Neurol 9:245
65. Rosen DE, Siddique T, Patterson D, et al (1993) Mutations in CuZn superoxide dismutase gene are associated with familial amyotrophic lateral sclerosis. Nature 362:59
66. Rowland LP (1991) Ten central themes in a decade of ALS research. Adv Neurol 56:3
67. Sakoda S, Azuma T, Mizuno R, Tsujino S, Kishimotor S, Adachi E, Arita N, Suzuki T (1991) A novel antineuronal antibody in serum and CSF of a patient with motor neuron disease. Eur Neurol 31:430
68. Salazar-Grueso EF, Routbort MJ, Martin J, Dawson G, Dawson G, Roos RP (1990) Polyclonal IgM anti-GM1 ganglioside antibody in patients with motor neuron disease and variants. Ann Neurol 27:558
69. Santoro M, Thomas FP, Fin ME, Lange DJ, Uncini A, Wadia NH, Latoov N, Hays AP (1990) IgM deposits at nodes of Ranvier in a patient with amyotrophic lateral sclerosis, anti-GM1 antibodies, and multifocal motor conduction block. Ann Neurol 28:373
70. Schauf CL, Antel JP, Arnason BGW, Davis A (1980) Neuroelectric blocking activity and plasmapheresis in amyotrophic lateral sclerosis. Neurology 30:1011
71. Sebire G, Emilie D, Wallon C, Hery C, Devergne O, Delfraissy JF, Galanaud P, Tardieu M (1993) In vitro production of IL-6, IL-1beta, and tumor necrosis factor-alpha by human embryonic microglial and neural cells. J Immunol 150:1517
72. Selmaj K, Raine CS, Farooq M, Norton WT, Brosnan CF (1991) Cytokine cytotoxicity against oligodendrocytes. Apoptosis induced by lymphotoxin. J Immunol 147:1522
73. Sharif SF, Hariri RJ, Change VA, Barie PS, Wang RS, Shajar JB (1993) Human astrocyte production of tumour necrosis factor-alpha, interleukin-1 beta, and interleukin-6 following exposure to lipopolysaccharide endotoxin. Neurol Res 15:109
74. Shaw IT, Cashman NR (1993) CNTF inhibits apoptosis in serum-deprived NSC-34 motor neuron hybrid cells. Soc Neurosci Abstr 19:416
75. Shaw IT, Boulet S, Wong E, Cashman NR (1990) Unique immunologic determinants of neuroblastoma-spinal cord (NSC) hybrid cells. Neurology 40[Suppl 1]:273
76. Shea TB (1994) Toxic and trophic effects of glial-derived factors on neuronal cultures. Neuroreport 5:797
77. Shy ME, Rowland LP, Smith T, Trojaborg W, Latov N, Sherman W, Pesce MA, Lovelace RE, Osserman EF (1986) Motor neuron disease and plasma cell dyscrasia. Neurology 36:1429
78. Smith RG, Hamilton S, Hofmann F, Schneider T, Nastainczyk W, Birnbaumer L, Stefani E, Appel SH (1992) Serum antibodies to L-type calcium channels in patients with amyotrophic lateral sclerosis. N Engl J Med 327:1721
79. Smith RG, Alexianu ME, Crawford G, Nyormoi O, Stefani E, Appel SH (1993) Cytotoxicity of immunoglobulins from amyotrophic lateral sclerosis patients on a hybrid motor neuron cell line. Proc Natl Acad Sci USA 91:3393
80. Stevens A, Wellwe M, Wietholter H (1993) A characteristic ganglioside antibody pattern in the CSF of patients with amyotrophic lateral sclerosis. J Neurol Neurosurg Psychiatry 56:361
81. Tan E, Lynn DJ, Amato AA, Kissel JT, Rammohan KW, Sahenk Z, Warmolts JR, Jackson CE, Barohn RJ, Mendell JR (1994) Immunosuppressive treatment of motor neuron syndromes. Attempts to distinguish treatable disorder. Arch Neurol 51:194

82. Thomas F (1995) Antibodies to GM1 and Gal (β1-3) GalNAc at the nodes of Ranvier in human and experimental autoimmune neuropathy. Microscop Res Tech (in press)
83. Troost D, Van den Oord JJ, Vianney de Jong J (1990) Immunohistochemical characterization of the inflammatory infiltrate in amyotrophic lateral sclerosis. Neuropathol Appl Neurobiol 16:401
84. Troost D, Das PK, van den Oord JJ, Louwerse ES (1992) Immunohistological alterations in muscle of patients with amyotrophic lateral sclerosis: mononuclear cell phenotypes and expression of MHC products. Clin Neuropathol 11:115
85. Troost D, Claessen N, van den Oord JJ, Swaab DF, De Jong JM (1993) Neuronophagia in the motor cortex in amyotrophic lateral sclerosis. Neuropathol Appl Neurobiol 19:397
86. Tsai CP, Lin KP, Liao KK, Wang SJ, Wang V, Kao KP, Wu ZA (1993) Immunosup-pressive treatment in lower motor neuron syndrome with autoantibodies against GM1 ganglioside. Eur Neurol 33:446
87. Ulvestad E, Williams K, Vedeler C, Antel J, Nyland H, Mork S, Matre R (1994) Reactive microglia in multiple sclerosis lesions have an increased expression of receptors for the Fc part of IgG. J Neurol Sci 121:125
88. Ulvestad E, Williams K, Matre R, Nyland H, Olivier A, Antel J (1994) Fc receptors for IgG on cultured human microglia mediate cytotoxicity and phagocytosis of antibody-coated targets. J Neuropathol Exp Neurol 53:27
89. Ulvestad E, Williams K, Mrøk S, Antel J, Nyland H (1994) Phenotypic differences between human monocytes/macrophages and microglial cells studied in situ and in vitro. J Neuropathol Exp Neurol 53:492
90. Uncini A, Santoro M, Corbo M, Lugaresi A, Latov N (1993) Conduction abnormalities induced by sera of patients with multifocal motor neuropathy and anti-GM1 antibodies. Muscle Nerve 16:610
91. Voumvourakis C, Rombos A, Konstantoulakis MM, Sedditsa I, Papageorgiou C (1992) Serum anti-GM1 and anti-GD1a antibodies in patients with motor neuron disease. Acta Neurol Scand 86:599
92. Weiner LP, Stohlman SA, Davis RL (1980) Attempts to demonstrate virus in amyotrophic lateral sclerosis. Neurology 30:1319
93. Werdelin L, Boysen G, Jensen TS, Morgensen O (1990) Immunosuppressive treatment of patients with amyiotrophic lateral sclerosis. Acta Neurol Scand 82:132
94. Westarp ME, Westphal KP, Clausen J, Rasmussen HB, Hoff-Jorgensen R, Fohring B, Kornhuber HH (1993) Antiretroviral therapy in sporadic adult amyotrophic lateral sclerosis. Neuroreport 4:819
95. Williams C, Kozlowski MA, Hinton DR (1990) Degeneration of spinocerebellar neurons in amyotrophic lateral sclerosis. Ann Neurol 27:215
96. Williams K, Dooley N, Ulvestad E, Blain M, Yong VW, Antel JP (1995) Antigen presentation by astrocytes: correction of the inability of astrocytes to initiate immune responses by the addition of microglia or the microglia-derived cytokine IL-1. J Neurosci 15:1869
97. Williams KC, Ulvestad E, Antel JP (1994) Immune regulatory and effector properties of human microglia in vitro and in situ. Adv Neuroimmunol 4:273
98. Williams K, Ulvestad E, Waage A, Antel JP, McLaurin J (1994) Activation of adult human derived microglia by myelin phagocytosis in vitro. J Neurosci Res 38:433
99. Willison HJ, Paterson G, Kennedy PG, Veitch J (1994) Cloning of human anti-GM1 antibodies from motor neuropathy patients. Ann Neurol 35:471
100. Wirgin I, Branner T, Argov Z, Steiner I (1992) Multifocal motor nerve conduction abnormalities in amyotrophic lateral sclerosis. J Neurol Sci 112:199
101. Wright S, Sternberg H, Bjornskov EK, Stephenson DT, Kushner PD (1990) ALS and Family of human neuronal external surface epitopes defined by Torpedo monoclonal antibodies. J Neurosci Res 25:486
102. Yoshino H, Miyatani N, Saito M, Ariga T, Lugaresi A, Latov N, Kushi Y, Kasama T, Yu RK (1992) Isolated bovine spinal motoneurons have specific ganglioside antigens recognized by sera from patients with motor neuron disease and motor neuropathy. 59:1684
103. Younger DS, Rowland LP, Latov N, et al (1991) Lymphoma, motor neuron diseases, and amyotrophic lateral sclerosis. Ann Neurol 29:78

Guillain-Barré syndrome: clinical and immunological aspects

A.M. Rostami

Department of Neurology, Hospital of the University of Pennsylvania, 3400 Spruce Street, Philadelphia, PN 19104, USA

Historical aspects

Guillain-Barré syndrome (GBS), as we recognize it today, was first described by Landry in 1859 [32] and for years was known as Landry's ascending paralysis. In 1916, Guillain, Barré and Strohl [23], using the newly developed technique of lumbar puncture, described albuminocytological dissociation, i.e., increased albumin with very few or no cells, in the cerebrospinal fluid of patients with Landry's ascending paralysis and helped to distinguish the disorder from other paralytic disorders especially poliomyelitis. Over the next several years, Landry's ascending paralysis with new diagnostic features in the cerebrospinal fluid (CSF) became kown as Guillain-Barré syndrome.

The first extensive clinicopathological study of GBS was made by Haymaker and Kemohan in 1949 [32]. They studied 50 fatal cases of the disorder from the U.S. armed forces during World War II. They reported that only edema was present for the first several days after the clinical disease, followed by breakdown of the myelin sheath and some axonal degeneration. These investigators thought that lymphocytic infiltrations in the nerve appeared after 9 days into the illness and believed that they were not involved in demyelination and acted as part of the reparative process.

The earliest evidence to suggest an autoimmune basis for demyelination in GBS came from the work of Waksman and Adams in 1955 [90]. They described experimental allergic neuritis (EAN), a paralytic illness in rodents induced by immunization with peripheral nerve tissue in complete Freund's adjuvant. The peripheral nerves of these animals had segmented demyelination associated with mononuclear cell infiltration. They pointed out the analogy between EAN and GBS and suggested an autoimmune basis for the latter as well.

In 1969, Asbury et al. [6] reported on the clinical and autopsy findings of 19 patients with GBS. They reported mononuclear cell infiltrations in the nerves of all patients as early as 30 h after the onset of the symptoms. They also noted close analogy with EAN. Electron microscopy of GBS by Prineas [60] demonstrated the role of activated macrophages in peripheral nerve system (PNS) demyelination. Recognition of humoral factors capable of inducing PNS demyelination in the serum of GBS patients [46, 47, 76, 78] and experimental animals [77] led to the clinical trial and acceptance of plasmapheresis as a treatment modality for GBS.

Understanding the immunology of EAN has helped elucidate some of the immune mechanisms in GBS. Future treatment modalities will most likely arise from studies on the treatment of this experimental model. Three recent monographs [35, 57, 64], and proceedings of a symposium [5] on GBS are valuable sources of information on this disease.

Clinical aspects

GBS is the most common acquired demyelinating neuropathy of man. The incidence varies from 0.6 to 1.9 per 100000 per year [82] and there appears to be a regular increase in incidence with age. The disease is widely distributed throughout the world. Children and adults of both sexes are affected with males being more susceptible [38].

The disease is characterized by acute progressive and symmetrical motor weakness of the extremities and of bulbar and facial musculature. Sensory symptoms are a reduction or absence of mild and deep tendon reflexes. GBS has features of demyelination on electrophysiological studies, with the pathological findings of mononuclear cell infiltration and segmental demyelination.

About two-thirds of patients report a viral-like illness, usually a respiratory or gastrointestinal tract infection within 8 weeks of the onset of GBS. Association of GBS with vaccinations, preceding surgery, Hodgkin's disease, lymphoma, and lupus erythematosus are well known. Infection with *Campylobacter jejuni* has been reported in association with GBS [10, 42] and GBS-like syndromes are seen in patients with AIDS and Lyme disease [22, 56].

Weakness usually develops rapidly but ceases to progress after 4 weeks in more than 90% of the patients. Facial weakness, usually bilateral, occurs in approximately 50% of patients. Other cranial nerves may be involved but to a much lower frequency. Recovery usually begins 2–4 weeks after the cessation of progression but may be delayed for months. Many patients are left with a residual deficit of various severity after 1 year. The median time of recovery of independent walking is about 85 days and for respiration is 169 days [82]. Autonomic dysfunction such as cardiac arrhythmia, tachycardia, postural hypotension, and hypertension are seen and at times can be fatal. CSF shows albuminocytological dissociation with increased protein and no or very few cells. These changes are usually seen after the 1st week of symptoms.

Approximately 80% of patients have electrophysiological abnormalities, including reduced conduction velocity, conduction block, temporal dispersion, prolonged distal latencies, and prolonged or absent F-waves and H-reflexes. However, up to 20% of patients have normal conduction studies. Electrophysiological parameters may stay normal until several weeks into the illness. Conditions such as hexacarbon abuse, porphyria, diphtheria, lead poisoning, botulism, toxic neuropathies, hysteria, tick paralysis, and poliomyelitis may occasionally be confused with GBS and need to be ruled out. We have reported three patients with GBS and magnetic resonance imaging evidence of gadolinium enhancement of the cauda equina and lumbar nerve roots. Similar enhancement has been observed in patients with chronic inflammatory demyelinating polyneuropathy and suggests proximal nerve inflammation. Magnetic resonance imaging in GBS and chronic inflammatory demyelinating polyneuropathy may have diagnostic utility [11, 12].

An ad hoc National Institute of Neurological Disorders and Stroke (NINDS) committee established guidelines for the diagnosis of GBS [4]. Although these guidelines are somewhat restrictive and exclude certain entities, such as Miller-Fisher syndrome of ophthalmoplegia, ataxia and areflexia, sensory loss and areflexia, polymyelitis cranialis, and pandysautonomia, they are every useful for diagnosing GBS, especially to exclude conditions that mimic this disorder.

Peripheral nervous system structure and antigenes

The PNS, compared to the central nervous system, (CNS), is a relatively simple structure. Myelinated and unmyelinated nerve fibers are surrounded by perineurium which form a nerve bundle or fascicle. Several nerve fascicles are wrapped in epineurium comprised primarily of collagen fibers and fibroblasts. The interstitial connective tissue within nerve fascicles is termed endoneurium. The normal endoneurium contains a few cell types including fibroblasts, myelinating and nonmyelinating Schwann cells, resident macrophages, and mast cells. The PNS myelin is an extension of Schwann cell plasma membrane. It is comprised of a lipid bilayer which constitutes 70–80% of its dry weight. The remaining 20–30% of dry weight is protein [54]. Galactocerebroside, sulfatides, and gangliosides are immunologically important lipids of the myelin. Major proteins of myelin are P_0, P_1, P_2, and myelin-associated glycoprotein (MAG) [49]. P_0 is a glycoprotein with a molecular mass of 28 000 Da and is capable of inducing EAN in experimental animals. P_1 and P_2 are basic proteins. P_1 is analogous to th CNS myelin basic protein with a molecular mass of 18 500 Da. P_2 protein is comprised of 131 amino acids and has a molecular weight of 14 500 Da [7]. It is the major protein responsible for the induction of EAN. MAG has a molecular mass of 100 000 Da and is a relatively minor glycoprotein comprising about 1% of myelin protein [61]. It is related structurally to the immunoglobulin super gene family and is present in both CNS and PNS myelin. The first few millimeters of the spinal nerve roots, after emerging from the spinal cord, are immersed in the spinal subarachnoid space and are in contact with the CSF. Therefore, inflammatory changes in the nerve roots may be reflected in the CSF.

Peripheral nerves are immunologically privileged sites. The rich blood supply of the nerve is impermeable to many macromolecules and constitutes the blood-nerve barrier (BNB). This barrier is formed by tight intercellular junctions of the capillary endothelium of the endoneurium and its surrounding basement membrane [55]. The BNB is more permeable in the region of the dorsal root ganglia. Peripheral nerves lack a lymphatic system; and resident macrophages, and to a lesser extent the Schwann cells, perform the phagocytic activities in the nerve.

Etiology and pathogenesis

The etiology and pathogenesis of GBS are not well understood. There is strong evidence that demyelination in the peripheral nerves is caused by an immunological attack on PNS myelin or Schwann cells. However, an infectious etiology for GBS, either direct destruction of myelin or Schwann cells by an infectious agent or a primary infection of the nerve followed by a secondary immune reaction leading to demyelination, is under investigation. Since many infectious agents such as cytomegalovirus (CMV), Epstein-Barr virus,. *C. jejuni*, and non-infectious conditions such as surgery, fever therapy, and malignancies may trigger an infectious reactivation, it is more probable that the final attack on the myelin is an immunological mechanism triggered by an infection or other conditions. The exact mechanism for how an infectious agent initiates an immune reaction that results in the disease is not understood. Antigenic mimicry between the infectious agent and myelin antigens is a popular hypothesis. Similarities to the animal model, EAN, are strong evidence for the immunological basis of myelin destruction in GBS.

Based on the observation that the herpes viruses, CMV, and Epstein-Barr virus are associated with GBS, viruses have been thought to be possible etiological agents for the disorder. Indeed, for some time the disorder was called acute infective polyneuritis [7]. A herpes virus of birds, known as Marek's disease virus, produces a lymphoproliferative disorder in chickens. This disease can be associated with paralysis of the wings and legs in chickens and a nerve pathology similar to GBS [52]. A condition similar to GBS in patients with HIV infection and high titer of antibodies to *C. jejuni* or mycoplasma is also taken as evidence that bacteria and viruses, directly or indirectly through activation of elements in the immune system, may be involved in the pathogenesis of GBS.

Association of *C. jejuni* infection and GBS was first reported by Kaldor and Speed in 1984 [42]. They reported antecedent *C. jejuni* infection in 38% of a group of the GBS patients in Australia. In a recent U.S. study, 17.2% of GBS patients had antibodies to *C. jejuni* compared to 7% of the controls [89]. In this study poor recovery was associated with serological evidence of recent *C. jejuni* infection. However, there was no correlation between *C. jejuni* infection and antibodies to GM1 or GD1b gangliosides. The authors concluded that antibodies to GM1 or GD1b do not necessarily mediate the extensive axonal damage seen in these severely affected patients. Other investigators have not found a poor prognosis in GBS associated with *C. jejuni* [13].

An epidemic of acute motor axonal neuropathy (AMAN) in China has recently been reported [53]. The patients were mainly children and young adults and mainly in rural arras. The clinical and CSF manifestations are consistent with a diagnosis of GBS; however, the electrodiagnostic features indicate a motor axonal neuropathy. Pathological studies on a few of these patients showed extensive motor nerve fiber degeneration and a striking chromatolysis of anterior horn cells without inflammation. In other cases, pathological changes were those of predominantly demyelinating process with focal lymphocytic infiltrates similar to classical demyelinating GBS. A similar paralytic disorder was reported in 1969 by Ramos-Alvarez et al. [62] in Mexico. A recent pathological examination of the materials from these Mexican patients showed striking similarity to those from the Chinese AMAN patients [33].

In one Japanese study [75] *C. jejuni* infection, especially of Penner serogroup 19, was associated with about half of the Japanese patients with GBS. Among those patients with *C. jejuni* infection, GM1 antibody titer was elevated in 68%. Molecular mimicry between the GM1 oligosaccharide and the carbohydrate structure of the LPS from *C. jejuni* (PEN 19) has been reported [93].

In the fall of 1976, a GBS epidemic occurred following immunization with the A/New Jersey A influenza vaccine. Peak incidence of GBS was seen 2–3 weeks post vaccination. By mid-December, 1100 cases of GBS had been reported to the centers for Disease Control, 10–20 times the rate of the disease in unvaccinated individuals.

The differences between the disease rate in vaccinated and unvaccinated groups remained significant for up to 8 weeks after vaccination. Originally, this observation was thought to be an artifact. Further review, however, indicated a true association between GBS and the vaccine [5].

The landmark work of Asbury et al. [6] on the pathology of GBS and the finding of mononuclear infiltration in the peripheral nerves in the areas of segmental demyelination similar to that of EAN, focused the attention of workers in the field on the possible immunological basis for myelin injury. A variety of immunological abnormalities have been reported in GBS. Cell-mediated immunity to P2 protein has been reported in GBS [1, 79] but not confirmed by others [40, 94]. Reports that GBS peripheral blood lymphocytes can demyelinate myelinated cultures in vitro [3] cannot be taken as evidence of the cytotoxic capability of these cells due to the differences in the major histocompatibility complex of effector cells and target tissues. Lymphokines or antibodies secreted by T and B cells or nonspecific killer cell activity may be the explanation for these observations. There are reports of changes in T cell subsets in the peripheral blood showing a decrease in either CD4 (T helper/inducer) or CD8 (T suppressor/cytotoxic) cells in GBS [37, 51], but others have reported no changes in the T cell subsets [31]. Serum interleukin-2 (IL-2) concentration is increased in GBS and correlates with the activity of the disease, suggesting an ongoing T cell activation and proliferation in these patients [29]. Serum levels of soluble IL-2 receptors and higher proportions of peripheral blood lymphocytes expressing transferrin and IL-2 receptors [28] are also indicative of the active role of T cells in GBS. These T cells can exert their action by helping B cells to secrete autoantibodies and to produce harmful cytokines to injure the myelin sheath and or Schwann cells either direcly or through the activation of the macrophages. Despite these observations, no clear cut role for T cells in the pathogenesis of GBS is documented.

The antigen-presenting cells in human peripheral nerves are thought to be Schwann cells [59]; however, data from EAN indicate that resident macrophages rather than to Schwann cells are the antigen-presenting cells.

Antibodies to various peripheral nerve components are represented in GBS. On the basis of the observation that anti-galactocerebroside antisera can demyelinate rat peripheral nerves in vivo, antibodies to galactocerebroside have been sought in serum and CSF from patients with various demyelinating disorders, including GBS, chronic demyelinating inflammatory polyneuropathy, multiple sclerosis, and various other neurological disorders using various assays, including radioimmunoassay and enzyme-linked immunosorbent assay (ELISA). No humoral immune response to galactocerebroside in human demyelinating disorders was observed [73]. Serum from acute GBS can produce demyelination when injected into the rat sciatic nerve [78]. Various antibodies to peripheral nerve components and circulating immune complexes have been reported in patients with GBS [16, 47]. Oligoclonal bands have been seen in the CSF of GBS, but were usually transient unlike those seen in cases of multiple sclerosis. High titers of antibodies including complement fixing antibodies to human peripheral nerve myelin have been demonstrated in the serum of GBS patients compared to normals and to those found in a variety of inflammatory neurological and non-neurological disorders. The target antigen(s) for these antibodies are not well defined, but neutral glycolipids cross-reactive with Forssman antigen [46], gangliosides [39], and sulfoglucoronyl glycolipids [92] have been suggested as possible candidates. These observations on the role of antibodies in GBS and EAN were the basis for the use of, and subsequent demonstration of efficacy of, plasmapheresis in GBS. Nearly all patients with Miller Fisher syndrome, as well as GBS patients with ophthalmoplegia, have IgG antibodies to GQ_{1b} ganglioside in their sera [9].

Various cytokines and adhesion molecules have been reported to be associated with GBS. Griffin and co-workers [20, 21] have used a combination of reverse transcriptase-polymerase chain reaction (RT-PCR) for cytokine and growth factor transcripts, immunocytochemistry, and standard pathological techniques to examine expression of selected cytokines and nerve growth factor in the nerve tissue from GBS patients. They reported prominent T lymphocyte infiltrates early in two cases of GBS with the expression of interferon-γ (IFN-γ) transcripts, as well as the tumor necrosis factor (TNF) and IL-1. In some demyelinated regions showing little Wallerian degeneration prominent nerve growth factor (NGF) transcripts were seen. Some of the early cases with an axonal pattern of neuropathy had only mild demyelination, prominent Wallerian degeneration, and very sparse lymphocytic infiltrates. There was an excellent correlation between IFN transcripts and the presence of T cells in the nerve segments.

In one study, serum levels of TNF-α were elevated in GBS compared to that found in other CSF polyneuropathies. TNF-α was not detected in the CSF of GBS or other neuropathies. TNF-α level correlated directly with disease severity, and these concentrations returned to normal in parallel with clinical recovery. It is believed that TNF-α may be important in the pathogenesis of demyelination in GBS [83]. Serum levels of E-selectin were increased in GBS [30].

Animal models

Experimental allergic neuritis

EAN has been induced in various laboratory animals by immunization with whole nerve, PNS myelin, myelin P_2 protein, some peptides of P_2 protein [8, 71, 84, 90] and galactocerebroside [77]. A synthetic peptide corresponding to residues 53–78 of P_2 protein (SP26) has also been shown to produce EAN in Lewis rats [74]. Lymph node cell populations from SP26 immunized rats elicited a proliferation response to the peptide and to the P_2 protein. These cells were mainly $CD4^+$ cells. About 2 weeks after active immunization, the animals developed weight loss, flaccid tails, ataxic paraplegia, tetraplegia. The majority of the animals recovered from the disease after about 1 weak.

Electrophysiological studies have shown evidence of a slowing in the sciatic motor conduction starting 10 days after immunization, with maximum slowing at about 24 days post immunization. Dispersion of action potentials, conduction block and prolongation of F-wave latencies indicative of demyelination were observed [71]. The pathological changes of EAN in Lewis rats consist of multifocal areas of primary demyelination, also seen in ganglia, roots and sciatic nerves. Demyelinative lesions are often associated with axonal degeneration. The presence of large diameter, thinly myelinated axons indicates recent remyelination. Multifocal mononuclear cell infiltration frequendy occurs in a perivenular distribution; in some cases these inflammatory changes appear to be unrelated to focal demyelination. Demyelinating and inflammatory lesions are less extensive and more variable in P_2-immunized rats, compared to myelin-immunized animals. In general, primary demyelination is most evident in dorsal and ventral roots and less in the sciatic nerves. No lesions are found in the spinal cord or the brain tissue. Collections of inflammatory cells are first identified in nerve roots 14 days after immunization. Ia-positive cells predominate in the evolving lesions and T helper cells are the dominant tissue type with T suppressor cytotoxic cells appearing later in the course of the disease. Vesiculation, the earliest change seen in the myelin sheath, and endoneural edema can be observed in the absence of mononuclear cell contact [65]. Macrophage-associated myelin stripping is detected on day 12 post-immunization. Macrophage infiltrations are extensive by day 14. $CD4^+$ and $CD8^+$ cells to start to appear from day 14 post immunization and by day 19 more $CD8^+$ than $CD4^+$ cells are observed. The pattern of EAN induction by SP26 is similar to that of the P_2 protein; however, the disease can be more severe.

EAN in Lewis rats can be transferred with P_2-reactive T cells [50, 70, 72]. These cell lines can induce EAN within 7 days following adoptive transfer to normal Lewis rats. The pathology has been shown to be very similar to actively induced EAN with mononuelear cell infiltration and demyelination. There was no anti-P_2 antibody response in the recipients of the P_2-incubated cells from donors with high anti-P_2 antibody levels, suggesting that transfer of EAN in this model is not simply a result of active immunization of the recipients by P_2 bound to the cultured cells. T cells reactive to the peptide 53–78 of P_2 protein (SP26) were also able to transfer EAN [68]. As few as 2×10^6 SP26-reactive T cells injected intraperitoneally, transfered server EAN. The disease appeared 7-8 days post inoculation of the cells and persisted feamms 5-10 days. The observed pathological features are indistinguishable from SP26-induced active EAN which appears 10-15 days after immunization. The cells capable of transfering the disease are mainly $CD4^+$, Ia^+ cells. These observations provided direct evidence that EAN is a cell-mediated immune phenomenon.

The nature of the antigen-presenting cells in the PNS is not well understood. Schwann cells from EAN rats have class II antigen on their surface and present antigen to myelin basic protein reactive T cells in vitro [91]. However, resident macrophages are present in normal nerves and are most likely the antigen-prresenting cell in the PNS. EAN can be successfully transferred with P_2-reactive T cells of Lewis origin into DA rats that have been lethally radiated and reconstituted with (Lewis x DA) F1 bone marrow cells [44]. This experiment suggests that Schwann cells are probably not necessary for antigen presentation in the peripheral nerves. Schwann cells in the peripheral nerves do not express Ia molecules on their surfaces. Double-labeled experiments using antibodies to the S-100 protein of Schwann cells and Ia molecules have clearly demonstrated that Ia^+ cells are not Schwann cells and are most likely $ED1^+$ macrophages [69]. Macrophages are believed to be the cells responsible for destruction of myelin sheath, either directly or by secreting proteins and inflammatory mediators [25, 80].

A paralytic disorder has been produced in rabbits by immunization with galactocerebroside [76, 77, 85]. This form of EAN had a distinct pathology compared to myelin or P_2-induced EAN. Perivenal infiltration of the small lymphocytes was not seen, but macrophages insinuated themselves between the myelin lamellae and phagocytized them. Rabbits with EAN have high titers of antibodies to galactocerebroside and their serum can transfer demyelination by intraneural injection of rat sciatic nerve, indicating that the disorder is most likel mediated by antibodies to galactocerebroside.

The role of complement in the pathogenesis of EAN is not well understood; however, terminal complement complex (C5B9) deposition on Schwann cells and along the myelin sheath has been observed 11 days after immunization which is prior to demyelination [86]. It has been shown that the cobra venom factor, which can decomplement animals, can suppress EAN [14].

Various cytokines are implicated in the padiogenesis of demyelination in CNS and PNS. TNF-α-positive macrophages are seen in EAN induced by active immunization or by adoptive transfer of autoreactive T cells. These macrophages pass through blood vessels or are adherent to nerve fibers. Large phagocytic macrophages are seen at later stages of demyelination and TNF-α-negative cells appear in the lesion [26]. Intraperitoneal application of anti-TNF-α, antibody to EAN rats has been shown to reduce significantly the degree of inflammatory demyelination, suggesting a pathogenic role for TNF-α. IFN-γ predominandy produced by the TH1 subtype of $CD4^+$ T lymphocytes and has various inflamatory effects. It has been shown that IFN-γ can upregulate intracellular adhesion molecule (ICAM) and major histocompatibility complex (MHC) class II on macrophages and microglia and enhance the release of oxygen radicals from macrophages [15], and IFN-γ has been carefully studied in EAN [81]. IFN-γ is observed in the nerves from EAN rats, mainly in the T cells, polymorphonuclear leukocytes, and ED_1^+ macrophages at inflammatory sites. They are detectable prior to onset of clinical disease but not thereafter. When recombinant IFN-γ is systemically injected into rats, it produces a more severe EAN with massive inflammatory infiltrate and MHC class II expression on macrophages and T cells. Monoclonal antibodies to IFN-γ suppresses EAN. These experiments suggest a pathogenic role for IFN-γ in EAN. [27]. Antibody to lymphocyte function-associated antigen-1 have also been shown to prevent or effectively suppress EAN [2].

Transforming growth factor-β (TGF-β1) is expressed in the reactive spinal cord microglia and infiltrating inflammatory cells in the nerve roots in EAN [43]. Systemic administration of TGF-β1 markedly suppresses the clinical and histological signs of EAN and reduces the delayed-type hypersensitivity response to the inducing antigen [19].

Coonhound paralysis

This is a paralytic disease of hunting dogs. Typically, dogs after 7–14 days of being bitten by a raccoon develop subacute paralysis that may progress to death. Pathological lesions are identical to those of GBS and EAN. Although a viral etiology has been postulated, no virus has been recovered from the saliva of the raccoons, and immunization of the rabbits or guinea pigs by raccoons saliva or salivary gland has not produced any paralysis [34, 45].

Cauda equina syndrome of the horse

This is a subacutely evolving paralysis of the hind legs and at times ptosis and vocal cord paralysis in the horse. The condition is mainly seen in the sacral and coccygeal nerves and has a demyelinating feature similar to that of EAN. Antibodies to P_2 protein have been observed in some of these horses [41].

Marek's disease

Chickens infected with Marek's disease virus, an oncogenic herpes-type virus of the chicken develop a paralytic disorder similar to EAN. In one form, weakness and paralysis associated with lymphoid infiltrates, enlargement and edema of peripheral nerves accur. There are lymphocytic and mononuclear cell inflammatory infiltrates and segmental demyelination in the nerves. The majority of the cells in the nerve are T cells and they do not appear to be tumor cells. The disease can be transferred with lymphoid cells to normal chicken's. Inoculation of cultures containing the Marek's disease. Wether the virus directly damages the peripheral nerve myelin or whether it elicits an immunological attack on the myelin sheath by molecular mimicry is not known [48].

Treatment

Although about 80% of GBS patients recover from the disease, the course of the disorder can be stormy. About 30% of patients in one study required intubation for an average of 51 days, with an average hospital stay of 61 days [63]. The mainstay of therapy is the management of paralyzed patients, which is the best accomplished in intensive care units.

Plasmapheresis was found to be effective in decreasing morbidity and shortening the course of GBS. In a United States multicenter controlled study performed on 245 patients with GBS [24], patients treated with plasmapheresis had better outcome when compared in terms of time needed to improve one clinical grade, outcome at 4 weeks and 6 months, and the time at which the patient could walk unaided. Plasmapheresis also reduced the number of days on the respirator. In severe cases, especially when ventilatory assistance is needed or when the patient has a rapidly progressive course, plasmapheresis is indicated. Once the decision to treat patients with plasmapheresis has been made, it should begin as early as possible.

In one randomized clinical trial of corticosteroid treatment for GBS, no beneficial effect was observed [36]. Another study using adrenocorticotropic hormone reported significant reduction's in duration of the disease [87]. A British multicenter study with high-dose methylprednisone as one treatment arm did not show efficacy for this therapy in GBS. There is no conclusive evidence that azathioprine and cyclophosphamide can be of any help in GBS [58].

Results from a multicenter study in the Netherlands comparing high-dose intravenous gamma-globulin (IVIG) with plasma exchange in 150 patients demonstrated a better outcome in patients receiving IVIG compared to those treated with plasma exchange. They defined beneficial outcome as improvement in at least one grade on a seven-point functional scale 4 weeks after randomization [88]. Patients were enrolled into the study within 2 weeks of the onset of the disease and received a median of 140 g IVIG or 12.5 l of plasma exchange. After 4 weeks of treatment, 52% of IVIG-treated patients and 34% of those in the plasmapheresis group had improved by one or more functional grade, a highly significant difference. The median time to reach this level of improvement was 27 days in the IVIG group and 41 days in patients assigned to plasma exchange. The median time to independent locomotion was 55 days with IVIG as compared with 69 days with plasma exchange. Another multicentered clinical trial to compare IVIG with plasmapheresis and plasmapheresis followed by IVIG in GBS is in progress.

Newer approaches to therapy of autoimmunc PNS disease have been tried in EAN. These studies are designed mainly to abrogate the immune attack against self antigens by inductions of immunological tolerance or suppression of abnomally heightened immune response. Antibodies to $CD4^+$ cells have been shown to decrease the severity of EAN in Lewis rats. Antibodies to cell adhesive molecules such as ICAM and to cytokines TNF-α and IFN-γ have decreased the severity of demyelination in experimental animals. Induction of immunological tolerance to the peptide SP26, which can induce EAN, has been accomplished. Oral administration of SP26 [70] or intravenous injections of spleen cells coupled with SP26 have resulted in marked reduction of clinical EAN in Lewis rats [17, 18]. These approaches to therapy may become applicable to human GBS in the future.

Summary and conclusions

Immune-mediated PNS disorders comprise a significant segment of diseases of the nervous system. Studies on GBS as a prototype of these disorders and its experimental model EAN have helped to elucidate some of the mechanisms responsible for myelin injury in the PNS. These mechanisms, although partially understood have been useful in implementing therapies such as plasmapheresis and IVIG and various other immunomodulators. The question of whether an infectious agent such as a virus can direcdy damage the myelin sheath and/or Schwann cells or whether the agent triggers an immune response against self through antigenic mimicry remains unanswered.

The association between *C. jejuni* infection and GBS has opened new areas of investigation in understanding the immunopathogenesis of the disease [20]. Similar observations with other environmental factors may be made in the future, pointing to the possibility that GBS may not be caused by a single agent but could be the result of an immunological attack on the PNS myelin assembly by a variety of agents or factors.

Regardless of the etiology, if the myelin injury is aggravated by product of the immune cells, such as various cytokines, neutralization of these factors could help lessen the burden of injury to the nerves.

Future research in autoimmune disorders of the PNS needs to focus on identifying environmental factors that directly, or indirectly through antigen mimicry, damage the PNS myelin. In parallel, further understanding of the immunopathogenesis by dissecting the immunological phenomenon at the systemic and local levels, especially the role of cytokines, growth factors, and adhesion molecules will pave the way for more rational therapies, even if the causative factors are not known.

Studies in laboratory animals have demonstrated the efficacy of selective immunotherapy through modulation of the trimolecular complex, i.e., T cell receptor, MHC/molecule, and antigen. Immunological tolerance, presumably through deletion of autoreactive clones, clonal anergy, or active suppression, has proven effective in animals. Other modes of immunotherapy such as nonspecific depletion of T or B cells or down-regulation of activited cells have also been shown to abolish or decrease the severity of experimental autoimmune neurological disorders, including EAN. These immunotherapeutic modalities may become applicable to human autoimmune neuropathies.

Acknowledgement. This work was supported in part by grants NS08075 and AR39489 from the National Institutes of Health.

References

1. Abramsky O, Webb C, Teitelbaum D, Amon R (1975) Cell-mediated immunity for neuronal antigens in idiopathic polyneuritis and myeloradiculitis. Neurology 25:1154
2. Archelos JJ, Maurer M, Jung S, Miyasaka M, Tamatani T, Toyka KV, Hartung H-P (1994) Inhibition of experimental autoimmune neuritis by an antibody to the lymphocyte function-associated antigen-1. Lab Invest 70:667
3. Arnason BGW, Winkler GJ, Hadler NM (1969) Cell-mediated demyelination of peripheral nerve in tissue culture. Lab Invest 1:1
4. Asbury AK (1981) Diagnostic considerations in Guillain-Barré syndrome. Ann Neurol 9 [Suppl]:1
5. Asbury AK, Gibbs CJ (1990) Autoimmune neuropathies: Guillain Barré syndrome. Ann Neruol [Suppl] 27:79
6. Asbury AK, Arnason BG, Adams RD (1969) The inflammatory lesion in idiopathic polyneuritis: its role in pathogensesis. Medicine 48:173
7. Bradford JR, Bashford EF, Wilson JA (1919) Acute infective polyneuritis. Q J Med 10:88
8. Brostoff SW, Burnett P, Lampert PW, Eylar EH (1972) Isolation and characterization of a protein from sciatic nerve myelin responsible for experimental allergic neuritis. Nature 235:210
9. Chiba A, Kusunoki S, Obata H, Machinami R, Kanazawa I (1993) Serum anti-GQ_{1b} IgG antibody is associated with ophthalmoplegia in Miller Fisher syndrome and Guillain-Barré syndrome: clinical and immunohistochemical studies. Neurology, 43:1911
10. Cook SD, Dowling PC, Blumberg BM (1987) Infection and autoimmunity in Guillain-Barré syndrome. In: Aarli AJ, Behan WMH, Behan PO (eds) Neuroimmunology. Blackwell, Oxford, pp 225–244
11. Crino PB, Grossman RI, Rostami AM (1993) Magnetic resonance imaging of the cauda equina in chronic inflammatory demyelinating polyneuropathy. Ann Neurology 33:311
12. Crino PB, Zimmerman R, Laskowitz D, Raps EC, Rostami AM (1994) Magnetic resonance imaging of the cauda equina in Guillain-Barré syndrome. Neurology 44:1334
13. Enders U, Karch H, Yoyka KV, Michels M, Zielasek J, Pette M, Heesemann J, Hartung H-P (1993) The spectrum of immune responses to *Campylobacter jejuni* and glycoconjugates in Guillain-Barré syndrome and in other neuroimmunological disorders. Ann Neurol 34:130
14. Feasby TE, Gilbert JJ, Hahn AF, Neilson M (1987) Complement depletion suppresses Lewis rat experimental allergic neuritis. Brain Res 419:97
15. Fierz W, Endler B, Reske I, Werkerle H, Fontana A (1992) Astrocytes as antigen-presenting cells. I. Induction of Ia antigen expression on astrocytes by T cells via immune interferon and its effect on antigen presentation. J Immunol 134:3785
16. Goust JM, Chenais F, Cames J. Hames CG, Fudenberg HH, Hogan EL (1978) Abnormal T cell subpopulations and circulating immune complexes in Guillain-Barré syndrome and multiple sclerosis. Neurology, 28:421
17. Gregorian SK, Rostami AM (1994) Delayed-type hypersensitivity response in experimental autoimmune neuritis treated with peptide-coupled spleen cells. J Neuroimmunol 51:69
18. Gregorian SK, Clark L, Heber-Katz E, Amento EP, Rostami AM (1993) Induction of peripheral tolerance with peptide-specific anergy in experimental autoimmune neuritis. Cell Immunol 150:298
19. Gregorian SK, Lee WP, Beck LS, Rostami A, Amento EP (1994) Regulation of autoimmune neuritis by transforming growth factor-beta 1. Cell Immunol 156:102
20. Griffin JW, Ho TW-H (1993) The Guillain-Barré syndrome at 75: the Campylobacter connection. Ann Neurol 34:125
21. Griffin JW, Macko C, Li CY, Ramos-Alvarez M, Ho T, Asbury AK (1994) New perspectives on the pathology of Guillain-Barre syndrome. Muscle Nerve [Suppl]1:S73
22. Gruenewald R, Ropper AH, Lior H, Chan J. Lee R, Molinaro VS (1991) Serologic evidence of *Campylobacter jejuni*/coli enteritis in patients with Barre syndrome. Arch Neurol 48:1080
23. Guillain G, Barre JA, Strohl A (1916) Sur un syndrome de radiculonevrite avec hyperalbuminose du liquide cephalorachidien sans reaction cellulaire: remarques sur les chracteres cliniques et graphiques des reflexes tendineux. Bull Soc Med Hop Paris 40:1462

24. Guillain-Barre Syndrome Study Group (1985) Plasmapheresis and acute Guillain-Barre syndrome. Neurology 35:1096
25. Hart NH, Hanks DT, Mackay A (1972) Ultrastructural observations in Guillain-Barre syndrome. Arch Pathol 93:552
26. Hartung HP, Heininger K, Schafers B. Fierz W, Toyka KV (1988) Immune mechanisms in inflammatory polyneuropathy. Advances in neruoimmunology. Ann NY Acad Sci 540:122
27. Hartung H-P, Schafer B, van der Meide PH, Fierz W, Heininger K, Toyka KV (1990) The role of interferon-gamma in the pathogenesis of experimental autoimmune disease of the peripheral nervous system. Ann Neurology 27(3):247
28. Hartung H-P, Hughes RAC, Taylor WA, Heininger K, Reiners K, Toyka KV (1990) T cell activation in Guillain-Barre syndrome and in MS: elevated serum levels of soluble IL-2 receptors. Neurology 40:215
29. Hartung H-P, Reiners K, Schmidt B, Stoll G, Toyka KV (1991) Serum interleukin-2 concentrations in Guillain-Barre syndrome and chronic idiopatic demyelinating polyradiculoneuropathy: comparison with other neurological diseases of presumed immunopathogenesis. Ann Neurology 30:48
30. Hartung H-P, Reiners J. Michels M, Hughes RAC, Heidenreich F, Zielasek J. Enders U, Toyka KV (1994) Serum levels of soluble E-selectin (ELAM-1) in immune-mediated neuropathies. Neurology 44:1153
31. Hauser SL, Ropper AH, Perio VP, Reinherz EL, Schlossmann SF, Weiner HL (1982) T cell subsets in human autoimmune disease. Neurology 32:1321
32. Hauser W, Kernohan JW (1949) The Landry-Guillain-Barre-syndrome: a clinical pathologic report of 50 fatal cases and a review of the literature. Medicine 29:59
33. Ho T. Griffin JW, Li CY, Ramos-Alvarez R, Lui Y, Sabin AB, Cornblath DR, McKhann GM Asbury AK (1993) Acute motor axonal neuropathy in China compared to acute neuropathy in Mexico. Prodeedings of the Peripheral Nerve Study Group, 11th Biennal Meeting, Jakobsberg, Boppard, Germany, p 2
34. Holmes DF, Lahunta A de (1974) Experimental allergic neuritis in the dog and its comparison with the naturally occurring disease, coonhound paralysis. Acta Neuropathol (Berl.) 30:329
35. Hughes RAC (1990) Guillain-Barre sydrome. Springer, Berlin Heidelberg New York
36. Hughes RAC, Newsom-Davis CJ, Perkins GD, Pierce JM (1978) Controlled trial of prednisone in acute polyneuropathy. Lancet II:750
37. Hughes RA, Asian CS, Gray IA (1983) Lymphocytic subpopulations and suppressor cell activity in acute polyradiculoneuritis (Guillain-Barre syndrome). Clin Exp Immunol 51:448
38. Hurwitz ES, Holman RC, Nelson DB, Schoenberger LB (1983) National surveillance for Guillain-Barre syndrome. Neurology, 33:156
39. Ilyas AA, Willison HJ, Quarles RH, Jungalwala FB, Cornblath DR, Trapp BD (1988) Serum antibodies to gangliosides in Guillain-Barre syndrome. Ann Neurol 23:440
40. Iqbal A, Oger J-F, Amason BGW (1981) Cell-mediated immunity in idiopathic polyneuritis. Ann Neurol 9 [Suppl]:65
41. S Kadlubowski M, Ingram PL (1981) Circulating antibodies to the neuritogenic myelin protein P2 in neuritis of the cauda equina of the horse. Nature 293:299
42. Kaldor J, Speed RB (1984) Guillain-Barre syndrome and *Campylobacter jejuni*: serological study. Br Med J Clin Res Ed 288(6434):1867
43. Kiefer R, Gold R, Gehrmann J. Lindholm D,.Wekerle H, Kreutzberg GW (1993) Transforming growth factor beta expression in reactive spinal cord microglia and meningeal inflammatory cells during experimental allergic neuritis. J Neurosci Res 36:391
44. Kimura H, Silvers WK, Venture E, Honda H, Rostami AM (1989) Perivascular macrophages as the antigen presenting cells in the peripheral nervous system. Neurology 39 [Suppl]:174
45. Kingma FJ, Catcott EJ (1954) A paralytic syndrome in coonhounds. N Am Vet 35:115
46. Koski CL (1990) Characterization of complement-fixing antibodies to peripheral nerve myelin in Guillain-Barre syndrome. Ann Neurol 27 [Suppl]:S44–S47
47. Koski CL, Gratz E, Sutherland J. Mayer RF (1986) Clinical correlation with antiperipheral myelin antibodies in Guillain-Barre syndrome. Ann Neurol 19:573
48. Lambert P, Garrett R, Powen H (1977) Demyelination in allergic and Marek's disease virus induced neuritis. Comparative electron microscopic studies. Acta Neuropathol (Berl) 40:103
49. Linington C, Brostoff SW (1993) Peripheral nerve antigens. In: Dyck PJ et al. (eds) Peripheral neuropathy. Saunders, Philadelphia, vol 25, pp 404–417

50. Linington C, Izumo S, Suzuki M, Uyemura K, Meyermann R, Wekerle H (1984) A permanent rat T cell line that mediates experimental allergic neuritis in the Iewis rat in vivo. J Immunol 133:1946
51. Lisak RP, Zweiman B, Guerrero F, Moskovitz AR (1985) Circulating T cell subsets in Guillain-Barre syndrome. J Neuroimmunol 8:93
52. Marek J (1907) Multiple Nerventzündung (Polyneuritis) bei Hühnern. DTW Dtsch Tierärztl Wochenschr 15:417
53. McKhann GM, Cornblath DR, Griffin JW, Ho TW, Li CY, Jiang Z, Wu HS, Zhaori G, Liu Y, Jou LP, Liu TC, Gao CY, Mao JY, Baluser MJ, Mishu B, Asbury AK (1993) Acute motor axonal neuropathy: a frequent cause of acute flaccid paralysis in China. Ann Neurol, 33:333
54. Norton WT, Cammer W (1984), Isolation and characterisation of myelin. In: Morrel P (ed) Myelin. Plenum New York, p 147
55. Olsson Y (1984) Vascular permeability in the peripheral nervous system. In: Dyck PJ, et al (eds) Peripheral neuropathy. Saunders, Philadelphia, pp 579–597
56. Parry GJ (1988) Peripheral neuropathies associated with human immunodeficiency virus infection. Ann Neurol 23 [Suppl]:S45
57. Parry GJ (1993) Guillain-Barre syndrome. Thieme, New York
58. Pollard JD (1987) A critical review of therapies in acute and chronic inflammatory demyelinating polyneuropathies. Muscle Nerve 10:214
59. Pollard JD, Baverstock J. McLeod JG (1987) Class II antigen expression and inflammatory cells in Guillain-Barre syndrome. Ann Neurol 21:337
60. Prineas JW (1969) Acute idiopathis polyneuritis: an electron microscope study. Lab Invest 26:133
61. Quarles RH (1984) Myelin associated glycoprotein in development and disease. Devel Neurosci 6:285
62. Ramon-Alvarez M, Bessudo L, Sabin AB (1969) Paralytic syndromes associated with noninflammatory cytoplasmic or nuclear neuropathy. JAMA 207:1481
63. Ropper AH, Shahani BT (1984), Diagnosis and management of acute areflexic paralysis with emphasis on Guillain-Barre syndrome in disorders. In: Asbury AK, Gilliat R (eds) Peripheral nerve disorder. Butterworths, London Boston, pp 21–45
64. Ropper AH, Wijdicks EFM, Truax BT (1991) Guillain-Barre syndrome. Davis, Philadelphia
65. Rosen JL, Brown MJ, Hickey WF, Rostami AM (1990) Early myelin lesions in experimental allergic neuritis. Muscle Nerve 13:629
66. Rosen JL, Brown MJ, Rostami AM (1992) Evolution of the cellular response in P_2-induced experimental allergic neuritis. Pathobiology 60:108
67. Rostami AM, Cronkhite RI (1992) Prevention of experimental allergic neuritis (EAN) by oral induction of immunological tolerance. Neurology 42 [Suppl 3]:171
68. Rostami AM, Gregorian SK (1991) peptide 53–78 of myelin P_2 protein is a T cell epitope for the induction of experimental autoimmune neuritis Cell Immunol 132:1
69. Rostamin AM, Honda H (1991) Expression of major histocompatibility complex class II antigen in experimental allergic neuritis. Neurology 41:188
70. Rostami AM, Burns JB, Brown MJ, Lisak RP, Pleasure DE (1983) Transfer of experimental allergic neuritis (EAN) with P_2-reactive T cells. Ann Neurol 14:108
71. Rostami AM, Brown MJ, Lisak RP, Sumner AJ, Zweiman B, Pleasure DE (1984) The role of myelin P_2 protein in the production of experimental allergic neuritis. Ann Neurol 16:680
72. Rostami AM, Burns JB, Brown MJ, Rosen J. Zweiman B, Lisak RP, Pleasure DE (1985) Transfer of experimental allergic neuritis with P_2-reactive T-cell lines. Cell Immunol 91:354
73. Rostami AM, Eccleston PA, Silberberg DH, Manning MC, Bums JB, Lisak RP (1987) Search for antibodies to galactocerebroside in the sera and cerebrospinal fluid in human demyelinating disorders. Ann Neurol 22:381
74. Rostami AM, Gregorian SK, Brown MJ, Pleasure DE (1990) Induction of severe experimental autoimmune neuritis with a synthetic peptide corresponding to the 53–78 amino acid sequence of the myelin P_2 protein. J Neuroimmunol 30:145
75. Saida K (1993) *Campylobacter jejuni* infection, GM1 antibody, and HLA antigens in Guillain-Barre syndrome. Proceedings of the Peripheral Nerve Study Group. 11th Biennal Meeting, Jakobsberg, Boppard, Germany, p 4
76. Saida K, Sumner AJ, Saida T (1980) Antiserum-mediated demyelination: relationship between remyelination and functional recovery. Ann Neurol 8:12
77. Saida T, Saida K, Silberberg DH, Brown MJ (1981) Experimental allergic neuritis induced by galactocerebroside. Ann Neurol 9 [Suppl]:87

78. Saida T, Saida K, Lisak RP, Brown MJ (1982) In vivo demyelinating activity of sera from patients with Guillain-Barre syndrome. Ann Neurol 11:69
79. Scheremata W, Colby S, Karkhansis Y, Eylar E (1975) Cellular hypersensitivity to basic myelin (P_2) protein in Guillain-Barre syndrom. Can J Neurol Sci 2:87
80. Schmidt B, Stoll G, Hartung HP, Heininger K, Schafer B, Toyka KV (1990) Macrophages but not Schwann cells express Ia antigen in experimental autoimmune neuritis. Ann Neurol 28:70
81. Schmidt B, Stoll G, van der Meide P, Jung S, Hartung H-P (1992) Transient cellular expression of gamma-interferon in myelin-induced and T-cell line-mediated experimental autoimmune neuritis. Brain 115:1633
82. Schoenberger LD, Hurwitz ES, Katona P, et al (1981) Guillain-Barre syndrome: its epidemiology and association with influenza vaccination. Ann Neurol 9 [Suppl]:31
83. Sharief MK, McLean B, Thompson EJ (1993) Elevated serum levels of tumor necrosis factor alpha in Guillain-Barre snydrome Ann Nurol 33:591
84. Smith ME, Hofmann WW (1979) Experimental allergic neuritis in the Lewis rat. J Neurophatol Exp Neurology 38:377
85. Stoll G, Schwendermann G, Heininger K, Kohne W, Hartung H-P, Seitz RJ, Toyka KV (1986) Relation of clinical, serological, morphological, and electrophysiological findings in galactocerebroside-induced experimental allergic neuritis. J Neurol Neurosurg Psychiatry 49:258
86. Stoll G, Schmidt B, Jander S, Toyka KV, Hartung H-P (1991) Presence of the terminal complement complex (C5b-9) precedes myelin degradation in immune-mediated demyelination of the rat peripheral nervous system. Ann. Neurol 30:147
87. Surik HM, McQuillen MP (1976) The use of steroids in the treatment of idiopathic polyneuritis. Neurology, 25:205
88. Van der Meche FGA, Schmitz PIM, Group S-BS (1992) A randomized trial comparing intravenous immune globulin and plasma exchange in Guillain-Barre syndrome. N Eng J Med 326:1123
89. Vriesendorp FJ, Mishu B, Blaser MJ, Koski CL (1993) Serum antibodies to GM1, GD1b, peripheral nerve myelin, and *Campylobacter jejuni* in patients with Guillain-Barre syndrome and controls: correlations and prognosis. Ann Neurol 34:130
90. Waksman BH, Adams RD (1955) Allergic neuritis: an experimental disease of rabbits induced by the injection of peripheral nervous tissue and adjuvants. J Exp Med 102:213
91. Wekerle H, Schwab M, Linington C, Meyermann R (1986) Antigen presentation in the peripheral nervous system: Schwann cells present endogenous myelin autoantigens to lymphocytes. Eur J Immunol 16:1551
92. Yu R, Ariga T, Kohriyama T, Kusunoki S, Maeda Y, Miyatani N (1990) Autoimmune mechanisms in peripheral neuropathies. Ann Neurol 27 [Suppl]:S30
93. Yuki N, Yoshino H, Sato S, Miyatake T (1990) Acute axonal polyneuropathy associated with anti-GM1 antibodies following Campylobacter enteritis. Neurology 40:1900
94. Zweiman B, Rostami AM, Lisak RP, Moskovitz AR, Pleasure DE (1983) Immune response to P_2 protein in the human inflammatory demyelinative neuropathies. Neurology 33:234

Systemic lupus erythematosus: immunopathogenesis of neurologic dysfunction

Patricia M. Moore, Robert P. Lisak

Wayne State University School of Medicine, 4201 St. Antoine, Detroit, MI 48201, USA

Summary. Neurologic complications of systemic lupus erythematous (neuro-SLE) are common. The most frequent manifestations of neuro-SLE are seizures, encephalopathy, and behavioral changes, but a wide variety of other neurologic abnormalities affecting the central and peripheral nervous system and muscle also occur. Although the prevalence of neuro-SLE is high, the diversity of clinical presentations, the multiple potential etiologies, and the absence of sensitive and specific diagnostic tests render diagnosis difficult. Recent advances in understanding mechanisms of neuronal dysfunction combined with advances in imaging techniques, including functional imaging, should help in diagnosis and management. The mechanisms of neurologic injury can be divided into three broad categories. First, neuronal dysfunction may result from direct effects of the immune system on brain cells such as autoantibody binding to cell surface, immune complex deposition with secondary inflammation, and effects of cytokines. Second, immune- mediated injury to supportive structures such as the vasculature may also affect the nervous system by producing ischemia. Finally, the neuraxis may be affected by any one of several immune and non- immune effects of infection, toxins, and metabolic disturbances.

Introduction

Systemic lupus erythematosus (SLE), a multisystem autoimmune disease, is characterized by persistent circulating autoantibodies and immune complexes. Autoantibodies occur to a wide variety of autoantigens. The current incidence of SLE is 7.4 per 100,000 with a prevalence of 50.8 per 100,000. Most common in young women, the non-white (blacks, Sioux Indians, and Orientals) population appears disproportionately affected. Nonetheless, the disease is wide spread [79]. The cause of SLE is unknown. A genetic predisposition is suggested by population, fâmily, and twin stud-

Correspondence to: P.M. Moore

ies, although SLE does not fit a classic Mendelian genetic model. The concordance rate for monogygotic twins is about 30%, while dizygotic twins have a rate of 5% which does not differ from the frequency among non-twin siblings. The predisposition to SLE is polygenic and as many as four genes appear involved in the expression of human SLE. The genes influencing the expression of SLE govern immune regulation, autoantibody responses, clearance of immune complexes, complement and hormones [33, 100, 106]. Environmental factors such as drugs, chemicals, UV radiation, and infections, are also implicated in triggering disease [38, 90].

SLE most frequently affects the skin, kidneys, musculoskeletal, and nervous systems [108]. The clinical course may be benign or severe, the tempo indolent or fulminant. Diagnosis and treatment may be difficult. The American Rheumatic Association criteria for SLE were revised in 1982 and these remain in use [110]. Primarily developed as a classification and research rather than for diagnosis, the low sensitivity and high specificity of the criteria may make definitive diagnosis difficult in individual patients. Recent multicenter studies have attempted to delineate features associated with a virulent clinical course [21]. The clinical spectrum of SLE has evolved over the past three to four decades. Although the involvement of skin and joints continues to be the most common feature, the incidence of lupus nephritis appears to be declining. The reasons are not clear but may be the result of successful treatment strategies. Neuropsychiatric manifestations of the disease have become more prominent over the past 30 years. Whether this is a true increase in incidence (possibly related to a decline in mortality from other causes) or represents improved recognition of neuro-SLE is not known.

Immunopathogenesis of SLE

Autoantibodies, self tolerance and immunoregulation

The hallmark of SLE is excessive autoantibody formation. These antibodies are directed against nuclear, cytoplasmic, and cell surface antigens. Certain families of autoantigens are over represented. Characteristically, elevated titers of antibodies directed against the native β-helical form of DNA, which is not immunogenic in normal subjects, occur. Other important antibodies are those to ribonucleoprotein particles and the associated Sm (Smith) antigens involved in processing of nuclear RNA [69]. Studies of the cross- reactivity of anti-DNA and anti-phospholipid antibodies provide further information on the nature of autoantigens in SLE. The high frequency of public idiotypes (antibodies to shared domains on immunoglobulin molecules) suggests immunologic cross-reactivity may be important in the pathogenesis of disease [43,97]. Autoantibodies that are associated with SLE are listed in Table 1.

Table 1. Autoantibodies associated with SLE

Anti-nuclear, anti-DNA, anti-ribosomal, anti-Ro, anti-La, anti-Sm, antiphospholipid, anti-lymphocytes, anti-collagen, anti-neuroblastoma, anti-brain cortex, anti-cerebellum

Central to the immunologic abnormalities in SLE is the loss of self tolerance. Numerous mechanisms could lead to loss of self tolerance among individuals with SLE and different animals strains [35, 42, 78, 96]. The stimuli for autoantibody formation, whether an antigen-driven process or a result of polyclonal activation, remains a matter of controversy. A plausible hypothesis is that initial polyclonal activation is later biased by antigen-driven mechanisms. Overactive B cells, both produce self-stimulatory factors and respond to excessive signals from T cells. The extent of the participation of T cells and macrophages in abnormal B cell activation in SLE is not clear. T cell abnormalities may be critical because synthesis of many of the pathogenic autoantibodies appears to be T cell dependent. An apparent reduction in T cell activity in vitro and increased activity of T cells in vivo remains unexplained. Defects in immune regulation, present in patients and relatives of patients with SLE, contribute to the breakdown of self tolerance and subsequent autoantibody formation [22]. Increasing evidence of a stem cell defect in murine and human SLE could explain the abnormalities in multiple immune cells [59]. Accelerated maturation of T and B cells during development could impair the usual efficient deletion of self-reactive clones in the thymus. A recent study demonstrating that viral infection may limit apoptosis (programmed cell death) of B cells [41] suggests that an acquired mechanism as well as a genetic mechanism [85] contribute to the numerous and persistent autoreactive B cells. Similarly, thymocytes escaping from apoptosis might result in persistent self-reactive T cells, providing T cell help for autoreactive B cells.

Cytokines play a prominent role in the pathogenesis of joint inflammation and destruction in rheumatoid arthritis, and may be similarly important in the pathogenesis of organ damage in SLE Multiple abnormalities in cytokines secretion have been are reported [4, 7, 44a, 70]. SLE B cells spontaneously produce interleukin-1 (IL-1), IL-4, and IL-6 which stimulate B cells to differentiate into immunoglobulin (Ig)-producing cells [111, 112]. Receptors for IL-1 and IL-4 are increased on B cells in patients with SLE when compared with controls. IL-6 plays a major role in the maturation of activated B cells into Ig-secreting cells. In SLE, transcripts for IL-6 are increased in T cells and monocytes, and IL-6 is detected in the renal lesions. In addition, increased levels of IL-6 occurs in blood and cerebrospinal fluid of patients with neuro-SLE [70]. Abnormalities in both the synthesis and response to IL-2 are also present. Serum IL-2 is high; the diminished T cell response to exogenous IL-2 in vitro might reflect postactivation refractoriness [119]. Increased serum levels of both interferon γ and α are also associated with disease activity in SLE.

Immunopathogenesis of organ/tissue damage

Although the events leading to persistent autoantibody formation are not clear, much is known about events leading to tissue injury in SLE. Established immunopathogenic mechanisms of disease include immune complex deposition and direct antibody-mediated effects. Clinically, the most clearly delineated immunopathogenic effect in SLE is that of immune complex deposition. Many of the target antigens of the autoantibodies in connective tissue diseases are intracellular antigens (such as double-stranded DNA, Sm, Ro and other small ribonucleoproteins) and not readily accessible to antibody binding in vivo. Thus, antibodies to these intracellular antigens are pathogenic largely by their participation in immune complexes deposition [88,97]. The body normally eliminates immune complexes through several mechanisms. Persistent immune complexes (with certain immunochemical characteristics)may result in tissue damage. Deposition of these antibody-antigen complexes in the vasculature results in recruitment of the inflammatory cascade via the Fc portion of the antibody molecule [16].

Immune complex-mediated inflammation results in much of the renal and cutaneous tissue damage in SLE.

The second major immunopathogenic mechanism in SLE is cellular dysfunction or reduction in cell numbers, resulting from antibodies binding directly to their cognate antigens. Anemia, leukopenia, thrombocytopenia, and some coagulopathies appear to result from direct antibody-mediated effects. In addition, many of the immunoregulatory abnormalities in patients with SLE may result from autoantibody-mediated effects on lymphocytes. In general, the titers of anti-lymphocyte antibodies (ALA) correlate with systemic disease activity. ALA react with a variety of plasma membrane antigens including surface Ig, major histocompatibility complex (MHC) class I and II molecules, IL-2 receptors, and different isoforms of CD45. This binding is reflected in vitro by several changes in lymphocyte function [121]. The association of disease flares, lymphopenia and various abnormalities of cellular immunity support the clinical relevance of the increased ALA titers.

Murine SLE

Autoantibody profiles similar to those seen in patients with SLE occur spontaneously in some animals [30, 117]. The (NZB/NZW)F1 hybrid, MRL/lpr, and BXSB mice and animals with graft versus host disease are examples. Studies of these models provides considerable information on the genetic control of autoantibody formation, structure of epitopes recognized by autoantibodies, pathogenicity of autoantibodies, and therapy of disease resulting from autoantibody formation [109].

Murine strains developing clinical and serologic abnormalities similar to human SLE include the frequently studied MRL-*lpr/lpr*, (NZB/W) F1, and BXSB mice. These animals illustrate the complex interplay of genetic and hormonal influences on disease. A stem cell defect appears present in all of the animals. MRL/lpr mice develop a generalized autoimmune disease which includes increased autoantibody production, glomerulonephritis, development of lymphadenopathy as well as systemic vasculitis and some elements of the sicca complex. Serologically, the spectrum of autoantibodies include anti-DNA antibodies, and anti-ribonucleo-protein antibodies. The MRL/lpr strain also develop the anti-Sm antibodies which are infrequent (about 30%) but highly specific for SLE in humans. The disease is accelerated by the *lpr* gene; disease occurs much later in the congenic MRL/++. The *lpr* gene is also disease accelerating when introduced on other genetic backgrounds. The *lpr* genetic defect is identified as a defect in the *Fas* apoptosis gene, resulting in low expression of *fas* mRNA. An intrinsic abnormality in lpr B cells as well as an associated abnormality in lpr T cell appear.

In addition, NZB and NZB/W F1 hybrid mice also spontaneously develop autoimmune manifestations. The NZB manifests hemolytic anemia, while the NZB/W F1 animals share the glomerulonephritis, hypertension, and cutaneous clinical disease of human SLE. The disease occurs predominantly in female mice and is accelerated by estrogen. In contrast to that seen in the MRL/lpr strain, IgG but not IgM production is T cell dependent in the NZB/W F1 mouse [78].

Autoimmune disease in the BXSB, which occurs in the male sex, depends, on the Y chromosome, rather than hormonal environment. These animals develop their circulating Ig and immune complex disease fairly early in life as do the MRL/lpr.

Neurologic features present in the (NZB/W) F1 and the MRL/lpr mice include behavioral changes, diminished learning of spatial information, and altered socialization. Focal features of CNS vascular disease are not present clinically or histologically. Neuropathologic features vary somewhat among the strains. Scattered foci of plasma cells and lymphocytes in the meninges appear in the MRL/lpr more often than in the other strains. Cerebrovascular disease has not been identified in several series [5, 93]. Autoantibodies can be identified in the brain substance and immune complex are deposited in the choroid plexus. The potential role of autoantibodies in the pathogenesis of neurologic abnormalities is being explored in several laboratories.

Clinical syndromes in neuro-SLE

The clinical features of Neuro-SLE range from overt seizures, encephalopathy, stroke or psychosis to more subtle deficits in cognition (Table 2). These protean abnormalities represent a major cause of morbidity and mortality in SLE. The neurologic abnormalities often represent a diagnostic and therapeutic challenge. This reflects our incomplete understanding of immunopathogenic mechanisms and our paucity of specific diagnostic studies. Most perplexing are the encephalopathies and behavioral disorders which have been frustratingly difficult to define. Physicians (and family members) recognize decreased cognition, poor attention span, diminished judgment, depression, flattened affect, emotional liability, and difficult (negative) personality features present in patients with SLE. Cognitive impairment appears more common than previously suspected particularly in patients with no overt neurologic abnormalities. These appear to be distinct from changes seen in other autoimmune or chronic diseases. Overt symptomatic neurologic disease may appear early in the course of SLE. Recent studies show that cognitive changes are present in 55% of patients with SLE [24]. Behavioral abnormalities are even more perplexing. Although some investigators minimize their relevance, other studies demonstrate a spectrum of changes [47, 50, 80, 102].

Table 2. Clinical features

Seizures
Encephalopathies
Behavioral changes
Focal motor/sensory abnormalities
Ataxias
Movement disorders
Myelopathies
Polyradiculopathies
Peripheral neuropathies
Cranial neuropathies
Increased intracranial pressure

Overlapping and perhaps contributing to these cognitive and behavioral derangements are the neuro-endocrine features which are emerging as a link between the brain and the immune system in SLE. Activation of the hypothalamic-pituitary-adrenal (HPA) axis [which may result from the action of cytokines, IL-1, IL-6, or tumor necrosis factor-α (TNF-α)] initiates release of corticotropin-releasing hormone (CRH) into the hypophysial-portal system and, thus, activates a counter regulation of inflammation through glucocorticoids. However, other neuropeptides and hormones which are released and participate in the stress response adversely impact learning, memory, and behavior particularly in chronic activation of the HPA axis [107]. In fact, CRH itself may be anxiogenic. In addition the HPA axis also influences neurotransmitter levels. Norepinephrine, which affects lymphocytic function (lymphoid organs are sympathetically innovated) as well as CNS and cardiovascular function, participates in a positive feedback loop with the CRH neurons. Other neurotramsmitters, including serotonin and acetylcholine, also respond to HPA axis stimulation, although the effects are less well studied. Thus, the chronic stimulation of the HPA axis which occurs in SLE may set the stage or contribute directly to the neuro-psychiatric manifestations.

Seizures are prominent in SLE and may occur transiently or as a lifelong but controllable abnormality. Both focal and generalized seizures occur. Strokes occur in SLE, but clinical and histologic evidence of ischemia or infarction is far less common than blood flow studies suggest. Cerebral venous thromboses, also manifest in patients with SLE, presents with pleomorphic clinical manifestations [120].

Seizures, encephalopathies, and psychiatric changes are the predominate syndromes. Headaches, although frequent, are nonspecific. Cranial neuropathies (including a painful ophthalmoplegia) may herald disease.

Strokes and transient ischemic episodes may occur at an time during the course of the disease and are not linked to systemic vascular disease. Ataxia and movement disorders (usually chorea) are distinctive, although less common manifestation of CNS involvement. Myelopathies occur in two patterns, an acute syndrome of traverse myelopathy and a slowly progressive vascular myelopathy, which simulates combined system degeneration peripheral neuropathies. Polyradiculopathies, mononeuropathology multiplex, autonomic neuropathies and myopathies in SLE are all well documented, although the true incidence has not yet been determined [56, 57, 60, 105].

Difficulties in diagnosis and therapy stem from our inability to distinguish specific pathogenic mechanisms and the absence of reliable sensitive and specific diagnostic tests. Encephalopathies may result from autoantibody-mediated neuronal changes, immune complex deposition, increased intracranial cytokines such as IL-2, but may also be the secondary features of infections and metabolic changes. Likewise, focal neurologic abnormalities can reflect a chronic vasculopathy, an acute coagulopathy, or a fungal meningoencephalitis. There are no immunochemical, radioimaging or neurophysiologic tests to unequivically distinguish among the immunopathogenic mechanisms. Computer axial tomography, magnetic resonance imaging, and positron emission tomography may demonstrate discrete abnormalities but the findings correlate poorly with clinical changes. Newer imaging studies and quantitative electrophysiologic tests, particularly those combined with measures of cognitive and behavioral function are promising techniques [91].

Immunopathogenesis of neuro-SLE

The immune manifestations of the neurologic syndromes in SLE are diverse. There are several ways to categorize the mechanisms of clinical injury to provide information of the temporal profile on the immunopathogenesis of disease and suggest therapeutic strategies. One useful scheme distinguishes immunologic effects directly affecting nervous system cells (neurons, glial, ependyma) from those immunologic effects on vascular tissue which secondarily affect the nervous system by ischemia. In each of the categories there are multiple immunologic effects, but the tempo of disease and acute therapy vary considerably. The vascular disease is typically a chronic process, resulting from years of sub-clinical injury. Table 3 summarizes the immunopathogenic mechanisms of neuro-SLE.

Table 3. Immunopathogenesis of neuro-SLE

Primary:	Immune-mediated *direct* effects
	immune complex
	autoantibodies
	cytokines
	? activated lymphocyte
	Immune mediated *indirect* effects
	vasculopathy
	coagulopathy
	emboli from cardiac disease
	hematologic disturbances
Secondary:	Infectious
	Metabolic
	Toxic

Direct immune-mediated effects

Immune-mediated direct effects result from the action of antibodies, cytokines, or immune complexes on neurons or, possibly, glia. Consequences of these interactions reflect the diversity of neuronal dysfunction. Effects ranging from increased or decreased membrane excitability, enhanced or diminished release of neurotransmitters, activation of neuro-endocrine circuits and release of cytokines are all implicated in clinical disease. Typically these effects are acute or sub-acute in onset.

Autoantibodies. Various autoantibodies are linked directly or indirectly to nervous system dysfunction. Antibodies may directly affect cell function without cytotoxicity by mechanisms such as agonist action [65, 113], mimicking interleukins [101], and receptor blockade [118] in different autoimmune diseases. In neuro-SLE three prominent groups of autoantibodies are implicated in the pathogenesis of disease: anti-neuronal antibodies, anti-ribosomal P antibodies, and anti-phospholipid antibodies.

A current hypothesis in neuro-SLE is that anti-neuronal antibodies bind to neuronal membranes and cause a transient alteration of cell function without cell death or inflammation. This would account for reversibility of symptoms and the poor correlation of CNS pathologic changes and clinical signs. Neuron-reactive antibodies are present in the serum and CSF of patients with neuro-SLE. Further, the presence of anti-neuronal antibodies correlates with certain varieties of neuro-SLE, particularly encephalopathies and seizures. Early studies identified a lymphocytotoxic antibody cross-reactive with brain tissue in patients with active neurologic disease including patients with cognitive impairment [73], although later studies failed to correlate a temporal profile of antibody with disease activity. Reports also described serum autoantibodies reactive with neuronal tissue sections of brain cortex, brain-derived plasma membranes and cultured neuroblastoma cells [14, 40, 45, 55]. A variety of target antigens including membrane proteins, membrane glycolipids are described. Access of anti-neuronal antibodies to the CNS may occur by alterations in the blood-brain barrier or by intrathecal synthesis. Evidence for both exists [54, 122]. CSF studies in patients with SLE reveal increased intrathecal Ig synthesis and/or oligoclonal bands [92] occurring in about half of the patients with encephalopathies, seizures and psychiatric abnormalities [39, 122].

Antibodies to ribosomal P proteins, particularly 60S ribosomal subunit phosphoproteins PO, P1 and P2 occur [15]. As intracellular antigens the access of antibodies to ribosomal P protein would be considered problematic, although antibodies to ribosomal P protein (38 kDa) were recently shown to also be immunoreactive with cell surfaces including those of neuroblastoma cells [67]. Recent studies have both confirmed [94] and refuted the association between psychoses and depression and anti-ribosomal antibodies [27, 115, 116]. The lower concentration of anti-ribosomal P antibodies in the CSF than in the serum raises some doubts about their role in disease [114].

Anti-phospholipid antibodies (APA), a large group of antibodies, include natural autoantibodies, which may have a physiologic role, as well as pathogenic antibodies such as those associated with hemolytic anemia, thrombocytopenia or venous occlusive disease. Even antibodies which are potentially pathogenic may not be so normally because of their association with natural anti-coagulants such as B2GP-1 [1]. In patients with SLE, only transverse myelitis appeared strongly associated with APA in one large series. Distinguishing primary anti-phospholipid syndrome in vascular diseases associated with APA from neurologic changes in SLE is under active investigation.

Limitations in interpreting studies correlating antibodies with neurologic disease include the facts that: (1) antibodies are also present in patients without demonstrable CNS involvement, (2) the specificity of the antigen(s) in neuronal tissue is unidentified and are likely heterogeneous, (3) no large, standardized, longitudinal studies exist correlating antibody levels and neuropsychiatric disease in unselected SLE patients [49], and (4) no in vivo effects of this population of antibodies have been demonstrated. Further studies are required to establish the pathogenicity of these antibodies in human disease.

The previously described autoimmune mice, which share antigen specificity with many human autoantibodies, have proved particularly useful in the investigations of neuronal autoantigens. In MRL/lpr and NZB/W F1 mice, serum antibodies bind to brain plasma membranes [82, 87]. Neuron-reactive antibodies are present in the brains of some animals and can be eluted from the parenchyma [82-84]. Early neuropathologic studies of murine SLE [5, 93] are joined by studies analyzing neurocognitive abnormalities, autoantibodies, and histopathology [25, 52].

Immune complex. Immune complex-mediated tissue injury is a hallmark of SLE, well established in the pathogenesis of renal and cutaneous disease. There are potentially at least two pathogenic processes attributed to immune complex-mediated injury to the nervous system in SLE: (1) acute inflammation in the choroid plexus or vasculature, and (2) chronic deposition in the intracranial vasculature. In addition, mononeuropathy multiplex is likely similarly mediated.

Much of the interest in immune complexes and neurologic function has developed from the observations that immune complexes appear in the CSF, are deposited in the choroid plexus (a tissue with similarities to the renal glomeruli), and can alter the blood-brain barrier when injected into experimental animals [104]. However, evidence for immune complex-triggered inflammation in the CNS in humans is scanty. Several patients showed IgG and complement deposits in the choroid plexus post mortem. Two histologic patterns of deposition (membranous and vascular) were identified, both associated with immune complexes in the CSF [95]. CSF DNA-anti-DNA complexes, detected in a patient with acute glomerulonephritis and hypocomplementemia, cleared with prednisone therapy [62]. CSF immune complexes of unknown antibody specificity were present in 33 out of 34 patients with CNS-SLE [98]. In most of these patients blood-brain barrier impairment was not present on the basis of Ig/protein ratios in the CSF, although focal or regional defects in the barrier could exist and not be detected by CSF analysis. In other patients loss of integrity of the blood-brain barrier exists but does not correlate with the presence of neurologic abnormalities or the presence of circulating immune complexes [122].

Against immune complex deposition serving as a major cause of neurologic disease in neuro-SLE are (1) the presence of immune complex deposition in the choroid plexus in patients both with and without neurologic symptoms [17], and (2) the paucity of histologic evidence of inflammation. Furthermore, neurologic abnormalities are usually not associated with evidence of active elevation of circulating immune complexes and active renal disease, as would be anticipated if immune complex deposition in the choroid plexus or vasculature were a prominent cause of encephalopathies or seizures.

Another type of immune complex-mediated injury is more indolent, not always associated with evidence of systemic inflammation, and more difficult to diagnose. Low-level deposition of immune complexes in blood vessels may produce a chronic degenerative vasculopathy rather than acute tissue inflammation. Some of these issues are discussed in indirect immune-mediated effects as discussed below.

Cytokines. Cytokines are prominent in the immunopathogenesis of neurologic disease. Cytokines, small proteins produced by lymphocytes, macrophages, and other cells such as fibroblasts, cause biologic affects in cells with receptors for that particular cytokine. The short distances over which cytokines traditionally manifest their effects and their normal regulatory role complicate research into their pathogenic consequences in the nervous system. Both physiologic and pathologic effects on neurologic function may result through several pathways: direct effect of cytokines on function of cells of the nervous system and increased direct effects on endothelial cells which could lead to ischemia.

Evidence that cytokines play a role in normal neuronal function (and therefore may be implicated in disease) begins with histologic localization of cytokines, their receptors, and their antagonists in the neuraxis. IL-1 and IL-1 receptor antagonist co-localize to the hippocampus, hypothalamus, and several other brain regions, forming an intrinsic paracrine system. IL-1 activates the hypothalamic-pituitary-adrenal axis, causing release of pituitary hormones and ultimately glucocorticoids [12]. Brain cytokines are synthesized in response to local tissue injury; IL-6 appears with nonspecific cerebral injury [77]. Production of CNS cytokines also elicit other cytokines, e.g., the production of IL-6 in response to injections of IL-1 [28]. Thus, cytokines may play a role in the behavioral, cognitive, and neuro-endocrine changes in neuro-SLE.

Activation of cytokines within the CNS reportedly induces profound changes in neural function. Behavioral and cognitive changes induced by cytokines are pertinent to the investigation of neuro-SLE. IL-1β inhibits long-term potentiation in mouse hippocampi with potential effects on memory [61]. Pharmacologic doses of IL-2 induce fever, slow wave sleep, anorexia, and, in mice, exploring behavior [26, 51]. The neuropathogenic effects of several infectious agents that do not invade the nervous system are attributed to cytokines produced by glial and immune cells or to serum cytokines on areas of the brain with a naturally incomplete blood-brain barrier. Neuropeptides such as substance P and substance K (neurotransmitters released at the peripheral terminals of afferent nerves) can stimulate immunocompetent cells as well as neuroglial cells to secrete inflammatory cytokines IL-1, TNF-α and IL-6 [74]. IL-1, TNF-α, and IL-6, in turn, act directly and indirectly on the hypothalamus to stimulate synthesis and secretion of CRH. This is a potent interaction between the neuraxis and the immune system [11, 89]. Recent studies on IL-6 suggest the cytokine itself may be associated with neurologic abnormalities.

Another major effect of cytokines in the nervous system is the effect on the vasculature. Despite the potential effects on brain cells and blood vessels, there has been scanty demonstration of increased cytokines in the brain or CSF of patients with SLE. The only reported of abnormal cytokine concentrations in neuro-SLE is for IL-6 which is increased in the CSF [53]. The possible role of cytokines in vascular pathology is discussed in indirect immune-mediated effects.

Activated lymphocytes. Activated lymphocytes may have a pathogenic effort on brain tissue in SLE. Activated immunocompetent cells such as lymphocytes, monocytes, and macrophages can penetrate the blood-brain barrier, lodge in the brain, and secrete their full repertoire of cytokines, leukotrienes, and prostaglandins. B cells present, albeit sparsely, in human and murine CNS may be the source of intrathecal synthesis of Ig. The B cells may be amenable to immunotherapy directed at activation markers. Cell-mediated cytotoxicity, another effector mechanism mediated by a subset of lymphocytes, has not been demonstrated in the nervous system of human or murine SLE.

Indirect immune-mediated effects

Immune-mediated indirect neurologic effects typically result from the action of autoantibodies, cytokines, immune complexes, or activated lymphocytes on the vasculature. Neurologic effects result from tissue ischemia. Tissue ischemia or infarction from occluded blood vessels may develop from several different, indirect immune-mediated effects of disease. This group is distinguished from the first not only in immunopathogenesis but also in clinical presentation and therapy. In these disorders, the immunologic effects on the blood vessel occur over a period of time ranging from months to years. The incidence of stroke in clinical series of SLE ranges from 5 to 20% [34, 66]. Most clearly denoted are thromboses (coagulopathies and hematologic disorders), vasculopathies, with or without hemorrhagic infarcts and emboli from the heart. Arteries, veins, and microvasculature are affected.

Neurologic abnormalities develop when a critical degree of local disease or a series of sub-clinical events (such as mild changes in hemostasis) in abnormal blood vessels occur. Compounding the vascular damage are additional, non-immune abnormalities such as hypertension. Furthermore, corticosteroid therapy may accelerate the premature atherosclerosis present, at least in the coronary arteries.

Vasculopathy. Vasculopathy refers to structural abnormalities of the blood vessel wall. In patients with neuro-SLE, these changes occur primarily in small but sometimes in medium-sized, vessels. Two recurrent observations are noteworthy. First, degenerative changes in the blood vessel wall far exceed evidence of inflammation. Second, histologic changes suggest a chronic rather than acute process. In addition, histologic evidence of vasculopathy is more widespread than clinical findings, and also exceeds histologic evidence of ischemia.

Pathologic studies show endothelial proliferation and medial hyperplasia of small intracortical arterioles and capillaries, petechial hemorrhages, and parenchymal necrosis with a variable glial reaction [46]. Histologic staining with Masson-Trichrome reveals irregular proliferation of collagen with occasional hyalinization. The frequent association of extravasated blood with occasional hemosiderin-laden macrophages suggests structural incompetence of vascular hemostasis in those vessels [31]. Abnormalities in microvessels appear in 40-80% of patients, although lesions associated with evidence of tissue infarction are more limited in distribution and extent [58]. Major vessel thromboses associated with atheroma are occasionally reported. Although CNS vasculitis occurs, it is much less frequent than cutaneous or visceral vasculitis, and is an infrequent cause of neurologic abnormalities.

Numerous mechanisms could be responsible for the vasculopathy but three are considered most likely (1) deposition of circulating immune complexes including DNA-anti-DNA complexes (2) effects of autoantibodies including APA on the vessel wall; and (3) consequences of chronic inflammation.

Clues to the development of an immune complex-mediated degenerative rather than inflammatory vasculopathy are found in animal studies. In murine SLE the duration, quantity, and types of circulating immune complexes, as well as the genetic background of the host determine the effects of immune complex deposition on the vasculature. Chronic, low levels of immune complexes are associated with endothelial and intimal proliferation in the vessel wall plus thrombosis [10]. Immune complex-mediated injury also results in the production or accrual of vasoactive substances with a net prothrombotic effect. Evidence that immune complex deposition may be clinically relevant is the association of cerebrovascular disease with renal involvement and circulating anti-DNA/DNA complexes [66].

Direct autoantibody effects on the vessel wall may also produce changes. Endothelial cells, known regulators of blood flow through their actions on vascular smooth muscle, mediate these changes through vasodilators such as prostaglandin (PG) I1 endothelium-derived relaxing factor and the vasoconstrictor peptide endothelin or polypeptide PDGF [18]. Both anti-endothelial antibodies and APA may result in mild but recurrent injury to the endothelium [19, 44]. Antibodies binding phospholipids in endothelial cell membranes may result in decreased levels of prostacyclin and increased platelet aggregation and thrombosis [13, 64]. Abnormalities in the endothelial cell in SLE is likely accompanied by changes in blood flow and vessel diameter.

Chronic inflammation may also be injurious. Recent studies suggest serum amyloid A proteins, acute-phase reactants associated with high-density lipoproteins (HDL), displace apolipoprotein A1. This potentially leads to greater deposition of fat in the blood vessel and may result in premature atherosclerosis [8]. Inappropriate activation of adhesion molecules may contribute to the consequences of chronic inflammation [6, 9, 37, 48, 76].

Structural changes in the wall are compounded by coagulation abnormalities, the result of the prothrombotic effect of many cytokines. Subsequent inflammation, transient or recurrent, is controlled by a variety of cell regulatory events, which influence a wide array of cell responses including chemotaxis, apoptosis, differentiation and growth arrest. Intravascular complement activation may also result in occlusive vasculopathy [44]. Leukocytes and platelets aggregate in vessel walls secondary to increased adhesiveness rendered by lipopolysaccharide, IL-1, or TNF.

Coagulopathies. The coagulation system is a highly regulated cascade of surface associated enzymes and cofactors, resulting in the generation of thrombin. The interactions of the coagulant and anticoagulant proteins and the fibrinolytic system occur at the vascular endothelial cell surface. Acquired hypercoaguable states occur with abnormalities of protein, platelet function or endothelial cell dysfunction. Disorders of fibrin metabolism potentially facilitate deposition of fibrin and platelets on endothelium of small vessels and cardiac valves [86].

APA, often present in patients with SLE and reportedly associated with an increased risk of stroke, are the most frequently studied potential cause of stroke. This family of antibodies, including anti-cardiolipin antibody, shares reactivity with negatively charged phosphodiester moieties. They are prothrombotic factors in occlusive vascular disease both in the brain and elsewhere. Potential targets of APA are the endothelial cell, prostacyclin, protein C, the protein C-S complex, and platelets. Both acute effects on coagulation and more chronic effects on the vessel wall are possible. The APA manifestations of stroke in SLE may, thus, involve synergism with the vasculopathy or other inflammatory effects.

Thus, APA potentially contribute to stroke by two pathways, interaction with the blood vessel wall and with the coagulation cascade. Numerous studies link APA with stroke [2, 3, 32, 72]. A direct pathogenic role of APA and stroke is, however, not established. The classic features of the anti-phospholipid syndrome are recurrent fetal loss, thrombophlebitis, and stroke. Numerous other neurologic abnormalities have been associated with APA. However, in one large series of SLE patients studied prospectively, only the occurrence of transverse myelopathies (not stroke) was statistically significantly associated with APA [2]. Also, APA were not found in the CSF of five SLE patients with acute strokes, despite high serum levels of the antibody [32]. In another study only three of the six stroke patients had a serum lupus anticoagulant [120]. Other abnormalities leading to a hypercoaguable state include decreased fibrinolysis, decreased plasminogen activation, and impaired formation of prostacyclin [108].

Cardiac emboli. In 1924 Libman and Sacks reported four patients with atypical verrucous endocarditis. The vegetations consisted of soft clumps of fibrinous material with plasma cells usually found on the chordae tendiae, papillary muscles, or on the valves most frequently the mitral valve. These vegetations seldom dislodged and emboli were considered rare. Recently, the potential role of emboli from valvular heart disease has been reevaluated.

In 1988, an autopsy study of 50 patients with SLE identified 10 people with embolic cerebral infarcts, 5 caused by identified Libman-Sacks endocarditis, and 4 from other cardiac sources [29]. Clinical studies, particularly in patients who survive, reveal fewer diagnosed cerebral emboli [36]. An association of mitral valve disease and APA is also noted [63]. The actual incidence of cerebral emboli from the heart in neuro-SLE is not known. As described, the incidence of Libman-Sacks lesions and other mild valvular abnormalities detected by echocardiography is 18% of patients with SLE. There was no occurrence of cerebrovascular disease in 74 patients over 5-period years [36].

Hematologic changes. Thrombotic thrombocytopenic purpura (TTP) presents with wide-spread small vessel lesions in association with the hematologic abnormalities.

Histologically the brain lesions in the vasculopathy of SLE resemble those of TTP, except those of TTP tend to be all of the same age [58]. Thrombocytopenia appears to result from an antibody-mediated increased destruction of platelets. Underscoring the overlap with the coagulopathies, one of the types of antibodies reported to cause thrombocytopenia is APA [75].

Non-immune mediated effects

Neurologic complications of SLE also result from abnormalities not directly related to the primary disease. Many of these systemic or generalized abnormalities result in neurologic dysfunction, including confusion and seizures, that are clinically indistinguishable from primary immune-mediated damage. The major etiologies in this group are infections, toxins (including medications), and metabolic derangements. Although survival in SLE has improved over the years, *infection* remains a leading cause of morbidity and mortality in the disease. Meningitis from both bacterial pathogens such as *Neisseria* and opportunistic infections including *Candida* and *Cryptococcus* occur [81]. The spectrum of infections includes bacterial, viral, fungal, and parasitic organisms [20, 23, 68, 81, 103]. Septic emboli continue to be a cause for concern in patients with fever. The frequency of infections results from the patients' immunosuppression both by medication and by their underlying disease. The importance of identifying infections was highlighted by work indicating that high-dose corticosteroids did not improve the outcome of neurologic disease in patients with SLE. In fact, patients on high-dose corticosteroids had a higher mortality rate than those on lower doses [99, 123]. Unfortunately, many potentially treatable infections are only diagnosed post mortem.

The *toxins* most often causing neurologic abnormalities in SLE are medications. Corticosteroids and anti-hypertensive medications can produce neurologic and psychiatric symptoms. *Metabolic* abnormalities likely to affect the nervous system include uremia, as well as disorders of calcium, sodium, and magnesium.

References

1. Alarcon-Segovia D, Cabral AR (1994) Antiphospholipid antibodies. Where do they come from? Where do they go? J Rheumatol 21:982
2. Alarcon-Segovia D, Deleze M, Oria CV, Sanchez-Guerrero J, Gomez-Pacheco L, Cabiedes J, Fernandez L, Ponce De Lion S (1989) Antiphospholipid antibodies and the antiphospholipid syndrome in systemic lupus erythematosus. A prospective analysis of 500 consecutive patients. Medicine 68:353
3. Alarcon-Segovia D, Perez-Vazquez ME, Villa AR, Drenkard C, Cabiedes J (1992) Preliminary classification criteria for the antiphospholipid syndrome within systemic lupus erythematosus. Semin Arthritis Rheum 21:275
4. Al-Janadi M, Raziuddin S (1993) B cell hyperactivity is a function of T cell derived cytokines in systemic lupus erythematosus. J Rheumatol 20:1885
5. Alexander EL, Murphy ED, Roths JB, Alexander GE (1983) Congenic autoimmune murine models of central nervous system disease in connective tissue disorders. Ann Neurol 14:242
6. Argenbright LW, Barton RW (1992) Interactions of leukocyte integrins with intercellular adhesion molecule 1 in the production of inflammatory vascular injury in vivo. J Clin Invest 89:259
7. Bankhurst AD (1987) Interferons and systemic lupus erythematosus. J Rheumatol [Suppl 13] 14:63
8. Beer MC de, Kindy MS, Lane WS, Beer FC de (1994) Mouse serum amyloid A protein (SAA5) structure and expression. J Biol Chem 269:4661

9. Belmont HM, Buyon J, Giorno R, Abramson S (1994) Up-regulation of endothelial cell adhestion molecules characterizes disease activity in systemic lupus erythematosus. Arthritis Rheum 37:376
10. Berden JHM, Hang L, McConahey PJ, Dixon FJ (1983) Analysis of vascular lesions in murine SLE. J Immunol 130:1699
11. Besedovsky HO, DelRey AE, Sorkin E (1985) Immune-neuroendocrine interactions. J Immunol 135:750s
12. Besedovsky HO, DelRey A, Sorkin E, Dinarello CA (1986) Immunoregulatory feedback between interleukin-1 and glucocorticoid hormones. Science 233:652
13. Bevilacqua MP, Pober JS, Majeau GR, Cotran RS, Gimbrone MA Jr (1984) Interleukin 1 (IL-1) induces biosynthesis and cell surface expression of procoagulant activity in human vascular endothelial cells. J Exp Med 160:618
14. Bluestein HG (1978) Neurocytotoxic antibodies in serum of patients with systemic lupus erythematosus. Proc Natl Acad Sci USA 75:3965
15. Bonfa E, Golombek SJ, Kaufman LD, Skelly S, Weissbach H, Brot N, Elkon KB (1987) Association between lupus psychosis and anti-ribosomal P protein antibodies. N Engl J Med 317:265
16. Boros P, Odin JA, Chen J, Unkeless JC (1994) Specificity and class distribution of FcγR-specific autoantibodies in patients with autoimmune disease. J Immunol 152:302
17. Boyer RS, Sun NCJ, Verity A, Nies KM, Louie JS (1980) Immunoperoxidase staining of the choroid plexus in systemic lupus erythematosus. J Rheumatol 7:645
18. Brenner BM, Troy JL, Ballermann BJ (1989) Endothelium-dependent vascular responses. J Clin Invest 84:1373
19. Cines DB (1989) Disorders associated with antibodies to endothelial cells. Rev Infect Dis 2 [Suppl 4]:S705
20. Collins DN, Oppenheim IA, Edwards MR (1971) Cryptococcosis associated with systemic lupus erythematosus. Arch Pathol Lab Med 91:78
21. Combardier C, Gladman DD, Urowitz MB, Caron D, Chang CH (1992) Derivation of the sledai. A disease activity index for lupus patients. Arthritis Rheum 35:630
22. DeHoratius RJ, Pillarisetty R, Messner R, Talal N (1975) Anti-nucleic acid antibodies in systemic lupus erythematosus patients and their families: incidence and correlation with lymphocytotoxic antibodies. J Clin Invest 56:1149
23. Deleze M, Mintz G, CarmenMejia M (1985) *Toxoplasma gondii* encephalitis in systemic lupus erythematosus: a neglected cause of treatable nervous system infection. J Rheumatol 12:5:994
24. Denburg SD, Carbotte RM, Denburg JA (1987) Cognitive impairment in systemic lupus erythematosus: a neuropsychological study of individual and group deficits. J Clin Exp Neuropsychol 9:323
25. Denenberg VH, Sherman GF, Morrison L, Schrott LM, Waters NS, Rosen GD, Behan PO, Galaburda AM (1992) Behavior, ectopias and immunity in BD/DB reciprocal crosses. Brain Res 571:323
26. Denicoff KD, Rubinow DR, Papa MZ, Simpson C, Seipp CA, Lotze M, Chang AE, Rosenstein D, Rosenberg SA (1987) The neuropsychiatric effects of treatment with interleukin-2 and lymphokine-activated killer cells. Ann Intern Med 107:293
27. Derksen RHWM, van Dam AP, Meyling FHJG, Bijlsma JWJ, Smeenk RJT (1990) A prospective study on antiribosomal P proteins in two cases of familial lupus and recurrent psychosis. Ann Rheum Dis 49:779
28. De Simoni MG, Sironi M, De Luigi A, Manfridi A, antovani A, Ghezzi P (1990) Intracerebroventricular injection of interleukin 1 induces high circulating levels of interleukin 6. J Exp Med 171:1773
29. Devinsky O, Petito CK, Alonso DR (1988) Clinical and neuropathological findings in systemic lupus erythematosus: the role of vasculitis, heart emboli, and thrombotic thrombocytopenic purpura. Ann Neurol 23:380
30. Dixon FJ (1987) Basic elements of murine systemic lupus erythematosus. J Rheumatol 14:3
31. Ellis SG, Verity MA (1979) Central nervous system involvement in systemic lupus erythematosus: a review of neuropathologic findings in 57 cases, 1955-1977. Semin Arthritis Rheum 8:212
32. Fields RA, Sibbitt WL, Toubbeh H, Bankhurst AD (1990) Neuropsychiatric lupus erythematosus, cerebral infarctions, and anticardiolipin antibodies. Ann Rheum Dis 49:114
33. Fronek Z, Timmerman LA, Alper CA, et.al. (1990) Major histocompatibilty complex genes and susceptibility to systemc lupus erythematosus. Arthritis Rheum 33:1542
34. Futrell N, Millikan C (1989) Frequency, etiology, and prevention of stroke in patients with systemic lupus erythematosus. Stroke 20:584

35. Gahring LC, Weigle WO (1990) The regulatory effects of cytokines on the induction of a peripheral immunologic tolerance in mice. J Immunol 145:1318
36. Galve E, Candell-Riera J, Pigrau C, Permanyer-Miralda G, Garcia-Del-Castillo H, Soler-Soler J (1968) Prevalence, morphologic types, and evolution of cardiac valvular disease in systemic lupus erythematosus. N Engl J Med 319:817
37. Gibbons GH, Dzau VJ (1994) The emerging concept of vascular remodeling. N Engl J Med 330:1431
38. Golan DT (1988) Exposure to UV light in the pathogenesis of systemic lupus erythematosus(SLE). 24:360
39. Golombek SJ, Elkon KB (1986) Variety of auto-antibodies in the cerebrospinal fluid of lupus patients. Arthritis Rheum 29:S8
40. Golombek SJ, Graus F, Elkon KB (1986) Autoantibodies in the cerebrospinal fluid of patients with systemic lupus erythematosus. Arthritis Rheum 29:1090
41. Gregory CD, Dive C, Henderson S, Smith CA, Williams GT, Gordon J, Rickinson AB (1991) Activation of Epstein-Barr virus latent genes protects human B cells from death by apoptosis. Nature 349:612
42. Gutierrez-Ramos JC, Andreu JL, Alboran IM de, Rodriguez J, Leonardo E, Kroemer G, Marcos MAR, Martinez-A C (1990) Insights into autoimmunity: from classical models to current perspectives. Immunol Rev 118:73
43. Hahn BH, Ebling FM (1984) A public idiotype determinant is present on spontaneous cationic IgG antibodies to DNA from mice of unrelated lupus-prone strains. J Immunol 133:3015
44. Hammad A, Tsukada Y, Torre N (1992) Cerebral occlusive vasculopathy in systemic lupus erythematosus and speculation on the part played by complement. Ann Rheum Dis 51:550
44a. Handwerger B, Rus V, daSilva L. Via C (1994) The role of cyotkines in the immunopathogenesis of lupus. Springer Seminars in Immunopathology 16:153-180
45. Hanly JG, Rajaraman S, Behmann S, Denburg JA (1988) A novel neuronal antigen identified by sera from patients with system lupus erythematosus. Arthritis Rheum 31:1492
46. Hanly JG, Walsh NMG, Sangalang V (1992) Brain pathology in systemic lupus erythematosus. J Rheumatol 19:732
47. Hanly JG, Walsh NM, Fisk JD, Eastwood B, Hong C, Sherwood G, Jones JV, Jones E, Elkon K (1993) Cognitive impairment and autoantibodies in systemic lupus erythematosus. Br J Rheumatol 32:291
48. Hart MN, Zsuzsanna F, Waldschmidt M, Sandor M (1990) Lymphocyte interacting adhesion molecules on brain microvascular cells. Mol Immunol 27:1355
49. Hay EM, Isenberg DA (1993) Autoantibodies in central nervous system lupus. Br J Rheumatol 32:329
50. Hay EM, Huddy A, Black D, Mbaya P, Tomenson B, Bernstein RM, Holt PJL, Creed F (1994) A prospective study of psychiatric disorder and cognitive function in systemic lupus erythematosus. Ann Rheum Dis 53:298
51. Hellerstein MK, Meydani SN, Meydani M, Wu K, Dinarello CA (1989) Interleukin-1-induced anorexia in the rat. Influence of prostaglandins. J Clin Invest 84:228
52. Hess DC, Taormina M, Thompson J, Sethi D, Diamond B, Rao R, Chamberlain CR, Feldman DS (1993) Cognitive and neurologic deficits in the MRL/lpr mouse: a clinicopathologic study. J Rheumatol 20:610
53. Hirohata S, Miyamoto T (1990) Elevated levels of interleukin-6 in cerebrospinal fluid from patients with systemic lupus erythematosus and central nervous system involvement. Arthritis Rheum 33:644
54. Hirohata S, Hirose S, Miyamoto T (1985) Cerebrospinal fluid IgM, IgA, IgG indexes in systemic lupus erythematosus. Arch Intern Med 145:1843
55. How A, Dent PB, Liao S-K, Denburg JA (1985) Antineuronal antibodies in neuropsychiatric system lupus erythematosus. Arthritis Rheum 28:789
56. Hoyle C, Ewing DJ, Parker AC (1985) Acute autonomic neuropathy in association with systemic lupus erythematosus. Ann Rheum Dis 44:420
57. Isenberg DA, Snaith ML (1981) Muscle disease in systemic lupus erythematosus: a study of its nature, frequency and cause. J Rheumatol 8:917
58. Johnson RT, Richardson EP (1968) The neurological manifestations of systemic lupus erythematosus: a clinical-pathological study of 24 cases and review of the literature. Medicine 47:337
59. Jyonouchi H, Kincade PW, Good RA, Fernandes G (1981) Reciprocal transfer of abnormalities in clonable B lymphocytes and myeloid progenitors between NZB and DBA/2 mice1. J Immunol 127:1232

60. Kaell AT, Shetty M, Lee BCP, Lockshin MD (1986) The diversity of neurologic events in systemic lupus erythematosus. Arch Neurol 43:273
61. Katsuki H, Nakai S, Hirai Y, Akaji K, Kiso Y, Satoh M (1990) Interleukin 1B inhibits long-term potentiation in the CA3 region of mouse hippocampal slices. Eur J Pharmacol 181:323
62. Keeffe EB, Bardama EJ, Harbeck RJ, Pirofsky B, Carr RI (1974) Antibody to deoxyribonucleic acid (DNA) and DNA:anti-DNA complexes in cerebrospinal fluid. Ann Intern Med 80:58
63. Khamashta MA, Hughes GRV (1993) Antiphospholipid antibodies and valve disease in patients with systemic lupus erythematosus. J Am Coll Cardiol 22:1268
64. Khamashta MA, Harris EN, Gharavi AE, Derue G, Gil A, Vazquez JJ, Hughes GRV (1988) Immune mediated mechanism for thrombosis: antiphospholipid antibody binding to platelet membranes. Ann Rheum Dis 47:849
65. Khokher MA, Dandona P, Janah S, Coulston GL (1981) Insulin-like stimulatory effect of human immunoglobulin G on adipocyte lipogenesis. Diabetes 30:1068
66. Kitagawa Y, Gotoh F, Koto A, Okayasu H (1990) Stroke in systemic lupus erythematosus. Stroke 21:1533
67. Koren E, Reichlin MW, Koscec M, Fugat RD, Reichlin M (1993) Autoantibodies to the ribosomal P proteins react with a plasma membrane-related target on human cells. J Clin Invest 89:1236
68. Lee P, Urowitz MB, Bookman AAM, Koehler BE, Smythe HA, Gordon DA, Ogryzlo MA (1977) Systemic lupus erythematosus. A review of 110 cases with reference to nephritis, the nervous system, infections, aseptic necrosis and prognosis. Q J Med 46:1
69. Lerner MR, Boyle JA, Mount SM, Wolin SL, Steitz JA (1980) Are snRNPs involved in splicing? Nature 283:220
70. Linker-Israeli M (1992) Cytokine abnormalities in human lupus. Clin Immunol Immunopathol 63:10
71. Linker-Israeli M, Deans RJ, Wallace DJ, Prehn J, Ozeri-chen T, Klinenberg JR (1991) Elevated levels of endogenous IL-6 in systemic lupus erythematosus. J Immunol 147:117
72. Lo S, Salem HH, Howard MA, Oldmeadow MJ, Firkin B (1990) Studies of natural anticoagulant proteins and anticardiolipin antibodies in patients with the lupus anticoagulant. Br J Haematol 76:380
73. Long AA, Denburg SD, Carbotte RM, Singal DP, Denburg JA (1990) Serum lymphocytotoxic antibodies and neurocognitive function in systemic lupus erythematosus. Ann Rheum Dis 49:249
74. Lotz M, Vaughan JH, Carson DA (1988) Effect of neuropeptides on production of inflammatory cytokines by human monocytes. Science 241:1218
75. Love PE, Santoro SA (1990) Antiphospholipid antibodies: anticardiolipin and the lupus anticoagulant in systemic lupus erythematosus (SLE) and in non-SLE disorders. Ann Intern Med 112:682
76. Male D, Pryce G, Hughes C, Lantos P (1990) Lymphocyte migration into brain modelled in vitro: control by lymphocyte actication, cytokines, and antigen. Cell Immunol 127:1
77. McClain C, Cohen D, Phillips R, Ott L, Young B (1993) Increased plasma and ventricular fluid interleukin-6 levels in patients with head injury. J Lab Clin Med 118:225
78. Merino R, Iwamoto M, Fossati L, Izui S (1993) Polyclonal B cell activation arises from different mechanisms in lupus-prone (NZB x NZW)F1 and MRL/MpJ-lpr/lpr mice. J Immunol 151:6509
79. Michet CJ, McKenna CH, Elveback LR, Kaslow RA, Kurland LT (1985) Epidemiology of systemic lupus erythematosus and other connective tissue diseases in Rochester, Minnesota, 1950 through 1979. Mayo Clin Proc 60:105
80. Miguel EC, Pereira RMR, Pereira CADB, Baer L, Gomes RE, Sa LCF de, Hirsch R, Barros NG de, Navarro JM de, Gentil V (1994) Psychiatric manifestations of systemic lupus erythematosus: clinical features, symptoms, and signs of central nervous system activity in 43 patients. Medicine 73:224
81. Mitchell SR, Nguyen PQ, Katz P (1990) Increased risk of neisserial infections in systemic lupus erythematosus. Semin Arthritis Rheum 20:174
82. Moore PM (1990) Immunoglobulin binding to neuronal cell surface epitopes in murine systemic lupus erythematosus. J Neuroimmunol 30:101
83. Moore PM (1992) Evidence for bound antineuronal antibodies in brains of NZB/w mice. J Neuroimmunol 38:147
84. Moore PM, Joshi I, Ghanekar SM (1994) Affinity isolation of neuron reactive antibodies in MRL/lpr Mice. J Neurosci Res 39:140
85. Mountz JD, Wu J, Cheng J, Zhou T (1994) Autoimmune disease. A problem of defective apoptosis. Arthritis Rheum 37:1415
86. Nachman RL, Silverstein R (1993) Hypercoagulable states. Ann Intern Med 119:819

87. Narendran A, Hoffman SA (1989) Characterization of brain-reactive autoantibodies in murine models of systemic lupus erythematosus. J Neuroimmunol 24:113
88. Pisetsky DS, Grudier JP, Gilkeson GS (1990) A role for immunogenic DNA in the pathogenesis of systemic lupus erythematosus. Arthritis Rheum 33:153
89. Reichlin S (1993) Neuroendocrine-immune interactions. N Engl J Med 329:1246
90. Reidenberg MM (1983) Aromatic amines and the pathogenesis of lupus erythmatosus. Am J Med 75:1037
91. Ritchlin CT, Chabot RJ, Alper K, Buyon J, Belmont HM, Roubey R, Abramson SB (1992) Quantitative electroencephalography. Arthritis Rheum 35:1330
92. Rudh I, Olsson T, Lindstrom F, Skogh T (1985) Cerebrospinal fluid immunoglobulin abnormalities in systemic lupus erythematosus. J Neurol 48:807
93. Rudick RA, Eskin TA (1983) Neuropathological features of a lupus like disorder in autoimmune mice. Ann Neurobiol 14:325
94. Schneebaum AB, Singleton JD, Sterling G, West MD, Blodgett JK, Allen LG, Cheronis JC, Kotzin BL (1991) Association of psychiatric manifestations with antibodies to ribosomal P proteins in systemic lupus erythematosus. Am J Med 90:54
95. Schwartz MM, Roberts JL (1983) Membranous and vascular choroidopathy: two patterns of immune deposits in systemic lupus erythematosus. Clin Immunol Immunopathol 29:369
96. Schwartz RH (1989) Acquisition of immunologic self-tolerance. Cell 57:1073
97. Schwartz RS (1986) Anti-DNA antibodies and the problem of autoimmunity. Cell Immunol 99:38
98. Seibold JR, Buckingham RB, Medsger TA Jr, Kelly RH (1982) Cerebrospinal fluid immune complexes in systemic lupus erythematosus involving the central nervous system. Semin Arthritis Rheum 12:68
99. Sergent JS, Lockshin MD, Klempner MS, Lipsky BA (1975) Central nervous system disease in systemic lupus erythematosus. Am J Med 58:644
100. Shen HH, Winchester RJ (1986) Susceptibility genetics of systemic lupus erythematosus. Springer Semin Immunopathol 9:143
101. Sibbitt WL Jr, Froelich CJ, Bankhurst AD (1984) Interferon-regulation of lymphocyte function in systemic lupus erythematosus. Clin Immunol Immunopathol 32:70
102. Sibley JT, Olszynski WP, Decoteau WE, Sundaram MB (1992) The incidence and prognosis of central nervous system disease in systemic lupus erythematosus. J Rheumatol 19:47
103. Sieving RR, Kauffman CA, Watanakunakorn C (1975) Deep fungal infection in systemic lupus erythematosus-three cases reported, literature reviewed. J Rheumatol 2:61
104. Simon J, Simon O (1975) Effect of passive transfer of anti-brain antibodies to a normal recipient. Exp Neurol 47:523
105. Singh RR, Prasad K, Kumar A, Misra A, Padmakumar K, Malaviya AN (1988) Cerebellar ataxia in systemic lupus erythematosus: three case reports. Ann Rheum Dis 47:954
106. Steinberg AD, Gourley MF, Klinman DM, Tsokos GC, Scott DE, Kreig AM (1991) Systemic lupus erythematosus. Ann Intern Med 115:548
107. Sternberg EM, Chrousos GP, Wilder RL, Gold PW (1992) The stress response and the regulation of inflammatory disease. Ann Intern Med 117:854
108. Stevens MB (1986) Systemic lupus erythematosus clinical issues. Springer Semin Immunopathol 9:251
109. Stott DI (1990) Lessons about autoantibody specificity in systemic lupus erythematosus from animal models (editorial review). Clin Exp Immunol 81:1
110. Tan EM, Cohen AS, Fries JF, Masi AT, McShane DJ, Rothfield NF, Schaller JG, Talal N, Winchester RJ (1982) The 1982 revised criteria for the classification of systemic lupus erythematosus. Arthritis Rheum 25:1271
111. Tanaka Y, Saito K, Shirakawa F, Ota T, Suzuki H, Eto S, Yamashita U (1988) Production of B cell-stimulating factors by B cells in patients with systemic lupus erythematosus. J Immunol 141:3043
112. Tanaka Y, Shirakawa F, Ota T, Suzuki H, Eto S, Yamashita U (1988) Mechanism of spontaneous activation of B cells in patients with systemic lupus erythematosus. J Immunol 140:761
113. Taylor SI, Grunberger G, Marcus-Samuels B, Underhill LH, Dons RF, Ryan J, Roddam RF, Rupe CE, Gorden P (1982) Hypoglycemia associated with antibodies to the insulin receptor. N Engl J Med 307:1422
114. Teh L, Isenberg DA (1994) Antiribosomal protein antibodies in systemic lupus erythematosus. Arthritis Rheum 37:307

115. Teh LS, Bedwell AE, Isenberg DA, Gordon C, Emery P, Charles PJ, Harper M, Amos N, Williams BD (1992) Antibodies to protein P in systemic lupus erythmatosus. Ann Rheum Dis 51:489
116. Teh LS, Hay EM, Amos N, Black D, Huddy A, Creed F, Bernstein RM, Holt PS, Williams BD (1993) Anti-P antibodies are associated with psychiatric and focal cerebral disorders in patients with systemic lupus erythematosus. Br J Rheumatol 32:287
117. Theofilopoulos AN, Dixon FJ (1981) Etiopathogenesis of murine SLE. Immunol Rev 55:179
118. Venter JC (1982) Monocolonal antibodies and autoantibodies in the isolation and characterization of neurotransmitter receptors: The future of receptor research. J Molec Cell Immunol 14:687
119. Via CS, Tsokos GC, Bermas B, Clerici M, Shearer GM (1993) T cell-antigen-presenting cell interactions in human systemic lupus erythematosus. J Immunol 151:3914
120. Vidailhet M, Piette J, Wechsler B, Bousser MG, Brunet P (1990) Cerebral venous thrombosis in systemic lupus erythematosus. Stroke 21:1226
121. Winfield JB, Mimura T (1992) Pathogenetic significance of anti-lymphocyte autoantibodies in systemic lupus erythematosus. Clin Immunol Immunopathol 63:13
122. Winfield JB, Shaw M, Silverman LM, Eisenberg RA, Wilson HA III, Koffler D (1983) Intrathecal IgG synthesis and blood-brain barrier impairment in patients with systemic lupus erythematosus and central nervous system dysfunction. Am J Med 74:837
123. Wysenbeek AJ, Leibovici L, Zoldan J (1990) Acute central nervous system complications after pulse steroid therapy in patients with systemic lupus erythematosus. J Rheumatol 17:1695

Antigen-specific therapies for the treatment of autoimmune diseases

David A. Hafler[1], **Howard L. Weiner**[2]

[1] Laboratory of Molecular Immunology, and
[2] The MS Clinical and Research Unit, Center for Neurologic Diseases, Brigham & Women's Hospital and Harvard Medical School, Boston, MA 02115, USA

Overview of human autoimmune disease

Disease initiation

There are two general hypotheses for the etiology of human autoimmune disorders. In the first instance, the organ becomes infected by a virus or other infectious agent and cells that infiltrate that organ are targeted to the infectious agent. These events would then begin the cascade, leading to the organ-specific autoimmunc disease. The alternative hypothesis is that the initial infiltrating cells are autoimmune in nature and recognize the organ-specific proteins that are presented by local antigen-presenting cells in the context of the major histocompatibility complex (MHC). At our present state of knowledge, either hypothesis is possible. However, once the autoimmune cascade begins, it is likely that "epitope spreading" will occur, and T cells recognizing other organ-specific proteins will be recruited. Perhaps viruses and infectious agents acting as superantigens initially trigger or drive the autoimmune process, as opposed to being the primary target of infiltrating cells. Understanding the mechanism for initiation of autoimmune disorders and the range of antigens recognized by self-reactive T cells is critical for developing antigen-specific therapies for these disorders.

Organ-specific proteins as targets for T cells

What are the characteristics of a tissue-specific autoantigen that will be recognized by T cells? Approximately 0.1% of a restricting class II MHC protein must be occupied with the antigen peptide for T cell activation to occur [21]. Considering that there is competition by a large variety of self and foreign peptides for binding to MHC class II proteins [1], it is possible that this requirement may be met only by peptides which are present in relatively large quantities in antigen-presenting cells, processed

Correspondence to: D.A. Hafler

efficiently and bound with a high affinity to class II molecules. Thus, the physical characteristics and processing of the antigen in the tissue may limit the number of potential autoantigens that the immune system is capable of recognizing.

Secondly, a T cell autoantigen must be processed in such a way as to be presented in the context of class II MHC by the antigen-presenting cells specific for that tissue site (Fig. 1). Using the central nervous system (CNS) as an example, microglia, which are specialized CNS macrophages, might phagocytose breakdown products of myelin and present peptide antigens to T cells [22, 74]. Alternatively, glial cells such as astrocytes, which after activation can express class II MHC and present antigen, might present CNS antigens in the context of class II MHC [14, 15].

Epitope mapping of autoantigen

Of the CNS antigens, myelin basic protein (MBP) is perhaps the most extensively characterized autoantigens and is a primary candidate autoantigen in multiple sclerosis (MS) because of its ability to induce experimental autoimmune encephalomyelitis (EAE). We will use MBP as an example in discussing epitope mapping. There have been many investigations of MBP reactivity in MS using 7-day proliferation assays of whole peripheral blood mononuclear cells [26, 32]. These investigations have shown a slight increase in T cell responses to human MBP in subjects with MS, as compared to normal subjects or other neurological disease patients, but the magnitude of the difference has generally been less than convincing. T cell cloning techniques have recently been applied to study autoreactive T cells both in the peripheral blood and spinal fluid. Antigen-specific T cell clones are generated by repeated stimulation and culture of T cells with antigen in the presence of T cell growth factors.

In the EAE model, T cell clones can be derived from animals injected with MBP in adjuvant that recognize immunodominant regions of MBP that are influenced by the MHC of the animal. These clones can transfer EAE when injected into naive animals [35, 37, 57, 67]. For example in Lewis rats, the MBP(68-88) peptide is immunodominant and is the epitope for the majority of encephalitogenic T cell clones [35, 37, 67]. Lewis rat T cell clones generated against MBP (residues 43–67) are only weakly able to cause disease.

In humans, immunodominant regions of MBP that are presented by defined MHC proteins have beeen identifed. Immunodominant regions of MBP are between residues 84–106 and 143–172 near the C terminus [25, 36, 49, 75, 77]. In $DR2^+$ individuals, a high proportion of T cell clones reactive with MBP recognized the 84–106 region, whether or not they had MS [49]. Both DR and DQ antigens were found to be restricting elements, indicating that immunodominant MBP peptides can be presented by different class II molecules. It has also been demonstrated that T cells from genetically diverse individuals responding to different epitopes of human MBP are associated with distinct MHC class II molecules [9]. Of interest was the finding that DR^+ possessed an unusual capacity to restrict all of the epitopes identified on MBP [9, 25, 77]. Moreover, different types of DR2 have the capacity to present a number of different MBP epitopes [50]. These data together indicate that disease-associated DR2 antigens can present a variety of MBP peptides to T cells, and this is likely to be associated with competition for peptide binding between different MBP regions. Thus, peptide competition for binding to class II molecules may in part determine immunodominant sites of an autoantigen.

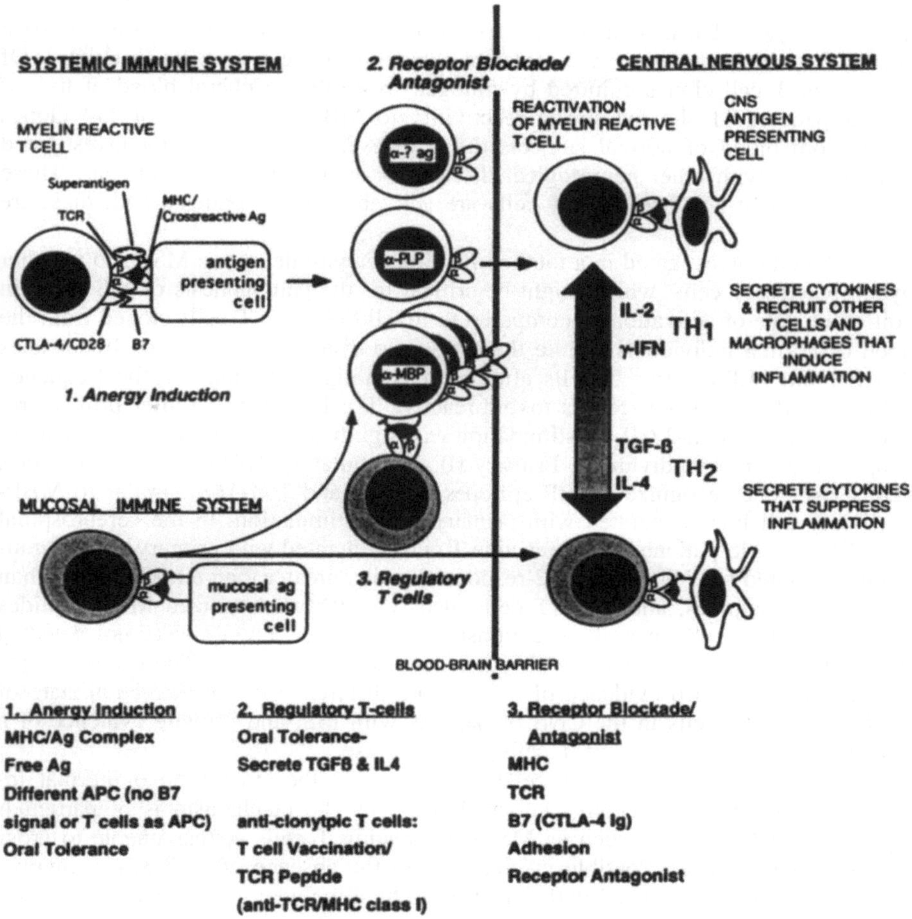

Fig. 1. Overview of proposed pathophysiology for multiple sclerosis (MS) and specific areas where immune system can be manipulated. The minimal requirements in mammals for inducing inflammatory autoimmune disease in the CNS white matter are activated $CD4^+$ T cells recognizing myelin basic protein (MBP) or proteolipid (PLP). In MS, we hypothesize that some activating event, such as a viral superantigen or molecular mimicry activates an autoreactive T cell. This allows the activate T cell to migrate into the CNS where it can now recognize CNS antigen in the context of NMC, recruiting effector cells leading to the inflammatory CNS response. The process may be blocked by: (1) anergy induction; (2) by regulatory T cells; or (3) by receptor blockade or antagonist (see text) (*Ag*, antigen; *TCR* T cell receptor; *APC* antigen-presenting cell; *TGF-β*, transforming growth factor-*β*; *IL*, interleukin; *IFN*, interferon)

Activation of autoreactive T cells

T cells recognizing MBP and proteolipid protein (PLP) exist in normal humans. From investigation of the EAE model, it has been learned that for these cells to be pathogenic they must be activated in vivo. To examine whether MBP-reactive T cells are activated in vivo, an $hprt^-$ mutant assay was used [2]. This assay is based on the observation that dividing cells acquire random mutations during DNA synthesis. Some of these mutations occur in the hprt gene, which results in inactivation of the hprt enzyme. Thus, mutant cells do not metabolize thioguanine to a cytotoxic metabolite, which allows a very effective selection of these mutants in culture. Of 258 mutant T cell clones cultured by mitogen from the peripheral blood of five of six MS patients 11 showed strong reactivitity to MBP, while none of 114 clones grown from blood of normal subjects did. However, clones were not investigated from subjects with other neurological diseases or with non-myelin antigens. These data suggest that MBP-reactive T cells are activated in MS patients and, thus, are pathogenic.

We directly investigated in a total of 72 subjects with definitive MS as to whether myelin-reactive T cells, which might be critical for the pathogenesis of MS, exist in a different state of activation as compared to myelin-reactive T cells cloned from the blood of normal individuals. While there were no differences in the the frequencies of MBP and PLP-reactive T cells after primary antigen stimulation, the frequency of MBP-or PLV- but not tetanus toxoid-reactive T cells generated after primary recombinant interleukin-2 (rIL-2) stimulation was significantly higher in MS patients as compared to control individuals. Primary rIL-2-stimulated MBP-reactive T cell lines were $CD4^+$, and recognized MBP epitopes 84–102 and 143–168, similar to MBP-reactive T cell lines generated with primary MBP stimulation. In the cerebrospinal fluid (CSF) of MS patients, MBP-reactive T cells generated with primary rIL-2 stimulation accounted for 7% of the IL-2-responsive cells, greater than tenfold higher than paired blood samples, and these T cells also selectively recognized MBP peptides 84–102 and 143–168. In striking contrast, MBP-reactive T cells were not detected in CSF obtained from patients with other neurological diseases. These results provide definitive in vitro evidence of an absolute difference in the activation state of myelin-reactive T cells in the CNS of patients with MS, and provide evidence of a pathogenic role of autoreactive T cells in the disease.

If circulating, autoreactive T cells are present in the circulation of normal individuals, how do they become activated (Fig. 1)? The mechanism is of particular interest in a CNS disease such as MS, since resting T cells appear unable to cross the blood-brain barrier. Possible mechanisms in the absence of autoantigen involve immune activation associated with infections which include: (1) molecular mimicry, (2) activation by superantigens, and (3) bystander CD2 activation.

Oral tolerance: antigen-specific immunotherapy

Overview of antigen-specific immunotherapy

In the section above, we presented an overview of antigen-specific recognition in autoimmune diseases. In total, these experiments suggest a number of manipulations that may be used to block the immune response. It is important to note the likelihood that no in vitro experiment can prove the autoinmune hypothesis for any of the presumed autoimmune diseases. Instead, only specific manipulations in vivo where antigen-reactive cells are targeted with associated amelioration of disease activity will allow the understanding of the disease's pathogenesis. Thus, these specific manipulations of the immune response represent both attempts to treat the disease as well as scientific experiments to undertand the disease.

There are two general aproaches to antigen-specific immunotherapy of autoimmune disease. One approach is to block either the initial activation or the subsequent recognition of autoantigen by autoreactive T cells (see Fig. 1). This approach requires the immune response against the self antigen to be restricted in terms of either the number of epitopes recognized or the T cell receptor (TCR) repertoire that is used. This includes MHC-blocking peptides, TCR antagonists, and TCR peptides which may inhibit specific T cell function. Considering MS as an example, it would appear that, although there are MBP and PLP dominant epitopes and somewhat restricted use of TCR in their recognition, the outbred human species contains too many myelin-reactive T cells with different TCR usage to sucessfully inhibit an of the spreading immune response to myelin antigens.

The second approach involves the antigen-specific targeting of T cells that down-regulate immune responses to the inflammatory sites. In this instance, it is not necessary to know the inciting antigen that elicites the immune response, nor is it required that the T cells be restricted in terms of antigen recognition or TCR usage. This approach may well be physiological in regulating the normal immune response. One such method which has attracted much attention recently is the use of oral tolerization, where the autoantigen is delivered orally with the generation of T cells secreting transforming growth factor-β (TGF-β), IL-4, and IL-10 that migrate to the site.of inflammation and suppress the immune response.

Mechanism of oral tolerance

Oral tolerance represents the exogenous administration of antigen to the peripheral immune system via the gut. As such, it is a form of antigen-driven peripheral immune tolerance. Immunological tolerance is not programmed into the germ line but is acquired during maturation of the immune system by mechanisms that delete or inactivate antigen-reactive clones. There are three basic mechanisms to explain antigen-driven tolerance: clonal deletion, clonal anergy, and active suppression [30, 42]. A large number of studies have shown that one of the primary mechanims associated with oral tolerance is the generation of active suppression [44]. More recently, clonal anergy has been demonstrated [38, 73]. There is little evidence that orally administered antigen induces clonal deletion. During the course of our investigations we have delineated two pathways by which oral tolerization results in systemic hyporesponsiveness by either active suppression or clonal anergy [16]. These pathways are described below in a schema which forms the theoretical basis for this review.

Based on recent findings in our laboratory and others, we believe that oral tolerance can be viewed as *anergy-driven oral tolerance* or *regulatory-cell-driven oral tolerance*. The primary factor which determines which form of peripheral tolerance develops following oral administration of antigen is the dose of antigen fed. Low doses of antigen favor the generation of active suppression or regulatory-cell-driven tolerance, whereas high doses of antigen favor anergy-driven tolerance (Fig. 2). Although these forms of oral tolerance are not mutually exclusive and may occur simultaneously, they are distinct and the use of oral tolerance to treat autoimmune diseases is critically dependent on which of these two mechanisms is triggered.

The delineation of these two mechanisms of oral tolerance was based on the following: (1) investigations in our laboratory using low doses of orally administered autoantigens were shown to suppress experimental autoimmune diseases via the generation of regulatory cells that suppressed in vitro and in vivo via the secretion of downregulatory cytokines such as TGF-β [41]; (2) investigations from other labora-

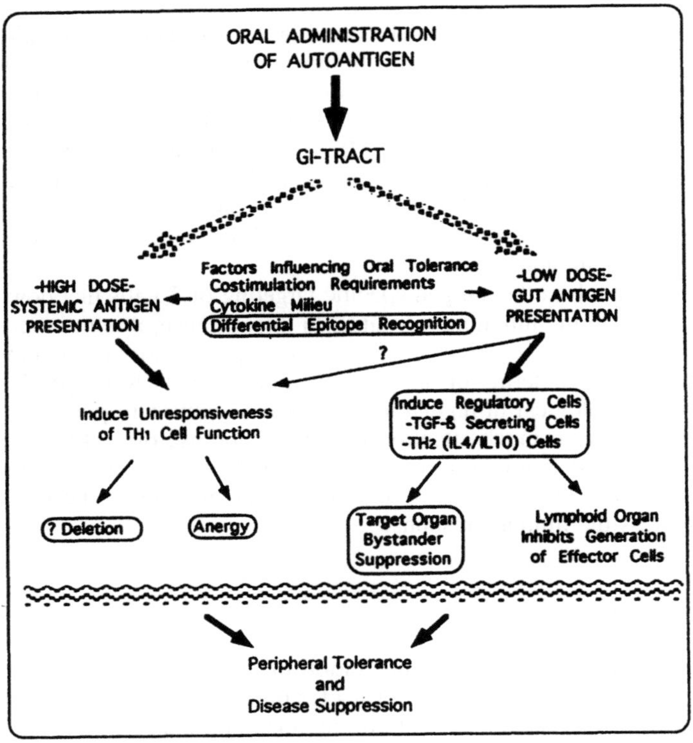

Fig. 2. Mechanisms of oral tolerance. High-dose antigen administration leads to the generation of anergy, while lower-dose oral antigen induces regulatory T cells that secrete TGF-β1 and/or IL-4/IL-10 which migrate to the CNS and down regulate the local immune response

tories which demonstrated clonal anergy following oral administration of large doses of antigen with no evidence of active suppression [38, 73]; (3) a large series of investigations demonstrated transferrable suppression following oral tolerance, including work which showed two components of oral tolerance, one that was abrogated by treatment with low-dose cyclophosphamide and one that was not, a difference that was dose dependent [45]; and (4) direct comparison in our laboratory demonstrating that the two mechanisms depend on the dose [16] (Fig. 2).

As shown in Fig. 2, *low doses* of antigen result in the generation of antigen-specific regulatory cells, and as such involve presentation of antigen by gut-associated antigen-presenting cells. Such presentation preferentially induces regulatory cells which upon subsequent recognition of antigen in vivo or in vitro, secrete the suppressive cytokine TGF-β. In addition, Th2 responses are preferentially generated in the gut, resulting in cells which secrete IL-4 and IL-10. These antigen-specific regulatory cells migrate to lymphoid organs and suppress immune responses by inhibiting the generation of effector cells, and to the target organ to suppress disease by releasing antigen-nonspecific cytokines (bystander suppression). Several factors can affect the generation of regulatory cells including co-stimulation requirements, the cytokine milieu in which the immune response is generated, and differential generation of epitopes which may preferentially trigger certain regulatory cells.

High doses of orally administered antigen result in systemic antigen presentation after antigen passes through the gut and enters the systemic circulation, either as intact protein or as antigen fragments. High doses of antigen induce unresponsiveness of Th1 cell function, primarily via clonal anergy. Whether, the degree to which clonal anergy occurs following high doses of antigen merely represents the direct passage of small amounts of antigen into the systemic or portal circulation, or is dependent on filtration by the gut is unknown. Why there is reduced active suppression with high doses of orally administered antigen is unclear, but could relate to anergizing cells involved in the generation of active suppression. In addition, it is not known to what degree costimulatory requirements, cytokine milieu, and differential epitope recognition may preferentially favor the generation of anergy in Th1 cells.

Active suppression

Many studies demonstrate that active suppression is an important mechanism for oral tolerance [10, 20, 34, 43, 44, 52, 61]. After feeding antigens such as ovalbumin or sheep red blood cells, transferrable suppression mediated by T cells from Peyer's patches, mesenteric lymph node, and spleen has been demonstrated. Investigators have also reported initial sensitization prior to the appearance of suppression [20]. Further characterization of active suppression as measured in these systems has not occurred, most probably related to the difficulties in defining the biology of suppression [6, 59]. Nonetheless, the demonstration of transferrable cellular suppression associated with oral tolerance is a recurrent theme reported by many investigators [44].

Our studies of oral tolerance in autoimmune models have found active suppression to be a primary mechanism, and have identified regulatory cells, generated following oral tolerance, which act via the secretion of antigen-nonspecific down-regulatory cytokines following triggering by the fed antigen [70]. These cells have been characterized both in the rat and murine model of EAE orally tolerized to MBP. In the Lewis rat model, regulatory cells were $CD8^+$ [31], and acted via the secretion of TGF-β following antigen-specific triggering. They transfer suppression in vivo and can suppress in vitro. The epitopes of guinea pig MBP triggering $CD8^+$ regulatory cells following orally administered MBP were different than the encephalitogenic determinant. In addition, TGF-β-secreting regulatory cells can be found in Peyer's patches 24–48 h after one feeding of MBP [54]. Of note is that cells from Peyer's patches removed after one feeding of MBP do not proliferate in response to MBP, even though they release TGF-β upon in vitro stimulation. The mechanism by which these regulatory cells are induced remains unknown. It is also not known to what degree to the generation of regulatory cells is related to unique antigen-presenting cells in the gut, the cytokine milieu, or other factors. Murine models of oral tolerance suggest that in addition to $CD8^+$ TGF-β secreting regulatory cells, $CD4^+$ regulatory cells are also induced. These $CD4^+$ cells secrete IL-4 and IL-10 in addition to TGF-β. Moreover, these TGF-β secreting T cells can adoptively transfer suppression of EAE to naive animals [8].

Anergy

Anergy has only recently been demonstrated to be a possible mechanism for oral tolerance [38, 73]. Anergy is defined as a state of T lymphocyte unresponsiveness characterized by absence of proliferation, IL-2 production and diminished expression of IL-2R [58]. Anergy may be experimentally differentiated from clonal deletion by demonstrating the presence of antigen-specific TCR clonotypes, or by release from the anergic state which is accomplished by pre-culture of cells in IL-2 [13]. Under these conditions we have shown that a single feeding of 20 mg ovalbumin (OVA) induced a state of anergy in OVA-specific T lymphocytes: cells did not respond to OVA by proliferation, OVA stimulation did not induce IL-2 production or IL2R expression, and the non-responsive state was reversed by pre-culture of tolerized cells in IL-2 [38]. One other study has indirectly demonstrated anergy as a mechanism for oral tolerance [73]. Whitacre et al. [73] reported diminished IL-2 and interferon-γ (IFN-γ) production in rats orally fed MBP in the presence of the soybean protease inhibitor; however, anergy was not confirmed in this study by TCR analysis or by IL-2-driven release. As discussed previously, the induction of anergy depends upon antigen dosage and frequency of feeding [16].

Treatment of organ-specific autoimmune diseases in animals

We have examined the manner in which oral tolerance may be applied to the treatment of autoimmune conditions in both humans and animals. Thompson and Staines [63] and Nagler-Anderson et al., [46] initially described suppression of collagen-induced arthritis by feeding type II collagen. Our laboratory [7, 23, 40] and that of Whitacre's [5, 17, 73] have studied suppression of EAE by orally administered myelin antigens. Additionally, we have investigated oral tolerance to suppress autoimmune models of uveitis [47], diabetes in the NOD mouse [4, 78] and adjuvant arthritis [3, 76], as well as orally-administered alloantigen or MHC peptide in transplantation models [28, 55, 56].

Investigators have also demonstrated suppression of other autoimmune models by orally administered antigen [72]. Thus, the ability to suppress autoimmunity in animal models via oral tolerance has been established, although the mechanisms responsible may differ depending on the laboratory and the models being studied.

Experimental autoimmune encephalomyelitis

Orally administered guinea pig MBP suppresses EAE in the Lewis rat model [5, 23]. As discussed above, this effect is mediated both by active suppression and anergy. An important question before entering human clinical trials was whether or not ongoing disease could be suppressed. We investigated chronic relapsing EAE in the Lewis rat and strain 13 guinea pig. Disease was also suppressed by oral administration of MBP or a bovine myelin preparation that is also being administered in human clinical trials. There was no exacerbation of disease in these animals, demonstrating that orally administered antigens do not appear to prime rather than suppress in an already inununized animal. Of note in the guinea pig model is that 10 mg of bovine myelin fed three times per week over a 3-month period suppressed disease and histological manifestations, whereas feeding 50 mg of bovine myelin did not. This may indicate that bystander suppression was responsible for the effect on chronic disease, and that in some instances, the oral administration of too high a dose will not suppress autoimmune models. As discussed later, loss of protection by orally administered antigen at higher doses was also seen with orally administered collagen in adjuvant arthritis (AA) and orally,administered insulin to suppress diabetes in the NOD mouse.

Lastly, we will discuss the immunhistology associated with oral tolerization to MBP and in animals recovering naturally from EAE [28]. Brains from OVA-fed animals at the peak of disease showed perivascular infiltration with activated mononuclear cells which secreted the inflammatory cytokines IL-1, IL-2, tumor necrosis factor-α IFN-γ, IL-6 and IL-8. Inhibitory cytokines, TGF-β and IL-4, and prostaglandin E2 (PGE_2) were absent. In MBP orally tolerized animals, there was a marked reduction of the perivascular infiltrate and down-regulation of all inflammatory cytokines. In addition, there was up-regulation of the inhibitory cytokine TGF-β. When bacterial lipopolysaccharide (LPS) was fed in addition to MBP, protection against EAE was enhanced and was associated with elevated IL-4 and PGE_2 in the brain [27, 28]. In control recovering animals (day 18) staining for inflammatory cytokines was diminished and there was up-regulation of TGF-β and IL-4. These results suggest that the suppression of EAE by oral tolerization and natural recovery is related to regulatory cells that secrete inhibitory cytokines at the target organ.

Collagen and adjuvant arthritis

Previous investigators have demonstrated suppression of collagen-induced arthritis by feeding collagen type II [46, 63]. We have studied AA in the rat, a well-characterized and more fulminant form of experimental arthritis [76]. Oral administration of chicken collagen type II (CII), given at a dose of 3 mg per feeding, consistently suppressed the development of AA. A decrease in delayed-type hypersensitivity responses to CII was observed that correlated with suppression to AA. Oral administration of collagen type I also suppressed AA; only minimal effects were seen with collagen type III. Suppression was antigen specific in that feeding collagen type II did not suppress EAE, and feeding MBP did not suppress AA. Suppression of AA could be adoptively transferred by T cells from CII-fed animals and was observed when CII was fed after disease onset. Of note is that suppression was observed at doses of 3 and 30 mg, but not at 300 or 1000 mg. These results suggest that oral collagen is suppressing AA via bystander suppression rather than clonal anergy since active suppression may be lost at higher doses. The effectiveness of such small amounts of oral collagen may be related to the fact that collagen has repeating amino acid subunits.

Uveitis

Oral administration of S-antigen (S-Ag), a retinal autoantigen that induces experimental autoimmune uveitis (EAU), prevented or markedly diminished the clinical appearance of S-Ag-induced disease as measured by ocular inflammation [19, 47, 60, 62, 64, 65, 68]. Furthermore, oral administration of S-Ag also markedly diminished uveitis induced by the uveitogenic M and N fragments of the S-Ag. Oral administration of S-Ag did not prevent MBP-induced EAE. In vitro studies demonstrated a significant decrease in proliferative responses to the S-Ag in lymph node cells draining the site of immunization from fed versus non-fed animals. Furthermore, the addition of splenocytes from S-Ag-fed animals to cultures of a $CD4^+$ S-Ag-specific lymphocyte line profoundly suppressed the line's response to the S-Ag, whereas these splenocytes had no effect on a PPD-specific lymphocyte line. The antigen-specific in vitro suppression was blocked by anti-CD8 antibody, demonstrating that suppression was dependent on $CD8^+$ T lymphocytes. As in EAE, EAU was also suppressed by feeding S-Ag-related peptides that were either uveitogenic, cross-reactive or synthetic [19, 60, 62, 65, 68]. Gregerson et al. [19], using high and low doses of S-Ag peptides, also found that low doses of antigen favor suppression, whereas high doses induce unresponsiveness or anergy.

Diabetes

NOD diabetic mice spontaneously develop an autoimmune form of diabetes associated with insulitis. To test oral tolerance in the NOD model, we administered porcine insulin at a dose of 1 mg orally twice a week for 5 weeks and then weekly until 1 year of age [69]. The severity of lymphocytic infiltration of pancreatic islets was reduced by oral administration of insulin, and there was a delay in the onset of diabetes. A decreased incidence of diabetes was seen in animals followed for 1 year. Suppression of insulitis as observed at a dose of 1 mg but not 5 mg. As expected, orally administered insulin had no metabolic affect on blood-glucose levels. Furthermore, splenic T cells from animals orally treated with insulin adoptively transferred protection against diabetes, demonstrating that oral insulin generates active cellular mechanisms that suppress disease. Ongoing studies have demonstrated the ability to suppress insulitis by administering insulin peptides, the B chain of insulin, or GAD. Immunohistochemical studies have demonstrated an increase of IL-4 in the islets of insulin fed animals. Given the mechanism of antigen-driven bystander suppression, our results do not implicate autoreactivity to insulin as a pathogenic mechanism in the NOD mouse. Indeed, initial experiments suggest that orally administered glucagon can suppress insulitis.

Treatment of autoimmune disease in humans

The first attempts at oral tolerization may have been utilized by Native Americans who we re thought to have fed their children *Rhus* leaves to prevent them from becoming sensitized to poison ivy [12]. Investigators have shown that exposure of a contact-sensitizing agent via the mucosa prior to subsequent skin challenge led to unresponsiveness in a portion of the subjects studied [33]. In another study on human volunteers serum antibodies to bovine serum albumin (BSA) were measured before and after feeding large amounts of this antigen (0.1–1.5 mg BSA per pound per day). Those subjects that had anti-BSA antibodies prior to eating BSA showed a rise in their serum anti-BSA titers. A similar response was observed when some subjects were given an iniection of BSA. Subjects who did not have anti-BSA antibodies before or after the test did not respond to subsequent intradermal immunization [29]. Oral desensitization has also been attemped in Rh disease [18]. In an attempt at oral immunization, human volunteers were given capsules containing killed streptococcus mutans and circulating IgA-producing cells were found in some subjects [11]. This suggests that some generation of a secretory immune response can occur following oral ingestion of microbial antigens. Orally administered keyhole limpet hemocyanin (KLH), 50 mg given daily for 2 weeks over a 3-week period, has been reported to decrease subsequent cell-mediated immune responses, although antibody responses were not affected [24]. In addition, preliminary experiments in subjects fed KLH have suggested that cell lines may be generated that suppress proliferative responses, although more work is needed in this area [51].

Multiple sclerosis

To determine whether orally ingested autoantigens could affect the clinical course and immune responses in patients with an autoimmune disease, 15 patients with relapsing remitting MS were fed a capsule containing 300 mg of bovine myelin or placebo daily for 1 year [71]. The results demonstrated a decrease in MBP-reactive cells in the bloodstream of MS patients as compared to controls. There was no evidence of sensitization either as measured by antibody levels to MBP or PLP, or by increased proliferative responses to the fed antigens at a 1-year period. Clinical responses demonstrated that 12 of 15 placebo-fed patients had major MS attacks, whereas only 6 of 15 in the control group had attacks ($P = 0.06$). It appeared that a subgroup of patients that were either males or $DR2^-$ preferentially responded to the oral tolerization. However, the sample size was small and the degree to which this subgroup response will occur in future studies is unknown. Based on these observations, a multicenter 520 patient double-blind placebo controlled trial of bovine myelin in $DR2^-$ relapsing remitting MS patients, both males and females, has begun. Of note is that trials of given MBP given by multiple subcutaneous injections resulted in prominent immune responses to the antigen [53].

Rheumatoid arthritis

A 60-patient double-blind trial of oral collagen administration to patients with rheuma demonstrated a decrease in joint swelling and disease index in patients fed collagen compared to placebo controls [66]. Patients were previously or currently on immunosuppressant drugs, such as methotrexate, that had failed such therapy. Patients were taken off these medications and treated for a 3-month period. In the 1st month, they received 100 mg oral collagen per day, and in the 2nd and 3rd months 500 mg per day. These doses were extrapolated from the small amounts of collagen used to suppress AA in the Lewis rat. There were no toxicities or evidence of sensitization to type II collagen in fed patients as measured by anti-collagen antibodies. There was no linkage to either DR type or sex in the patients that responded. In patients treated with the collagen there was less need for narcotic use during the course of the study, and four patients in the collagen-treated group apparently had complete remission of their rheumatoid arthritis. Given the biological effects seen, future trials will focus on other disease categories, more prolonged administration, and dosing studies. A multicenter trial is currently being planned. Given what is known of the mechanism of oral tolerization, these studies do not establish that type II collagen is a target autoantigen in the disease. Indeed, given the low doses fed, it is possible that the effect may have been mediated by regulatory cells that migrated to the joint and released anti-inflammatory cytokines such as TGF-β or IL-4. Whether patients that went into complete remission represent a separate category remains to be determined.

Uveitis

An open-label pilot study has been performed on two patients with uveitis, one with Bechet's disease [48] and the second with pars planitis. In this open label trial, patients had required steroids and/or cyclosporin to maintain visual function. Patients were started on 30 mg of bovine S-Ag three times a week, and then tapered from steroids and immunosuppressive mediation. A positive therapeutic response was observed in both patients, in that over a 2-year period they were able to reduce their previous medication without worsening of vision and with decrease in S-Ag responsiveness. A double-masked placebo-controled trial of 45 patients is currently in progress.

Future directions

Given the results in animal models of autoimmunity and initial studies in human disease states, it appears that orally administered autoantigens may find a place for the treatment of human organ-specific inflammatory autoimmune diseases. Such therapy would have the advantages of being orally administered, non-toxic, and antigenspecific. The mechanism of bystander suppression solves a major problem related to designing antigen- or T cell-specific therapy of inflammatory autoimmune diseases, since one need not necessarily identify the target autoantigen for oral tolerance to be effective. As discussed above, it is likely that in human autoimmune disease states, there are reactivities to multiple autoantigens from the target organ and multiple epitopes given that humans are an outbred population. Dosing appears to be important for stimulating the active suppression component of oral tolerance, and identification of regulatory cells in humans following oral tolerization is critical for demonstrating the immunological effects of oral tolerization. Given the results in animal studies, one would predict that homologous protein and the use of synergist or enhancers would increase the biological efficiency of oral tolerance, In this regard, recombinant human proteins and the concommitant administration of immune adjuvants to enhance generation of regulatory cells would be required.

References

1. Adorini L, Muller, S, Cardineaux F, Lehmann P, Falcioni F, Nagy ZA (1988) In vivo competition between self peptides and foreign antigens in T-cell activation. Nature 334:623
2. Allegretta M, Nickals JA, Sriram S, Albertini RJ (1990) T cells responsive to myelin basic protein in patients with multiple sclerosis. Science 247:717
3. Birnbaum G, Lackovic V, Kotilinek L, Tobolt D (1990) A comparison of regulatory cells in spinal fluid and blood in patients with multiple sclerosis and other neurologic diseases. Neurology 40:1785
4. Bisaccia G, Caputo D, Landoni AM, Macchi GP (1978) Heterogeneity of human T lymphocytes to bind sheep red blood cells in multiple sclerosis patients and controls. Boll Ist Sieroter Milan 56:603
5. Bitar D, Whitacre CC (1988) Suppression of experimental autoimmune encephalomyelitis by the oral administration of myelin basic protein. Cell Immunol 112:364

6. Bloom BR, Modlin RL, Salgame P (1992) Stigma variations: observations on suppressor T cells and leprosy. Annu Rev Immunol 10:453
7. Brod SA, Al-Sabbagh A, Sobel RA, Hater DA, Weiner HL (1995) Suppression of experimental autoimmune encephalomyelitis by oral administration of myelin antigens. IV. Suppression of chronic relapsing disease in the Lewis rat and strain 13 guinea pig. Ann Neurol (in press)
8. Chen Y, Kuchroo V, Hafler DA, Weiner HL (1994) Myelin basic protein specific regulatory T cell u clones from orally tolerized mice suppress autoimmune encephalomyelitis. Science 265:1737
9. Chou YK, Vainiene M, Whitham R, Bourdette D, Chou CH, Hashim G, Offner H, Vandenbark AA (1989) Response of human T lymphocyte lines to myelin basic protein: association of dominant epitopes with HLA class II restriction molecules. J Neurosci Res 23:207
10. Cowdery JS, Johlin BJ (1984) Regulation of te primary in vitro response to TNP-polymerized ovalbumin by T suppressor cells induced by ovalbumin feeding. J Immunol 132:2783
11. Czerinsky C, Prince SJ, Michalek SM, Jackson S, Russell MW, Moldoveanu Z, McGhee JR, Mestecky J (1987) IgA antibody-producing cells in peripheral blood after antigen intesation: evidence for a common mucosal immune system in humans. Proc Natl Acad Sci USA 84:2449
12. Dakin R (1829) Remarks on a cutaneous affection produced by certain poisonous vegetables. Am J Med Sci 4:98
13. DeSilva DR, Urdahl KB, Jenkins MK (1991) Clonal anergy is induced in vitro by T cell receptor occupancy in the absence of proliferation. J Immunol 147:3261
14. Fierz W, Endler B, Reske K, Wekerle HAF (1985) Astrocytes as antigen-presenting cells. I. Induction of Ia antigen expression on astrocytes by T cells via immune interferon and its effect on atigen presentation. J Immunol 134:3785
15. Fontana A, Fierz W, Wekerle H (1984) Astrocytes present myelin basic protein to encephalitogenic T-cell lines. Nature 307:273
16. Friedman A, Weiner HI (1993) Induction of anergy and/or active suppression in oral tolerance is determined by frequency of feeding and antigen dosage. FASEB nn:nn
17. Fuller KA, Pearl D, Whitacre CC. Oral tolerance in experimental autoimmunc encephalomyelitis: serum and salivary antibody responses. J Neuroimmunol 28:15
18. Gold WRJ, Queenan FT, Woody J. Sacher RA (1983) Am J Obstet Gyecol 146:980
19. Gregeerson D, Obritsch W, Donoso LA (1993) Suppression of clonal anergy play roles in oral tolerance and EAU. Invest Opthalmol Vis Sci 34 [Suppl]:902
20. Guatam SC, Chikkala NF, Battisto JR (1990) Oral administration of the contact sensitizer trinitrochlorobenzene: initial sensitization and subsequent appearance of a suppressor population. Cell Immunol 125:437
21. Herding CV, Unanue ER (1990) Quantitation of antigen-presentinc, cell MHC class II/peptide complexes necessary for T-cell stimulation. Nature 346:574
22. Hickey WF, Kimura HP (1988) Perivascular microglial cells of the CNS are bone marrow-derived and present antigen in vivo. Science 239:290
23. Higgins PJ, Weiner HL (1988) Suppression of experimental autoimmune encephalomyelitis by oral administration of myelin basic protein and its fragments. J Immunol 140:440
24. Husby S, Elson CO, Moldoveanu Z, Mestecky J (1992) Oral tolerance in humans. T cell but not B cell tolerance to a soluble protein antigen. 7th International Congress on Mucosal Immunology Prague, Czechoslavakia, 1992
25. Jaraquemada D, Martin R, Rosen Bronson S, Flerlage M, M Farland H, Long EO (1990) HLA-DR2a is the dominant restriction molecule for the cytotoxic T cell response to myelin basic protein in DR2Dw2 individuals. J Immunol 145:2880
26. Johnson D, Hafler DA, Fallis RJ, Lees MB, Brady RO, Quarles RH, Weiner HL (1986) Cell-mediated immunity to myelin-associated glycoprotein, proteolipid protein, and myelin basic protein in multiple sclerosis. J Neuroimmunol 13:99
27. Khoury SJ, Lider 0, Al-Sabbagh A, Weiner HL (1990) Suppression of experimental autoimmune encephalomyelitis by oral administration of myelin basic protein. III. Synergistic effect of lipopolysaccharide. Cell Immunol 88:4O4
28. Khoury SJ, Hancock WW, Weiner HL (1992) Oral tolerance to myelin basic protein and natural recovery from experimental autoimmune encephalomyelitis are associated with down-regulation of inflammatory cytokines and differential up-regulation of TGF-β IL-4 and PGE expression in the brain. J Exp Med 46:1355

29. Korenblatt PE, Rothberg RM, Minden P, Farr RS (1968) Immune responses of human adults after oral and parenteral exposure to bovine serum albumin. J Allergy 41:226
30. Kroemer G, Martinez AC (1992) Mechanisms of cell tolerance. Immunol Today 13:401
31. Lider O, Santos LMB, Lee CSY, Higgins PJ, Weiner HL (1989) Suppression of experimental autoimmune encephalomyelitis by oral administration of myelin basic protein. II. Suppression of disease and in vitro immune reponses is mediated by antigen-specific $CD8^+$ T lymphocytes. J Immunol 174:791
32. Lisak RP, Zweiman B (1977) In vitro cell-mediated immunity of cerebrospinal-fluid lymphocytes to myelin basic protein in primary demyelinating diseases. N Engl Med 297:850
33. Lowney ED (1968) Immunologic unresponsiveness to a contact sensitizer in man. J Invest Dermatol 51:411
34. MacDonald TT (1982) Immunosuppression caused by antigen feeding. I. Evidence for the activation of a feedback suppressor pathway in the spleens of antigen-fed mice. Eur J Immunol 12:767
35. Martenson RE, Levine S, Sowinski R (1975) The location of regions in guinea pig and bovine myelin basic proteins which induce experimental allergic encephalomyelitis in Lewis rats. J Immunol 114:592
36. Martin R, Howell MD, Jaraquemada D (1991) A myelin basic protein peptide is recognized by cytotoxic T cells in the context of four HLA-DR types associated with multiple sclerosis. J Exp Med 173:19
37. McFarlin DE, Blank SE, Kibler RF, (1973) Experimental allergic encephalomyelitis in the rat: response to encephalitogenic proteins and peptides. Science 179:478
38. Melamed D, Friedman A (1993) Direct evidence for anergy in T lymphocytes tolerized by oral administration of ovalbumin. Eur J Immunol 23:931
39. Miller A, Prabhu-Das M, Weiner HL (1992) Epitopes of myelin basic protein (MBP) that trigger TGF-β release following oral tolerization to MBP are different from encephalitogenic epitopes. FASEB J 6:1686
40. Miller A, Lider 0, Al-Sabbagh A, Weiner HL (1992) Suppression of experimental autoimmune encephalomyelitis by oral administration of myelin basic protein. V. Hierarchy of suppression by myelin basic protein from different species. J Neuroimmunol 39:243
41. Miller A, Lider O, Roberts AB, Spom NW, Weiner HI (1992) Suppressor T cells generated by oral tolerization to myelin basic protein suppresses both in vitro and in vivo immune responses by the release of TGF-β following antigen-specific triggering. Proc Natl Acad Sci USA 89:421
42. Miller JFAP, Moraham G (1979) Peripheral T cell tolerance. Annu Rev Immunol 10:51
43. Miller S, Hanson D (1979) Inhibition of specific immune responses by feeding protein antigens. IV. Evidence for tolerance and specific active suppression of cell-mediated immune responses to ovalbumin. J Immunol 123:2344
44. Mowat AM (1987) The regulation of immune reponses to dietary protein antigens. Inmunol Today 8:93
45. Mowat AM, Strobel S, Drummond HE, Ferguson A (1982) Immunological response to fed protein antigens in mice. I. Reversal of oral tolerance to ovalbumin by cyclophosphamide. Immunology 45:105
46. Nagler-Anderson C, Bober LA, Robinson ME, Siskind GW, Thorbeke FJ (19869) Suppression of type II collagen-induced arthritis by intragastric administration of soluble type II collagen. Proc Natl Acad Sci USA 83:7443
47. Nussenblatt RB, Caspi RR, Mahdi R, Chan CC, Roberge F, Lider O, Weiner HL (1990) Inhibition of S-antigen induced experimental autoimmune uveoretinitis by oral induction of tolerance with S-antigen. J Immunol 144:1689
48. Nussenblatt RB, Smet D, Weiner HL, Grey I (1993) The treatment of the ocular complications of Behcet's disease with oral tolerization with S-antigen. Abstracts of the 6th International Conference on Behcet's Disease. Elsevier Press, Amsterdam
49. Ota K, Matsui M, Milford EL, Mackin GA, Weiner HL, Hafler DA (1990) T-cell recognition of immunodominant myelin basic protein epitope in multiple sclerosis. Nature 346:183
50. Pette M, Fujita K, Wilkinson D, Altmann DM, Trowsdale J. Giegrich G, Hinkkanen A, Epplen JT, Kappos L, Wekerle H (1990) Myelin autoreactivity in multiple sclerosis: recognition of myelin basic protein in the context of HLA-DR2 products by T lymphocytes of multiple sclerosis patients and healthy donors. Proc Natl Acad Sci USA 87:7968
51. Polanski M, Matsui M, Khoury S, Weiner HL (1993) Oral tolerization to KLH in humans: generation of antigen-specific lines that suppress proliferative responses. J Immunol 150:114
52. Richman LK, Chiller JM, Brown WR, Hanson DG, Vaz NM (1978) Enterically induced immunological tolerance. I. Induction of suppressor T lymphocytes by intragastric administration of soluble proteins. J Immunol 121:2429

53. Salk RJS (1983) A study of myelin basic protein as a therapeutic probe in patients with multiple sclerosis. In: Hallpike JK, Adams CWM, Toutellotte n (eds) Multiple sclerosis. University Press, pp 621–630
54. Santos LMB, Al-Sabbagh A, Londono A, Weiner HL (1993) Oral tolerance to myelin basic protein induces TGF Beta secreting T cells in Peyer's patches. J Immunol 150:115A
55. Sayegh MH, Zhang ZJ, Hancock WW, Kwok CA, Carpenter CR Weiner HL (1992) Down-regulation of the immune response to histocompatabilitys antigen and preve sensitization by skin allografts by orally administered alloantigen. Transplantation 53:163
56. Sayegh MH, Khoury SJ, Hancock WH, Weiner HL, Carpenter CB (1992) Induction of immunity and oral tolerance with polymorphic class H major histocompatability complex allopeptides in the rat. Proc Natl Acad Sci USA 89:7762
57. Schluesener HJ, Wekerle H (1985) Autoaggressive T lymphocytic lines recognozing the enecephaliogenic region of myelin basic protein: in vitro selection from unprimed rat T lymphocyte populations. J Immunol 135:3128
58. Schwartz RH (1990) A cell culture model for T lymphocyte clonal anergy. Science 248:1349
59. Sercarz E, Krzych U (1991)The distinctive specificity of antigen-specific suppressor T cells. Immunol Today 12:111
60. Singh VK, Kalra HK, Yamaki K, Shinohara T (1992) Suppression of experimental autoimmune uveitis in rats by the oral administration of the uveitopathogenic S-antigen fragment at a cross-reactive homologous peptide. Cell Immunol 139:81
61. Strobel S, Mowat AM, Drummond HE, Pickering MG, Ferguson A (1983) Immunological responses to fed protein antigen s in mice. II. Oral tolerance for CMI is due to activation of cyclophosphamide-sensitive cells by gut-processed antigen. Immunology 49:451
62. Suh EDW, Vistica B, Chan CC, Raber JM, Gery 1, Nussenblatt RB (1995) Splenectomy abrogatees the induction of oral tolerance in experimental autoimmune uveoretinit Curr Eye Res (in press)
63. Thompson HSG, Staines NA (1986) Gastric administration of type II collagen delays the onset and severity of collagen-induced arthritis in rats. Clin Exp Immunol 64:581
64. Thurau SR, Caspi RR, Chan CC, Weiner HL, Nussenblatt RB (1991) Immunological suppression of experimental autoimmune uveitis. Fort Opthalmol 88:404
65. Thurau SR, Chan CC, Suh E, Nussenblatt RB (1991) Induction of oral tolerance to S-antigen induced experimental autoimmune uveitis by a uveitogenic 20mer peptide. J Autoimmun 4:507
66. Trentham DE, Dynesius-Trentham RA, Orav EJ, Combitchi D, Lorenzo C, Sewell KL, Hafler DA, Weiner HL /1993) Effects of oral administration of type II collagen on rheumatoid arthritis. Science 261:1727
67. Vandenbark AA, Hashim GA, Celnik B (1989) Determinants of human myelin basic protein that inducee encephalitogenic T cells in Lewis rats. J. Immunol 143:3512
68. Vrabec TR, Gregerson DS, Dua HS, Donoso LA (1992) Inhibition of experimental autoimmune uveoretinitis by oral administration of s-antigen and synthetic peptides. Autoimmunity 12:175
69. Weiner HL, Zhang ZJ, Khoury SJ, Miller A, Al-Sabbagh A, Brod SA, Lider O, Higgins P, Sobel R,. Nussenblatt RB, Hafler Da (1991) Antigen-driven peripheral immune tolerance. Suppression of organ-specific autoimmune diseases by oral administration of autoantigens. Ann NY Acad Sci 636:227
70. Weiner HL, Miller A, Khoury SJ, Al-Sabbagh A, Brod SA, Lider O, Higgins P, Sobel R, Matsui M, Sayegh M, Carpenter D, Eisenbarth G, Nussenblatt RB, Hafler DA (1992) Suppression of organ-specific autoimmune diseases by oral administration of autoantigens. Prog Immunol 8:627
71. Weiner HL, Mackin GA, Matsui M, Irav EJ, Khoury SJ, Dawson DM, Hafler DA (1993) Double-blind pilot of oral tolerization with myelin antigens in multiple sclerosis. Science 259:1321
72. Weiner HL, Friedman A, Miller A, Khoury SJ (1994) Al-Sabbagh A, Santos L, Sayegh M, Nussenblatt RB, Trentham DE, Hafler DA (1994) Oral tolerance: Immunologic mechanism and treatment of animal and human organ specific autoimmune diseases by oral administration of autoantigens. Annu Rev Immunol 12:809
73. Whitacre Cc, Gienapp IE, Orosz CG, Bitar D (1991) Oral tolerance in experimental autoimmune encephalomyelitis. II. Evidence for clonal anergy. J Immunol 147:2155
74. Woodroofe MN, Bellamy AS, Feldmann M, Davison AN, Cuzner ML (1986) Immunocytochemical characterization of the immune reaction in the central nervous system in multiple sclerosis. Possible role for microglia in lesion growth. J Neurol Sci 74:135
75. Wucherpfennig KW, Weiner HL, Hafler DA (1991) T cell recognition of myelin basic protein. Immunol Today 12:277

76. Zhang JZ, Lee CSY, Lider O, Weiner Hl (1990) Suppression of adjuvant arthritis in Lewis rats by oral administration of type II collagen. J Immunol 145:2489
77. Zhang JW, Chou CH, Hashim G, Medaer R, Raus JC (1990) Preferential peptide specificity and HLA restriction of myelin basic protein-specific T cell clones derived from MS patients. Cell Immunol 129:189
78. Zhang JA, Davidson K, Eisenbarth G, Weiner HL (1991) Suppression of diabetes in NOD mice by oral administration of porcine insulin. Proc Natl Acad Sci USA 88:10 252

Are current immunological concepts of multiple sclerosis reflected by the immunopathology of its lesions?

H. Lassmann[1], **K. Vass**[2]

[1] Neurological Institute, University of Vienna, Schwarzspanierstrasse 17, A-1090 Wien, Austria
[2] Department of Neurorehabilitation, University of Vienna, Austria

Abstract. Immunopathological studies on multiple sclerosis (MS) brain clearly indicate that a T cell-mediated immune response is the driving force in the induction of the lesions. This T cell-mediated response alone, however, is not sufficient to explain the widespread and selective destruction of myelin sheaths. According to present evidence, it is likely that antibodies directed against surface components of myelin sheaths are at least one factor involved in the demyelinating process. The patterns of inflammation, demyelination and oligodendrocyte destruction, however, suggest that the pathogenesis of the lesions may be fundamentally different in individual MS patients and that autoimmunity may not be the sole cause. In the case of autoimmune reactions various different proteins of the nervous system may become targets and it appears unlikely, that myelin basic protein is a major candidate for a pathogenetic role in MS.

Introduction

The pathology of multiple sclerosis (MS) was defined more than a hundred years ago [3, 6, 29]. The disease is characterized by plaques of demyelination and sclerosis, randomly distributed throughout the central nervous system (CNS). The nature of these lesions is thought to be selective, involving mainly myelin and its supporting cells, the oligodendrocytes, but sparing other tissue elements of the nervous system, such as axons, neurons or astrocytes. During active disease, demyelination is associated with an inflammatory reaction, which is predominantly accomplished by lymphocytes and macrophages. It is generally believed, although not formally proven, that the chronic inflammatory process in the nervous system is responsible for the destruction of myelin sheaths. Infectious agents have not been reproducibly isolated from MS lesions and immunosuppression has a significant beneficial effect on the development

Correspondence to: H. Lassmann

of the disease. Furthermore, autosensitization in humans may lead to a disease which, at least on the basis of the pathology of its lesions, is indistinguishable from MS [47, 57].

For these reasons MS is believed to be an autoimmune disease, triggered in genetically susceptible individuals by an exogenous factor, possibly a virus infection during childhood. However, the nature of the autoantigen, and the mechanism by which the immune system exerts its deleterious effects on the nervous system, are still subjects of controversy. In the present review, recent observations on the immunopathology of MS will be summarized and critically discussed in the light of current immunopathogenetic concepts of the disease.

Immunopathology of brain inflammation in MS

The inflammatory reaction in MS is dominated by lymphocytes and macrophages. More detailed characterization of lymphocytes in MS lesions revealed controversial data on the relative numbers of $CD4^+$ and $CD8^+$ cells and their subsets [9, 12, 53]. This, however, is not surprising due to the extensive recruitment of secondary, antigen-nonspecific inflammatory cells into established inflammatory foci [5, 51]. Yet the bulk of evidence suggests that in early active lesions $CD4^+$ cells of the "helper/inducer" phenotype dominate, whereas there may be a significantly lower incidence of $CD45R^+$ putative "suppressor/inducer" cells compared to other inflammatory diseases of the CNS [12, 53]. T cell receptor analysis suggests oligoclonal expansion of both α/β [33] and γ/δ T lymphocytes [63] within MS lesions.

Although much fewer in numbers, B lymphocytes and plasma cells contribute to the inflammatory reaction [8, 32]. Interestingly, the relative frequency of immunoglobulin-producing B cells and plasma cells is significantly higher in the lesions of MS patients in the late chronic stage of the disease, compared to lesions formed in Marburg's type of acute MS or during the first or second bout of chronic MS [22, 35]. A considerable proportion of plasma cells within chronic MS lesions have been shown to locally produce antibodies against myelin basic protein (MBP) [11].

The dominant leukocyte population in MS lesions are macrophages. Macrophage activation and phagocytosis of myelin proteins in the lesions are reliable indicators of ongoing demyelinating activity [18, 23, 35, 46].

The inflammatory reaction in MS brains is associated with the up-regulation of immune-associated molecules and cytokines on infiltrating leukocytes as well as on resident cells of the nervous system. As an example, activated endothelial cells in active lesions may express adhesion molecules [7, 54], fibronectin [52], urokinase plasmin activator receptor [7], major histocompatibility complex molecules [7, 55], and stress proteins, and may be dressed with activated complement [4]. Undoubtedly, these molecules are important for the passage of inflammatory cells through the blood-brain-barrier.

Data on the expression of cytokines in MS lesions are incomplete and controversial. By polymerase chain reaction (PCR) amplification interleukin (IL)-1 mRNA has consistently been found in active lesions [64], whereas mRNAs for IL-2, IL-4 and IL-10 were present only in a small minority of plaques or completely absent [43, 64]. In addition, mRNAs for certain intercrines (IL-8 and membrane cofactor protein) were detected. In situ hybridization revealed mRNAs for a variety of cytokines [IL-1, 2, 4, 6 10, interferon-γ (IFN-γ), tumor necrosis factor-α (TNF-α) and transforming growth factor (TGF)-β1 and 2] exclusively expressed in perivascular inflammatory cells [62]. On the other hand, immunocytochemistry demonstrated a reactivity for IL-1, IL-2 and lymphotoxin in inflammatory cells and for TNF-α and IFN in macrophages and astrocytes [13, 50, 56]. For technical reasons that reside in the poor suitability of anti-cytokine antibodies for immunocytochemistry as well as in problems of sensitivity of the in-situ hybridization techniques, the significance of these results is at present difficult to assess.

A major aspect of the discussion on the immunopathology of MS is, that all the above-described features are not specific for this disease, but can be found, with some variation due to the stage or the etiological background, in other inflammatory diseases of the nervous system, in particular in chronic virus infections [14, 31, 54]. Furthermore, similar studies performed in the well-defined models of autoimmune encephalomyelitis do not, in the absence of other experimental evidence, allow major conclusions on the immunopathogenesis of the disease. Thus, from all these results, it is safe to conclude that a T cell-mediated immune reaction, with secondary macrophage activation, is the driving force in the pathogenesis of the lesions and that locally produced antibodies may also be involved in the chronic stage (Fig. 1). However, all further reaching concepts on the immunopathogenesis of the lesions depend upon analogies to well-defined animal models rather than on hard data on the disease itself.

Demyelination and oligodendrocyte destruction

Besides inflammation, demyelination is the second most characteristic feature of MS pathology. As already proposed, at the beginning of this century [29], the demyelinating process is relatively selective, due to segmental destruction of myelin sheaths with sparing of axons. However, emphasis has to be placed on the term "relative", since a variable extent of axonal destruction is present in all demyelinated MS plaque, that, in extreme situations – especially in acute MS –, may affect up to 80% of the axonal population.

Myelin sheaths are destroyed at any location along their internode in a process that has been described as "melting away the myelin" along intact axonal profiles [29]. Macrophages are intimately engaged in the demyelinating process. They are closely attached to acutely demyelinating fibers, and sometimes an interaction between the coated pits and vesicles of macrophages and degenerating myelin sheaths can be seen [36]. In addition, invasion by macrophages and their cell processes of the widened periaxonal space and between myelin lamellae can sometimes be found [18, 37]. Myelin is further dissected into small fragments that are taken up and further degraded by macrophages.

Oligodendrocytes, the cells responsible for formation and maintenance of myelin sheaths, are lost to a large degree, or even completely, in most cases of typical chronic MS [18, 28, 36, 42]. However, in some cases, especially in early stages of the disease, a considerable number of oligodendrocytes may be present in the lesions, and remyelination in these cases may be rapid and complete [1, 18, 35, 38–41]. This raises the question of whether oligodendrocytes are preserved during the demyelinating process or destroyed and newly recruited from undifferentiated precursor cells.

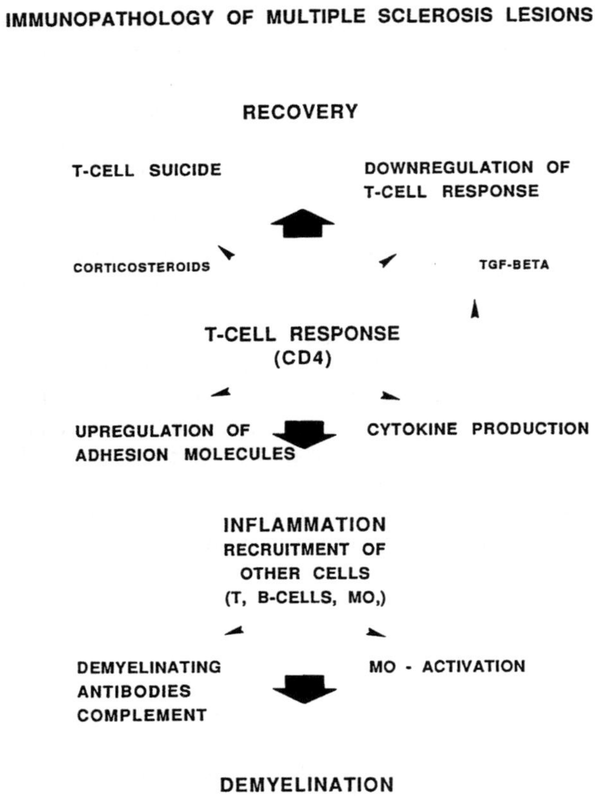

Fig. 1

Recently to address the problem of oligodendrocyte pathology in MS we used new techniques that allowed the positive identification of oligodendrocytes in the lesions and to differentiate between mature oligodendrocytes and precursor cells. In the majority of MS cases of typical long duration (3–20 years), oligodendrocytes were acutely destroyed during active demyelination and completely lost in the demyelinated plaques [22, 35]. However, in biopsy samples obtained during the first or second bout of the disease, oligodendrocytes survived the acute demyelinating episode, and remyelination was rapid and complete [1, 35]. In addition, we found evidence that in some cases of chronic MS oligodendrocyte destruction was secondary to demyelination, whereas in others a primary oligodendrocyte destruction led to secondary demyelination [35]. These data indicate a heterogeneity of the mechanisms that lead to demyelination in different cases and, in a single case, even in early and late stages of disease development.

What immunological mechanisms could be responsible for these patterns of demyelination and oligodendrocyte destruction (Table 1)? For immunologists, the favorite explanation would be that the T lymphocytes responsible for the induction of brain inflammation simultaneously mediate the demyelinating process. However, at present, only few models of pure T cell-mediated acute CNS inflammation lead to

Table 1. Mechanisms of demyelination in inflammatory demyelinating diseases

Mechanism in vitro	Evidence in MS
Virus-induced degeneration of oligodendrocytes	Few cases with primary oligodendrocyte destruction [35]
T cell cytotoxicity (γ/δ T cells)	Abundance of γ/δ T cells in some cases [49]
Demyelinating antibodies	Intrathecal synthesis of anti-MOG antibodies [34], receptor-mediated phagocytosis of myelin [36]
Bystander damage of myelin (enzymes, reactive oxygen species, nitric oxide, TNF-α)	Abundance of activated macrophages in lesions [23]
Complement-mediated lysis of myelin and oligodendrocytes	Presence of immunoglobulins and complement in lesions [4]

MS, Multiple sclerosis; MOG, myelin oligodendrocyte glycoprotein; TNF, tumor necrosis factor

significant demyelination [18, 21]. When more significant demyelination is present in putatively pure T cell models of CNS inflammation, it is found exclusively in chronic variants, where involvement of other, additional immunological mechanisms has not been excluded. In vitro, oligodendrocytes can be destroyed by TNF-α [48], complement or perforin [44, 45]. However, in vivo effects are minor [2] and possibly counteracted by other cytokines [27].

Another mechanism that has been suspected of being involved in demyelination is cytotoxicity of γ/δ T lymphocytes against oligodendrocytes that express the stress protein hsp65. In vitro, oligodendrocytes can be preferentially lysed by γ/δ T lymphocytes [10] and a co-localization of γ/δ T lymphocytes with oligodendrocytes that express hsp65 hase been described in MS lesions [49]. Clonal expansion of γ/δ cells within MS plaques [63] further argues in favor of their pathogenetic role in the disease. Yet, the essential questions on the role of γ/δ T cells in MS are still unresolved. It is not clear which antigen is recognized by γ/δ T cells within MS lesions and whether heat-shock proteins are involved at all. No experimental evidence is available which shows that γ/δ T lymphocytes home to the CNS after transfer and lyse oligodendrocytes in vivo. Heat-shock proteins, including hsp 65, in MS lesions are not restricted to oligodendrocytes, nor is infiltration by γ/δ T lymphocytes restricted to lesions with oligodendroglia destruction ([63] and Lassmann, unpublished). Finally, studies up to now have failed to show any evidence that oligodendrocytes degenerate in MS lesions in the course of contact with γ/δ T lymphocytes.

Extensive demyelination can be induced in vivo in models of autoimmune encephalomyelitis when, in addition to the encephalitogenic T cell response, antibodies against surface components of the myelin sheaths are present [19, 24, 25]. Indeed, several observations indicate that in MS patients such demyelinating antibodies may be involved in the pathogenesis of demyelination in the lesions. In chronic MS lesions a significant proportion of inflammatory cells in the lesions are B lymphocytes or plasma cells. B cells and plasma cells that produce antibodies against myelin oligodendrocyte glycoprotein (MOG) can be found in the CSF of chronic MS patients [34]. Since MOG is expressed on the surface of human myelin sheaths, local production of antibodies against this antigen is a potential risk for myelin sheaths when activated effector cells are present in the course of T cell-mediated brain inflammation [20, 25]. The interaction of myelin with coated pits of macrophages during demyelination indicates a receptor-mediated endocytosis of myelin, possibly through antibody opsonization [36]. On the other hand, to date no consistent pattern of anti-MOG or demyelinating antibodies has been found in serum and CSF of MS patients. Absence from the CSF is not surprising, since such antibodies, directed against antigens on the surface of myelin, have to be absorbed by the adjacent myelinated CNS tissue. Furthermore, it is not yet clear whether MOG is the only immunodominant myelin surface antigen in the induction of demyelinating antibodies.

It has to be emphasized that in our study on demyelination and oligodendroglia destruction in MS, the patterns were variable from case to case. The spectrum ranged from selective destruction of myelin with almost complete preservation of oligodendrocytes to substantial destruction of oligodendrocytes in the course of demyelination and even to primary oligodendroglia destruction with secondary demyelination [35]. These data suggest that the pathogenetic mechanisms leading to demyelination may be fundamentally different in different MS patients.

Lesional topography as a clue to identify target autoantigens in MS

MS, at least in its chronic variants is a disease that affects the CNS, but not the peripheral nervous system. Although inflammatory demyelination of peripheral nerves is a typical feature of Marburg's type of acute MS [29], its incidence in chronic MS is very low [15]. MS lesions may appear in any region of the CNS, yet certain areas of the brain are more frequently affected than others. The highest numbers of lesions are found either in the periventricular white matter or in the periphery of cerebral gyri, in the latter case being either contiguous to or wholly located within the gray matter [28]. Other regions with high density of lesions are the optic nerve and chiasm, pons, cerebellar peduncles, medulla oblongata and spinal cord. However, in early stages of MS different variants exist, predominantly affecting either the spinal cord (myelitis), the optic system (optic neuritis) or the periventricular white matter. In addition, in some MS patients inflammation can be found in the retina and in the uvea, in regions where myelin is absent [30].

This pattern of lesional distribution is in contrast to that found in experimental models induced by autoimmunity against MBP. In these models, CNS lesions are concentrated in the spinal cord and brain stem with very little affection of the periventricular white matter and other forebrain regions [17]. In addition, MBP-induced experimental autoimmune encephalomyelitis (EAE) is associated with inflammatory lesions in the peripheral nervous system, a feature that is rarely encountered in chronic MS [26]. Thus, it is unlikely that MBP is the major target antigen for pathogenic autoimmune reactions in MS.

However, other CNS antigens have been identified that could be targets of a T cell-mediated autoimmune encephalitis. They include proteolipid protein (PLP) [58], MOG [26], myelin-associated glycoprotein (MAG) [59], S-100 protein [17] and glia fibrillary acidic protein (GFAP) (Linington unpublished). The topography of lesions, induced by T cell reactions against these various nervous system antigens is quite variable. Obviously, inflammation is restricted to the CNS, when the T cells recognize an antigen, such as MOG or PLP, that is exclusively present there. When antigens are recognized that are present in the compact myelin, such as MBP or PLP, the highest incidence of lesions is present in areas with the thickest myelin sheaths: the spinal cord and the brain stem. Conversely, an antigen exclusively localized on the myelin surface e.g. MOG, is present in high concentrations in areas with many thin myelin sheaths. Thus, inflammation after transfer of MOG-reactive T cells severely affects the

Table 2. Target antigens for autoimmunity in the nervous system

Antigen	Histological localization	Topography of inflammatory lesions in experimental animals		Postulated human disease
MBP	Central myelin	Spinal cord	+++	Myelitis
	Compact myelin	Medulla	++	Neuromyelitis optica
		Optic nerve	+	
		Periventr.	+	
	Peripheral myelin	PNS	+	PNS involvement in acute MS
PLP	Central myelin	Spinal cord	+++	Myelitis
	Compact myelin	Medulla	++	Neuromyelitis optica
		Optic nerve	+	
		Periventr.	+	
MOG	Central myelin	Spinal cord	++	Classical chronic MS
	Myelin surface	Medulla	++	
		Optic nerve	+++	
		Cerebellum	+++	
		Periventr.	+++	
S100	Astrocytes	Spinal cord	+++	Chronic MS with eye involvement
	Eye	Medulla	++	
		Optic nerve	+	
		Cerebellum	+	
		Periventr.	+	
		Cortex	++	
		Retina	++	
		Uvea	++	
	Peripheral nerve	PNS	+	PNS involvement

MBP, Myelin basic protein; PLP, proteolipid protein; MOG; myelin oligodendrocyte glycoprotein; S100, S-100 protein; PNS, peripheral nervous system

brain, including the periventricular and cerebellar white matter. Inflammation induced by T cells against an astrocytic antigen, such as S-100 protein, in addition to the typical white matter lesions, also leads to severe involvement of the cerebral cortex, the retina and the uvea.

When the patterns of lesional topography are compared between the models of autoimmune encephalitis and MS, it is evident that MBP is in the vast majority of MS cases not a suitable candidate for a pathogenic autoimmune response. In the future it will have to be determined whether different forms of MS – pure spinal forms, neuromyelitis optica, cases with dominant periventricular involvement, cases with severe eye involvement – reflect dominant autoimmune reactions against different CNS components (Table 2). In addition, it has to be considered that the inflammatory reaction in MS may be induced or propagated by simultaneous immune reactions against a variety of different autoantigens [34]. In this case the final pathological outcome and lesional distribution has to depend upon the aggregate of the individual autoimmune responses.

In conclusion, the review of the immunopathology of MS suggests that the common denominator in its pathogenesis is a chronic immune reaction against one or more antigen(s) presented within the CNS compartment to peripherally predetermined T lymphocytes. A number of different autoantigens and foreign antigens, the latter present in the CNS in the course of latent non-cytopathic infections, may be the primary target of the immune response. With the chronicity of the disease, it is likely that the immune response broadens by recruiting additional autoimmune reactions. The pronounced and selective demyelination apparently requires immunological mechanisms in addition to the primary T cell response. One of these mechanisms appears to be the cooperation of antibodies in the pathogenesis of demyelination.

Consequence for diagnosis and therapy of MS

From the present experimental evidence, the question arises whether MS is a neurological syndrome with different immunopathological mechanisms triggering a final common pathway rather than a single disease. This may also explain the highly variable clinical course of MS [61] and raises major problems for the evaluation of new therapeutic strategies. The present state of the art is to perform large scale randomized double-blind clinical studies. These studies may be able to measure the efficacy of global immune suppression or detect an interference with a final immunopathological pathway. Yet they often do not reflect the fact that a therapy that is useful in the early stage of MS may be ineffective or even deleterious at later stages. In addition, the usefulness of more delicate therapeutic interventions, such as antigen-specific or immunomodulatory treatments, is probably difficult to prove by these methods.

Most antigen-specific therapeutic trials are based on the assumption that MBP is the immunodominant T cell antigen in MS. However, other candidates, such as MOG [16] or, in some instances, S-100 protein, may be more important, and additional new immunodominant encephalitogenic antigens will certainly be identified in the near future. Furthermore, individual MS patients may react to more than one epitope of known autoantigens. As long as we have no means of pinpointing the responsible encephalitogenic antigen(s) or even the responsible epitope(s) for a given individual MS patient, antigen-specific therapy is unlikely to be a successful option in this disease.

A recently published strategy for antigen-specific therapy is the paradigm of oral tolerization [60]. This is believed to transfer an antigen-specific Th1 response (which is known to induce T cell-mediated encephalitis) to a Th2 response. T cells of the Th2 type produce cytokines that drive mainly B cells and stimulate antibody production. Thus, oral tolerization may be effective in a stage of the disease in which T cell-mediated delayed-type hypersenstitivity is prominent, but in stages or patients with abundance of demyelinating antibodies, such a therapy could be deleterious for the outcome. At least in EAE a delicate interplay between low numbers of encephalitogenic T cells and demyelinating antibodies results in severe disease of the animals with widespread demyelinating lesions [20].

Thus, our growing knowledge of the multitude of different immunological mechanism, that can lead to inflammatory demyelinating lesions in the nervous system, encourages the development of techniques that allow their identification in individual MS patients. This should lead to therapeutic trials in specified subgroups of MS patients with exactly defined uniform patterns and stages of disease.

References

1. Brück W, Schmied M, Suchanek G, Brück Y, Breitschopf H, Poser S, Piddlesden S, Lassmann H (1994) Oligodendrocytes in the early course of MS. Ann Neurol 35:65
2. Butt AM, Jenkins HG (1994) Morphological changes in oligodendrocytes in the intact mouse optic nerve following intravitreal injection of tumor necrosis factor. J Neuroimmunol 51:27

3. Charcot JM (1868) Histologie de la sclerose en plaque. Gaz Hopital (Paris) 41:554
4. Compston DA, Morgan BP, Campbell AK, Wilkinsw P, Cole G, Thomas ND, Jasani B (1989) Immunocytochemical localization of the terminal complement complex in MS. Neuropathol Appl Neurobiol 15:307
5. Cross AH, O'Mara T, Raine CS (1993) Chronologic localization of myelin reactive cells in the lesions of relapsing EAE: implications for the study of MS. Neurology 43:1028
6. Dawson JW (1916) The histology of disseminated sclerosis. Trans R Soc Lond 50:517
7. Dore-Duffy P, Washington R, Dragovic L (1993) Expression of endothelial cell activation antigens in microvessels from patients with multiple sclerosis. Adv Exp Med Biol 331:243
8. Esiri MM (1980) Multiple sclerosis: a quantitative and qualitative study on immunoglobulin-containing cells in the central nervous system. Neuropath Appl Neurobiol 6:9
9. Estes ML, Rudick RA, Barnett GH, Ransohoff RM (1990) Stereotactic biopsy of an active multiple sclerosis lesion. Immunocytochemical analysis and neuropathologic correlation with magnetic resonance imaging. Arch Neurol 47:1299
10. Freedman MS, Ruijs TC, Selin LK, Antel JP (1991) Peripheral blood $\gamma\delta$ T cells lyse fresh human brain derived oligodendrocytes. Ann Neurol 30:794
11. Gerritse K, Deen C, Fasbender M, Ravid R, Boersma W, Claassen E (1994) The involvement of specific anti myelin basic protein antibody forming cells in multiple sclerosis immunopathology. J Neuroimmunol 49:153
12. Hayashi T, Morimoto C, Burks JS, Kerr C, Hauser SL (1988) Dual label immunocytochemistry of the active multiple sclerosis lesion: major histocompatibility complex and activation antigens. Ann Neurol 24:523
13. Hofman FM, von Hahnwehr RI, Dinarello CA, Mizel SB, Hinton D, Merrill JE (1986) Immunoregulatory molecules and IL2 receptors identified in multiple sclerosis lesions. J Immunol 136:3239
14. Iwasaki Y, Sako K, Tsunoda I, Ohara Y (1993) Phenotypes of mononuclear cell infiltrates in human central nervous system. Acta Neuropathol 85:653
15. Jellinger K (1969) Einige morphologische Aspekte der multiplen Sklerose. Wien Z Nervenheilk [Suppl] II:12
16. Kerlero de Rosbo N, Milo R, Lees MB, Burger D, Bernard CC, Ben-Nun A (1994) Reactivity to myelin antigens in multiple sclerosis. Peripheral blood lymphocytes respond predominantly to myelin oligodendroglia glycoprotein. J Clin Invest 92:2602
17. Kojima K, Berger T, Lassmann H, Hinze-Selch D, Zhang Y, Germann J, Reske K, Wekerle H, Linington C (1994) Experimental autoimmune panencephalitis and uveoretinitis transferred to the Lewis rat by T lymphocytes specific for the S100β molecule, a calcium binding protein of astroglia. J Exp Med 118:817
18. Lassmann H (1983) Comparative neuropathology of chronic experimental allergic encephalomyelitis and multiple sclerosis. Springer, Berlin Heidelberg New York Tokyo
19. Lassmann H, Kitz K, Wisniewski HM (1981) In vivo effect of sera from animals with chronic relapsing experimental allergic encephalomyelitis on central and peripheral myelin. Acta Neuropathol (Berl) 55:297
20. Lassmann H, Brunner C, Bradl M, Linington C (1988) Experimental allergic encephalomyelitis: the balance between encephalitogenic T lymphocytes and demyelinating antibodies determines size and structure of demyelinated lesions. Acta Neuropathol (Berl) 75:566
21. Lassmann H, Zimprich F, Rössler K, Vass K (1991) Inflammation in the nervous system. Basic mechanisms and immunological concepts. Rev Neurol (Paris) 147:763
22. Lassmann H, Suchanek G, Ozawa K (1994) Histopathology and the blood-cerebrospinal fluid barrier in multiple sclerosis. Ann Neurol [Suppl] 36:S42
23. Li H, Newcombe J, Groome NP, Cuzner ML (1993) Characterization and distribution of phagocytic macrophages in multiple sclerosis plaques. Neuropathol Appl Neurobiol 19:214
24. Linington C, Lassmann H (1987) Antibody responses in chronic relapsing experimental allergic encephalomyelitis: correlation of serum demyelinating activity with antibody titre to the myelin/oligodendroglia glycoprotein (MOG). J Neuroimmunol 17:61
25. Linington C, Bradl M, Lassmann H, Brunner Ch, Vass K (1988) Augmentation of demyelination in rat acute allergic encephalomyelitis by circulating mouse monoclonal antibodies against myelin/oligodendroglia glycoprotein. Am J Pathol 130:135

26. Linington C, Berger T, Perry L, Weerth S, Hinze-Selch D, Lu H, Lassmann H, Wekerle H (1993) T cells specific for the myelin oligodendrocyte glycoprotein mediate an unusual autoimmune inflammatory response in the central nervous system. Eur J Immunol 23:1364
27. Louis JC, Magal E, Takayama S, Varon S (1993) CNTF protection of oligodendrocytes against natural and tumor necrosis factor-induced death. Science 259:689
28. Lumsden CE (1970) The neuropathology of multiple sclerosis. Handb Clin Neurol 9:217
29. Marburg O (1906) Die sogenannte "akute Multiple Sklerose". Jahrb Psychiatr 27:211
30. Matthews WB (1991) Clinical aspects. In: Matthews WB, Compston A, Allen I, Martyn CN (eds) McAlpine's multiple sclerosis. Churchill Livingstone, Edinburgh, 107-138
31. Merrill JE (1992) Proinflammatory and antiinflammatory cytokines in multiple sclerosis and central nervous system acquired immunodeficiency syndrome. J Immunother 12:167
32. Mussini JM, Hauw JJ, Escourolle R (1977) Immunofluorescence studies of intracytoplasmic immunoglobulin binding lymphoid cells (CICL) in the central nervous system. Report of 32 cases including 19 MS. Acta Neuropathol (Berl) 40:227
33. Oksenberg JR, Panzara MA, Begovich AB, Mitchell D, Erlich HA, Murray RS, Shimonkevitz R, Sherritt M, Rothbard J, Bernard CC (1993) Selection for T cell receptor $V\beta$-$D\beta$-$J\beta$ gene rearrangements with specificity for a myelin basic protein peptide in brain lesions of multiple sclerosis. Nature 362:68
34. Olsson T (1992) Immunology of MS. Curr Opin Neurol Neurosurg 5:195
35. Ozawa K, Suchanek G, Breitschopf H, Brück W, Budka H, Jellinger K, Lassmann H (1994) Patterns of oligodendroglia pathology in multiple sclerosis. Brain 117:1311
36. Prineas JW (1985) The neuropathology of multiple sclerosis. Handb Clin Neurol 47:213
37. Prineas JW, Raine CS (1976) Electron microscopy and immunoperoxidase studies in early multiple sclerosis lesions. Neurology 26:29
38. Prineas JW, Kwoon EE, Goldenberg PZ, Ilyas AA, Quarles RH, Benjamins JA, Sprinkle TJ (1989) MS. Oligodendrocyte proliferation and differentiation in fresh lesions. Lab Invest 61:489
39. Prineas JW, Barnard RO, Kwon EE, Sharer LR, Cho ES (1993) Multiple sclerosis: remyelination of nascent lesions. Ann Neurol 33:137
40. Prineas JW, Barnard RO, Revesz T, Kwon EE, Sharer L, Cho ES (1993) Multiple sclerosis. Pathology of recurrent lesions. Brain 116:681
41. Raine CS, Scheinberg L, Waltz JM (1981) Multiple sclerosis: oligodendroglia survival and proliferation in an active established lesion. Lab Invest 45:534
42. Rodriguez M, Scheithauer BW, Forbes G, Kelly PJ (1993) Oligodendrocyte injury is an early event in lesions of multiple sclerosis. Mayo Clin Proc 68:627
43. Schlüsener HJ, Meyermann R (1993) Intercrines in brain pathology. Expression of intercrines in a multiple sclerosis and Morbus Creutzfeldt-Jakob lesion. Acta Neuropathol 86:393
44. Scolding N, Linington C, Compston A (1989) Immune mechanisms in the pathogenesis of demyelinating diseases. Autoimmunity 4:131
45. Scolding NJ, Jones J, Compston DA, Morgan BP (1990) Oligodendrocyte susceptibility to injury by T cell perforin. Immunology 70:6
46. Seitelberger F (1960) Histochemistry of demyelinating diseases proper including allergic encephalomyelitis and Pelizaeus-Merzbacher's disease. In: Cummings JN (ed) Modern scientific aspects of neurology. Arnold, London, pp 146
47. Seitelberger F, Jellinger K, Tschabitscher H (1958) Zur Genese der akuten Entmarkungsenzephalitis. Wien Klin Wochenschr 70:453
48. Selmaj KW, Raine CS (1988) Tumor necrosis factor mediates myelin and oligodendrocyte damage in vitro. Ann Neurol 23:339
49. Selmaj K, Brosnan CF, Raine CS (1991) Colocalization of lymphocytes bearing $\gamma\,\delta$ T cell receptor and heat-shock protein $hsp65^{+}$ oligodendrocytes in multiple sclerosis. Proc Natl Acad Sci USA 88:6452
50. Selmaj K, Raine CS, Cannella B, Brosnan CF (1991) Identification of lymphotoxin and tumor necrosis factor in multiple sclerosis lesions. J Clin Invest 87:949
51. Skundric DS, Kim C, Tse HY, Raine CS (1993) Homing of T cells to the central nervous system throughout the course of relapsing experimental autoimmune encephalomyelitis in Thy-1 congenic mice. J Neuroimmunol 46:113
52. Sobel RA, Mitchell ME (1989) Fibronectin in multiple sclerosis lesions. Am J Pathol 135:161
53. Sobel RA, Hafler DA, Castro EE, Morimoto C, Weiner HL (1988) The 2H4 (CD45R) antigen is selectively decreased in multiple sclerosis lesions. J Immunol 140:2210

54. Sobel RA, Mitchell ME, Fondren G (1990) Intercellular adhesion molecule-1 (ICAM-1) in cellular immune reactions in the human central nervous system. Am J Pathol 136:1309
55. Traugott U (1987) Multiple sclerosis: relevance of class I and class II MHC-expressing cells to lesion development. J Neuroimmunol 16:283
56. Traugott U, Lebon P (1988) Multiple sclerosis: involvement of interferons in lesion pathogenesis. Ann Neurol 24:243
57. Uchimura I, Shiraki H (1957) A contribution to the classification and the pathogenesis of demyelinating encephalomyelitis. J Neuropathol Exp Neurol 16:139
58. Waksman BH, Porter H, Lees MB, Adams RD (1954) A study of the chemical nature and components of bovine white matter effective in producing allergic encephalomyelitis in the rabbit. J Exp Med 100:451
59. Weerth S, Berger T, Lassmann H, Wekerle H, Linington C (1994) Analysis of the autoimmune response to myelin-associated glycoprotein (MAG) in the Lewis rat. Abstracts of the First Eur Meeting on Glia Cell Function, Unversity of Heidelberg, Heidelberg 24–27 March 1994, p 184
60. Weiner HL, Mackin GA, Matsui M, Orav EJ, Khoury SJ, Dawson DM, Hafler DA (1993) Double-blind pilot trial of oral tolerization with myelin antigens in multiple sclerosis. Science 259:1321
61. Weinshenker BG, Ebers GC (1987) The natural history of multiple sclerosis. Can J Neurol Sci 14:255
62. Woodroofe MN, Cuzner ML (1993) Cytokine mRNA expression in inflammatory multiple sclerosis lesions: detection by non-radioactive in situ hybridization. Cytokine 5:583
63. Wucherpfennig KW, Newcombe J, Li H, Keddy C, Cuzner ML, Hafler DA (1992) $\gamma\ \delta$ T cell receptor repertoire in acute multiple sclerosis lesions. Proc Natl Acad Sci USA 89:4588
64. Wucherpfennig KW, Newcombe J, Li H, Keddy C, Cuzner ML, Hafler DA (1992) T cell receptor V α-Vβ repretoire and cytokine gene expression in active multiple sclerosis lesions. J Exp Med 175:993

Multiple sclerosis: a unique immunopathological syndrome of the central nervous system

Samuel F. Hunter[1], **Moses Rodriguez**[1,2]

Departments of [1] Neurology and [2] Immunology, Mayo Clinic and Mayo Foundation, Rochester, MN 55905, USA

Multiple sclerosis: a clinical neurological syndrome mediated by immune phenomena

Multiple sclerosis (MS) is the most common, disabling neurological syndrome of young adults. The etiology is idiopathic, yet most authorities agree that the immune system is crucial in the pathogenesis of the recurrent and chronic injury to the central nervous system (CNS). Nearly identical histopathological findings occur in MS and related clinical syndromes, designated as "demyelinating diseases." Clinical presentation differentiates MS from the monophasic syndromes (optic neuritis, acute transverse myelitis, and Marburg disease) and the less-common relapsing syndromes (Devic's disease, Balo's concentric sclerosis). A chronic progressive variant of MS is also well recognized.

The primary pathological feature of demyelinating diseases is a focal loss of CNS myelin in the white matter, the areas of concentrated, parallel myelinated axons within the CNS. Oligodendroglia produce and maintain the myelin. This multilamellar lipid structure permits rapid, saltatory conduction of action potentials and mechanically separates CNS axons. The focal nature of demyelinating disease contrasts with the diffuse myelin loss of toxic or metabolic injuries, such as occurs in leukodystrophy, myelinolysis, radiation, and nutritional compromise. The great magnitude of myelin loss relative to the perivascular inflammatory infiltrate differentiates MS from the more cellular inflammatory response of acute disseminated encephalomyelitis.

Demyelinating disease afflicts 0.2% of individuals in high prevalence areas, and incidence is increasing. MS is not associated with other immune diseases [104]. Genetic studies reveal weak associations with major histocompatibility complex (MHC) and T cell receptor (TCR) genes and only moderate concordance in monozygous twins, supporting a polygenic inheritance for susceptibility and severity, as well as environmental factors.

Correspondence to: M. Rodriguez

Diagnosis is dependent upon clinical findings often in association with typical magnetic resonance (MR) imaging or immunological findings of cerebrospinal fluid (CSF). Less commonly biopsy or post-mortem examination are required to demonstrate typical histopathology. The characteristic focal inflammatory events in brain and spinal cord leave areas of increased water and decreased lipids visible with MR imaging. Paramagnetic contrast enhances these areas in acute lesions where inflammation perturbs the blood-brain barrier (BBB) [55]. Chronic inflammation in the brain and meninges also results in characteristic changes in the CSF.

Clinical neurological deficits depend primarily upon the location of the inflammation. Impairment is due to multifactorial axonal dysfunction, in part due to inflammation, edema, disordered axonal sodium channels, BBB abnormalities, demyelination, and axonal loss. The most common clinical deficits include impairment of visual acuity, double vision, abnormal sensation, incoordination, spastic weakness, tremor, mood disorders, and fatigue. Significant intellectual impairment is uncommon except in advanced stages of the disease, and most cerebral lesions are silent clinically. Lesions tend to cluster in the periventricular areas, but can occur almost anywhere within the CNS white matter.

The temporal pattern and progression of the disease are unpredictable. About half of cases have minimal progression, whereas about 10% of cases have a severe course which results in death due to MS or complications of the marked disability [76]. The most typical pattern is one of a relapsing-remitting deficit, but with time the attack rate decreases, the inter-attack disability increases, and eventually a progressive disease emerges. The primary chronic progressive variant has few or no distinct exacerbations. Large, fulminating and frequently life-threatening demyelinating lesions [55] are the least common presentation. In all of these clinical syndromes, the essential pathological lesion is similar.

Therapies for MS have targeted the immune response, mostly with limited benefit. Anti-inflammatory treatment with short courses of high-dose intravenous corticosteroids shorten exacerbations, may improve immune and inflammatory parameters, and prolong the interval to further neurological symptoms following an attack of optic neuritis [7]. Whereas clinical trials with interferon (IFN-γ) resulted in increased exacerbations [57], IFN-α suppresses the severity of exacerbations but does not alter disease progression [1]. IFN-β has similar actions to IFN-α, and, although larger active lesions are fewer on MR images [58], no difference in disability results over 3 years.

A hypothesized failure of self-tolerance, which could underlie autoimmunity, has spurred ongoing attempts to promote immunological tolerance by administration of proteins mimicking myelin antigens. Anti-proliferative and immunosuppressive treatments have been used with limited benefit and with significant toxicities. Attempts to modulate the disease through manipulation of blood components have had limited efficacy. Plasmapheresis (removal of the plasma) has a short-term benefit of less than 3 months [100], but may have an important role in the treatment of acute fulminating demyelinating disease [74]. Immunoglobulin (Ig) may improve function in optic neuritis [96] and is currently being studied as a way of promoting remyelination in double-blinded controlled trial.

The tissue lesion of MS

MS lesions are characterized by focal demyelination and inflammation. Areas of CNS myelin loss are widespread and easily detectable using myelin-specific stains. These "plaques" are the pathological hallmark of the gross and microscopic pathology of MS. These areas contain hypertrophied astroglia, and in more recent lesions "foamy" phagocytic cells with moderate mononuclear infiltrates predominate, mostly in perivascular areas. Older and less active lesions have many plasma cells.

During the course of MS the recurrent and chronic CNS inflammation damages myelin and axons. It is not clear whether demyelination represents oligodendroglial injury or one facet of the spectrum of inflammation, aiding the entry of the cellular immune response into the CNS. Myelin forms the greatest part of the white matter and it is connected to and maintained ultrastructurally by oligodendrocytes. Since myelin can often be repaired and axons cannot, one could view the myelin compartment as the most expendable of the white matter elements. The inflammatory pattern of the demyelinating response is different from the intense focal mononuclear infiltrates of acute disseminated encephalomyelitis, the diffuse infiltrates in viral encephalitis, and the granulocytic infiltrates in bacterial encephalitis or brain abscess. Loss of myelin is a minor feature of these other conditions, even though all the cellular and humoral components of the immune response are present.

An adequate framework for understanding the nature of the immune and glial functions in demyelinating disease might be as follows. After an as yet unclear inciting event of the inflammatory cascade, the vasculature attracts mononuclear cells. Astroglia regulate the BBB, express important adhesion molecules, produce cytokines, and regulate hypertrophy to provide a scaffold; pathway for movement of phagocytes and effector lymphocytes within the parenchyma. The microglia and infiltrating phagocytic cells present antigens, ingest the large amount of myelin and cellular debris produced, and may also serve immunoregulatory and cytotoxic roles. As inflammation subsides, attempts at myelin repair usually result, but renewed or continued inflammation shifts the inflammatory response to a chronic type, surveilled by plasma cells and T lymphocytes. The establishment of chronic inflammation within the CNS seems to be the hallmark of MS.

Actively demyelinating MS lesions

Acute lesions are those which have new perivascular inflammatory foci and recent onset of demyelination; these are probably the pathological basis of "attacks." Generally, these lesions have a paucity of intact myelin, an overwhelming number of phagocytic cells with large amounts of lipid and myelin debris [40], relative sparing of axons, disrupted BBB function [92], and a limited perivascular and disseminated mononuclear infiltrate of phagocytic cells and $CD3^+$ T lymphocytes [10]. Most phagocytes arise from normally ramified and quiescent resident microglia and develop into "foamy" macrophages by activation and engulfment of myelin debris.

Although both $CD4^+$ and $CD8^+$ T lymphocytes are present, the latter predominate throughout the lesion [29, 49, 92]. Most T cells in the lesion also express interleukin-2 receptor (IL-2R) [92]. Pan-T lymphocyte markers do not label all of the $CD8^+$ population, and the ratio of $CD4^+$ to $CD8^+$ phenotypes within active plaques is lower than the lesion edge and adjacent white matter [49]. The cells bearing CD45R (suppressor-inducer-associated antigen) are not seen frequently in MS lesions, unlike viral encephalitis, whereas the $4B4^+$ phenotypes (helper-inducer) are similar between the two conditions [85].

The character of T lymphocytes within the acute MS lesion has been further characterized on the basis of TCR subtypes. Many $\alpha\beta$ TCR-bearing T lymphocytes are present in acute lesions, and fewer are found in chronic lesions. Analysis of α- and β-chain TCR gene rearrangement indicates that many genes are used. Clonal T lymphocytes within active lesions are, therefore, probably a minor fraction [103]. Variable results regarding $\gamma\delta$-TCR-bearing cells have been reported: some investigators find them more prominently in chronic plaques [82], whereas others note more in acute plaques [102]. In either case they usually represent only a small percentage of total T cells.

MHC molecule and cytokine expression suggests immunological roles for inflammatory and parenchymal cells. Class I expression in astrocytes correlates with the lesion activity [92]. Surprisingly, oligodendroglia within lesions and nearby lack both immunoreactive class I [92] and class II [39]. A quantitative difference in class I antigen expression may explain this finding, since immunoreactive class I and class I-dependent cytotoxicity against human oligodendroglia can be demonstrated in vitro [77]. Class II is expressed around vessels in active lesions, especially near edges, and located ultrastructurally only on the pericytes, perivascular macrophages, microglia and astrocytes [39, 40, 92]. Perivascular lymphocytes lack immunoreactive class II and may represent unstimulated lymphocytes recruited from the circulation [40]. The presence of phagocytosed myelin in both astroglia and microglia in the setting of class II expression has suggested a possible role in autoantigen presentation. Cytokines are coexpressed with class II antigens. Astroglia within and near active lesions coexpress IFN-γ, and microglia coexpress IFN-α [93]. Roles for astroglia in promoting inflammation and microglia in containing the inflammatory response are thus inferred.

Membrane-associated adhesion molecules are also important in immune function, most likely for migration of cells into and movement within the CNS. Near the active lesion edges the microvasculature labels for intercellular adhesion molecule-1 (ICAM-1); microglia and astroglia throughout the lesion also express ICAM-1. However, the corresponding immune adhesion molecule, lymphocyte function adhesion molecule-1, is present only on infiltrating immune cells [86]. This pattern suggests that ICAM-1 assists in targeting of inflammatory cells to the developing edge of MS lesions and for movement of inflammatory cells within the brain parenchyma; however, these changes are not specific for MS and occur in all other CNS inflammatory diseases. Astrocytic hypertrophy and expression of adhesion molecules would seem to assist parenchymal migration, in an analogous fashion to the manner in which specialized astroglia (Bergmann and radial glia) guide neuronal migration during CNS architectonics.

Tumor necrosis factors (TNF), important cytokines in systemic inflammation, have been proposed to be responsible in part for demyelination due to their action in vitro in producing demyelination, oligodendrocyte apoptosis, and up-regulation of class II on astroglia. TNF is present in astrocytes throughout the lesion, foamy macrophages within the lesion center, and endothelial cells near the lesion. TNF-β is associated with $CD3^+$ T lymphocytes and microglia at the edges of acute or chronically active MS lesions, but absent in inactive chronic lesions [83].

Demyelination appears to be an early event during the acute attack. The earliest histological form of the lesion is unclear, but a few perivascular lymphocytes and lipid-laden macrophages without demyelination are seen frequently in otherwise normal MS tissue [90]. In smaller lesions the lymphocytes are almost completely $CD8^+$ but $CD4^-$ [29]. Whether this represents the primary inflammatory event or the secondary recruitment of inflammatory cells to oligodendrocyte injury is unknown.

Demyelination appears to occur as a disturbance of the oligodendrocyte myelinating function rather than as a primary attack or phagocytosis of the myelin sheath, since myelin first degenerates from the most distal part of the oligodendrocyte (inner glial loop) [72, 75]. Despite the loss of myelin, intact oligodendroglia persist in the lesion [10]. Although MS has always been cited as a "relatively axon-sparing" disease, axonal injury, as evidenced by swelling, is present in many acute lesions [72], and axonal density is lower than in surrounding normal white matter [10]. Nevertheless, the myelin degeneration is not usually secondary to axonal injury, as vesicular dissolution of myelin occurs from intact axons [40].

Reduction in the number of oligodendroglia is variable even in the early stages of the disease, ranging from near normal numbers to a virtual absence of oligodendroglia [10]. Oligodendroglia in MS lesions express heat-shock proteins, possibly indicative of stress [82]. Persisting oligodendrocytes in fresh lesions express cytoplasmic markers of early myelin formation, such as myelin-associated glycoprotein (MAG) [39, 40, 66].

Chronic lesions in MS

Chronic lesions may be "active" or "inactive" depending on the presence of perivascular lymphocytic infiltrates and evidence of ongoing myelin degeneration; both may coexist in separate areas of the parenchyma at a given time. The more active lesions are similar to the acute lesions described above, but myelin vesiculation is absent. Chronic inflammation with lymphocytes, plasma cells, and frequent lipid-laden macrophages are often found in demyelinated areas, even without perivascular infiltrates [40, 90]. Perivascular phagocytes are similar ultrastructurally to free tissue macrophages elsewhere, but have activated lysosomes. In contrast, microglia in MS brain show myelin debris in the cytoplasm, even outside of obviously lesioned areas. Whereas acute lesions have predominantly a T cell infiltrate, B-lineage plasma cells are increased both perivascularly and markedly within chronic lesions ($> 1700/mm^2$) [65]. Chronic lesions without inflammation have a relatively intact BBB [92], and the $CD4^+/CD8^+$ ratio is lower than in active regions [49].

Axonal changes also occur within chronic MS lesions, and signs of axonal degeneration (spheroids) accompany most areas with active inflammation [40]. Within inactive lesions, demyelinated axons lie either directly opposed or are separated by the fibrous processes of reactive astroglia [63]. Axons have a smaller diameter in demyelinated segments [63, 64]. Axonal loss probably occurs in chronic lesions since increased extracellular space can be seen even with an intact BBB and absence of inflammation [5].

Most astrocytes within inactive chronic lesions do not have immunoreactive class II or IFN-γ [93]. In contrast, reactive astrocytes as well as small numbers of IL-2R-bearing T cells near the edges of chronic lesions frequently express class II [92]. $CD8^+$ T cells continue to predominate [9].

Myelin repair

Myelinated areas of lesser density in the characteristic shape of lesions have been called "shadow plaques" and felt to be evidence for myelin repair. New myelin formation occurs in the CNS as manifested by the formation of new internodes at the edges of demyelinated lesions [67].

Various stages of this repair process have been documented even in the presence of an ongoing inflammatory response. Oligodendrocyte processes can be seen to make several loops of myelin in the presence of ongoing phagocytosis of degenerating myelin [22]. Acute lesions demonstrate substantial remyelination, as recognized by pale, abnormally thin myelin [72]. Although T cells and macrophages can coexist with ongoing remyelination, completely remyelinated areas are usually devoid of lymphocytes but have a few macrophages [10]. On the basis of careful study of completely demyelinated areas, it has been hypothesized that permanent myelin loss is in part due to recurrent demyelination in previously completely remyelinated zones [68].

Pathogens in MS

Certain viruses can cause chronic CNS disease with demyelination, as is the case for subacute sclerosing panencephalitis (SSPE; defective measles virus), Theiler's virus-induced demyelination (picornavirus in mice), visna (retrovirus in sheep), and progressive multifocal leukoencephalopathy (JC papilloma virus). These and similar viral pathogens have been proposed as the cause of MS, but no viral inclusions are seen in MS lesions. Herpes simplex virus (HSV) type-1 antigen is present in CSF immune complexes of some individuals with MS [15]. HSV type-2 antigen has been detected in nuclei of glia near demyelinated areas [45]. The presence of "paramyxovirus-like" bodies in MS tissues is nonspecific and probably represent degenerative nuclear chromatin [30]. Properly controlled polymerase chain reaction studies on CSF, brain, and blood from MS patients have failed to find evidence for HTLV-I [62] and HTLV-II [12].

Cellular immune abnormalities of the blood and CSF in MS

Because brain tissue itself can rarely be sampled in living patients with MS, most studies of the immunopathology of MS have concentrated on specimens of peripheral blood and CSF. The CSF circulates through the ventricular system and subarachnoid space, acquiring cells and humoral factors from the brain parenchyma, ependymal surfaces, and the leptomeninges. CSF normally turns over within 4 h, can be routinely sampled from the lumbar cistern, contains only a few cells and little protein, and differs in constituent concentrations from the extracellular parenchymal spaces of the brain due to this rapid turnover and contribution of the meninges to its contents.

Cellular immune activation in MS

Peripheral and CSF abnormalities characteristic of active immune states are present in MS. Peripheral blood has decreased numbers of $CD4^+CD45R^+$ [13] (suppressor-inducer) T lymphocytes but increased levels of T lymphocyte activation markers are seen [25]. In active MS CSF, minor increases in the numbers of lymphocytes are present, but the frequency of CD69-expressing lymphocytes is also increased, indicating activation [60].

Normally a physiological increase in the $CD4^+/CD8^+$ ratio is present in the CSF relative to the periphery [52]. During active MS, changes in phenotypes occur in both compartments. Variations in findings between laboratories likely result from different patient populations, varying immune activity, and even daily fluctuation of circulating phenotypes. Compared to patients with noninflammatory neurological diseases, an increased CSF $CD4^+/CD8^+$ ratio has been detected in patients with progressive MS even when pleocytosis is absent [38]. Similar changes occur in other types of subacute inflammation, such as cerebral vasculitis and inflammatory polyradiculopathy [61].

A similar physiological enrichment in CSF relative to blood is seen normally with $CD4^+CDw29^+CD45R^-$ cells (putative helper-inducer) [31] and class II expression [52]. Class II-bearing T lymphocytes increase in MS CSF during active disease, but are less commonly detected in MS CSF than in viral meningitis [18]. $CD4^+CD45R^+$ (suppressor-inducer) phenotypes are relatively decreased in MS CSF [14]. MS CSF shows elevated cytotoxic precursors and natural killer precursors, many of which are $CD4^+$ [99].

Autoreactive T cells

Autoreactive T lymphocytes have been extensively investigated in an effort to substantiate an autoimmune hypothesis for demyelinating disease. Oligoclonal usage of V-chain alleles is uncommon, and a wide variety of class II and V-chain alleles have been demonstrated. The response of peripheral T cells to myelin basic protein (MBP) is diverse with respect to $V\beta$ gene usage, MBP epitope, and class II alleles; however, individuals can consistently employ the same DRB alleles [50]. MBP-specific T cell lines have been derived from MS blood, but the precursor cells are not different in frequency or phenotype from control peripheral blood. Autoreactive cells typically are $CD3^+CD4^+CD8^-$ and DR-restricted, mostly functionally cytotoxic, recognizing mid- and C-terminal domains of MBP [36]. T cells detected by IFN-γ secretion in response to MBP are a small subset (1/5000) of the CSF T cells and far fewer in blood [87].

Other myelin protein-reactive cells have been an important focus of study. Myelin oligodendrocyte glycoprotein (MOG)-reactive lymphocytes are present in 50% of MS patients, but control patients also have reactive cells at equivalent frequencies [37]. In contrast to peripheral blood, IFN-γ-secreting MOG-reactive CSF cells represent 1/450 T cells [88]. Increased numbers of proteolipid protein (PLP)-reactive cells are found in MS patients, and these represent 1/500 T cells [89].

Cytokine activity and production in MS

The pivotal role of cytokines in regulation and amplification of the immune response has been suggested by changes in serum and CSF cytokines, cytokine receptors (both soluble and cell-associated), and cytokine-producing cells. Perturbations in cytokines may be transient, which may partially explain variability in results between laboratories.

Serum IL-1α (but not IL-1β) is increased threefold during attacks in relapsing-remitting MS. Quiescent disease and chronic progressive MS do not show this abnormality [48]. MBP stimulates both IL-1 and IL-2 production by peripheral blood mononuclear cells in MS patients but not controls [81]. In CSF, IL-1β can be increased both with MS and other diseases [11].

IL-2 and its receptor may have significant correlation with disease activity. Elevated serum IL-2 or soluble IL-2R (sIL-2R) is common in MS (89% of patients in multiple samples) but not usually in CSF [21]. Serum IL-2 levels are usually low in MS, but sIL-2R correlates with disease activity [19]. Serum IL-2 correlates with serum TNF-α, fluctuating in the course of chronic progressive MS with disability [94]. In relapsing-remitting MS, serum IL-2 correlates with disease as it does in inflammatory demyelinating polyradiculopathies [28]. IL-2-secreting, myelin protein-reactive peripheral blood lymphocytes are more frequent in MS patients than in controls in response to MAG, MOG, and PLP (but not MBP) stimulation but are similar in frequency to that found in acute viral meningitis [23].

Other lymphokines have been investigated to a lesser extent. IL-3 production by MS peripheral blood lymphocytes is decreased during exacerbation, whereas IFN-γ and TNF production are increased; these changes are not present when the disease is inactive [34]. IL-4 is rarely detected in MS sera and never in CSF [21]. IL-6 in MS plasma is higher than in viral meningitis and controls, but the inverse is true for the CSF [20].

In a prospective study of TNF-α in MS CSF, 53% of chronic-progressive MS and none of the clinically stable patients showed detectable levels. Intrathecal TNF synthesis correlated with severity of disease and a progressive phenotype [84] but not with IgG synthesis rate [44] or attack timing. TNF-α-producing cells in MS CSF are both $CD4^+$ and $CD8^+$ and often produce IFN-γ and IL-2 [8].

IFN represent important inflammatory and antiviral cytokines and have disease-modifying activity in MS. IFN-γ has been proposed to have an important role in promoting the inflammation in MS. Immunoreactive circulating IFN-γ can be absent in MS patients [21], but sensitive bioassay techniques can detect circulating IFN-γ even in remission [27]. IFN-α is frequently detectable in MS sera [33].

Transforming growth factor-β (TGF-β) may have a role in down-regulating of $CD4^+$ T cell activity, and TGF-β is elevated in patients with MS, and to a greater extent between relapses [6]. Isoforms of α_2-macroglobulin, a serum protease inhibitor, correlate with whether individuals have relapsing-remitting or chronic-progressive MS [3]. One could hypothesize a role for this protease inhibitor in limiting the size of MS lesions by interactions with phagocytic proteases or cytokines.

Humoral immunity and autoantibodies in MS

In contrast to the unclear clinical significance of changes in cellular immunity, the changes in humoral immunity are very helpful for diagnosis of MS. The presence of oligoclonal bands (OCB) of IgG and increased intrathecal (within the BBB) production of IgG are characteristic changes seen in more than 90% of clinically definite MS patients [91]. Evidence of typical lesions on MR imaging correlates highly with abnormalities of CSF IgG [54]. The Ig is produced by plasma cells within the brain parenchyma and the meninges [17], and detection of intrathecal Ig synthesis correlates with chronic inflammation within the CNS.

Intrathecal IgG synthesis in MS

Intrathecal Ig production is agreed to be a hallmark of clinically established MS, although it lacks sensitivity early in the course of the syndrome. These changes can be seen in other chronic CNS inflammatory disorders and also in some other diseases. Systemic inflammatory disorders such as connective tissue diseases [98], neoplasia and plasma cell dyscrasias [105], and viral infections (SSPE) can produce OCB and intrathecal Ig synthesis. However, the typical clinical presentation of MS and the characteristic Ig production are important to the diagnosis of MS. Those patients with clinical MS which lack the Ig changes also lack meningeal and parenchymal plasma cells [17].

The increased synthesis occurs in all Ig classes and isotypes even between attacks, but a predominance of IgG_1 (and to a lesser extent IgG_3) over other subclasses is similar to that seen in viral and chronic autoimmune diseases [42]. Intrathecal production of IgM and IgA are present in about half of the cases and IgD in 20% [41]. MS CSF B cells appear to be already maximally active in synthesizing Ig [32].

Reactivity of intrathecal antibodies in MS

The nature of the antibodies produced within the CNS has been extensively studied. These antibodies react with pathogens, self antigens, chemical haptens, and oligodendroglia. These antibodies are related to disease presence rather than genetic susceptibility, since twins discordant for MS are also discordant for CSF antiviral titers to many pathogens [101]. Most work with antibodies in MS has centered on potential pathological reactivity with myelin proteins. "Natural" autoantibodies reactive with chemical haptens have been detected, which may contribute either to pathological or to favorable immunoglial interactions.

Antibodies to a variety of viruses are present in MS CSF, unlike in most more acute diseases of the CNS [101], although mumps meningitis also causes a variety of specificities [78]. Antibodies to various bacterial antigens have also been described. Whether these antibodies represent specific immune reaction is a major issue. B lymphocytes secreting IgM antibody to HSV or measles virus represent 0.2% of total cells in MS CSF; no IgG- or IgA-secreting cells for these pathogens are found in blood from MS patients [4]. The great diversity of reactivities reported for MS CSF antibody fails to support activity of a single pathogen in the CNS.

The most extensively studied autoreactivity is that with MBP, the major myelin protein which can induce immune encephalomyelitis in animal models. However, OCB are not removed by adsorption with myelin [56], indicating that most of the OCB-associated Ig is not anti-myelin IgG. Anti-MBP titer in CSF varies with disease activity. MS CSF frequently has this antibody at the first attack, unlike controls, and with disease progression anti-MBP is increasingly antigen bound [98]. Although peripheral blood in MS patients contains anti-MBP-secreting cells, the frequency of these cells is not different from control subjects; in contrast, these cells are sixfold more frequent in the CSF of MS patients but represent only 1/380 CSF cells [46].

Other autoreactive antibodies have been described. A weakly elevated titer to anti-MAG carbohydrate is present in MS CSF but not in blood [53]. Anti-MOG-secreting cells are detectable in 80% of MS patients and preferentially present in CSF (1/265 cells) [88]. Anti-galactocerebroside in CSF is detectable in more than 50% of patients with MS or acute optic neuritis but in none of controls; however, this was seen also in other neurological diseases [35]. Anti-endothelial and anti-smooth muscle IgG antibodies are increased in MS patients with actively relapsing disease [96].

Oligodendroglial loss in MS has prompted a search for specific Ig reacting with oligodendrocytes but not myelin; such antibodies are present in MS sera at low titer, principally IgM, and some IgG [59]. Quantitative determinations of CSF anti-oligodendrocyte IgG show twofold increases in both MS and other neurological diseases as compared to controls [24]. When adjusted for CSF IgG concentration, Ig reacting to galactocerebroside-positive human oligodendroglia is not more prevalent in MS specimens [43]. These antibodies may, therefore, be increased along with general intrathecal antibody production. While nonspecific for MS, these Ig are reactive with oligodendroglia. Such antibodies could be pathogenic in the setting of BBB breakdown, complement, and inflammatory infiltrates which occur in association with acute lesions.

Complement activation has also been studied. As in inflammatory polyradiculopathy, MS CSF contains terminal complement complexes [79], and vesicles enriched in these complexes are present in MS CSF and are possibly oligodendrocyte derived [80]. C9 has been reported to be depleted in MS CSF, but not to the extent seen in acute meningitis [26], and the intrathecal production index is unaltered [73]. MS CSF frequently contains IgG and IgM immune complexes, but these occur only in relapsing-remitting MS. In sera, these complexes are of intermediate size, contain complement, and are increased in active disease [69].

In the absence of a proven detriment, the autoreactivity and polyreactivity of the above-described Ig may potentially represent a beneficial rather than pathogenic role. As regards humoral immunity, cross-reacting autoantibodies to infectious agents or idiotypes might serve to clear an infectious agent or to limit the activation of immune responses against self antigens exposed during the process of oligodendroglial and axonal degeneration.

The lack of "specificity" of the afore-mentioned immunoreactive antibodies to pathogens and autoantigens may be due, in part, to the presence of natural polyreactive antibodies. Apparently "nonspecific" in their purpose, these Ig may subserve an important immunological role in normal individuals. Natural autoantibodies reactive against a variety of serum antigens, cytoskeletal proteins as well as chemical haptens have been described in the sera of normal individuals. These antibodies are encoded by unmutated germ-line-derived sequences and probably serve an important first line of defense in humoral immunity in the immunologically naive host. In support of this interpretation, investigations have revealed increased polyspecific IgG antibodies in MS CSF but not serum [47].

Potentially beneficial actions of autoantibodies have been suggested by animal experiments. Ig against spinal cord antigens can promote remyelination in chronic infection of mice with Theiler's virus [71]. Remyelination can be enhanced in immune encephalomyelitis following treatment with CNS autoantigens in incomplete Freund's adjuvant [70]. An IgM monoclonal antibody, which promotes remyelination, reacts with CNS glial cells and has the properties of natural autoantibodies, including reactivity with multiple chemical haptens and autoantigens, use of germ-line variable Ig sequences, and homology to a known natural autoantibody [51].

Whether these natural autoantibodies normally have a special anti-inflammatory function can only be speculated. Although high-titer autoantibodies are strongly associated with autoimmune connective tissue diseases in both humans and animals, arguments have been set forth for beneficial actions of some autoantibodies [2]. The therapeutic modulation of the immune response by intravenous administration of Ig may be, in part, due to natural autoantibodies, as these represent a substantial part of circulating Ig [16]. Since humoral immune activation occurs in MS, some Ig may be the result of activation of polyspecific B cell clones as an epiphenomenon to CNS injury. Thus, the presence of autoreactive antibodies should not be presumed to represent primary pathogenic autoimmunity.

In summary, polyspecific production of antibodies occurs in MS which is sequestered within the CNS. This response is not unique to MS, since these changes occur in other inflammatory CNS diseases. The beneficial or deleterious nature of various autoreactive antibodies awaits experimental testing in demyelinating models.

Conclusions

Whether a microbial or environmental pathogen is involved in the initiation, recurrence or sustenance of the CNS inflammation in MS has not yet been determined. The inflammatory response could be characteristic of a persistent viral infection, autoimmunity, or both. The different phenotypes of demyelinating diseases may be due either to different initiating environmental pathogens or to varying factors of the host immune response. Indeed, the occurrence of demyelination itself may represent an extreme response during nonspecific CNS inflammation. The roles of specific genetic or environmental factors remain unclear in determining lesion size, chronicity of inflammation, axon survival, and remyelination.

Regardless of etiology, the recurrent and chronic leukoencephalitis that occurs in MS results in multifocal injury to glioaxonal units, causing axonal dysfunction. Axonal injury is apparent in established MS and is likely a principal factor in progressive disability. The lack of consistent correlation between abnormalities in immune function and clinical deficits may be explained by differing degrees of axonal injury. Inflammation in the midst of naked axons may result in persistent neuronal injury and permanent neurological deficit.

The CNS immunopathological process of MS is schematically shown in Fig. 1. The immune infiltrates in MS are consistent with the nature and temporal profile of the inflammatory response. Degeneration of myelin is an early event which is related to a disturbance of myelinating function. This event may be due either to direct oligodendroglial injury or to cytokines which induce the disintegration of the myelin sheath.

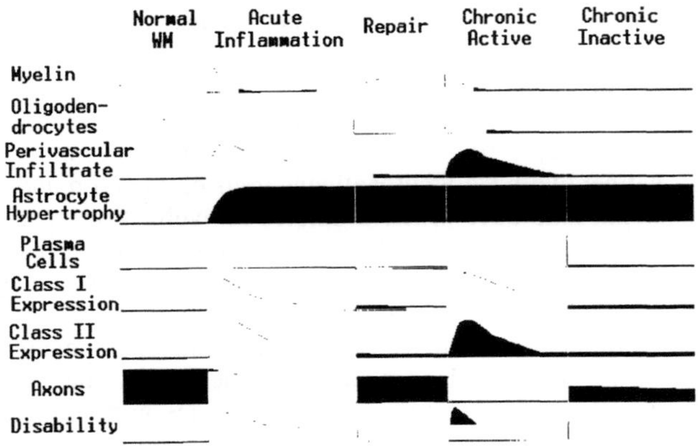

Fig. 1. The immunopathology of multiple sclerosis is presented schematically in a chronological fashion, indicating progression of the disease from the left to the right. Normal white matter (WM) is depicted at the left and the completely demyelinated chronic plaque at the right. The character of each stage is depicted in a relative fashion according to the amplitude of each bar

Myelin loss may not necessarily reflect permanent loss of oligodendrocytes. Nonlethal injury may interfere with the "luxury" function of the cells, i.e., myelination. Repair of myelin occurs but is interrupted by recurrent or persistent inflammation. CNS parenchymal cells are multifaceted, serving accessory functions to the immune response in initiating, targeting, and containing the immune response through ongoing surveillance. The magnitude of the perivascular inflammatory infiltrate is less dramatic than that of acute infections or other immune processes which do not result in demyelination. The character of the immune response is one of polyclonal and polyspecific activation of both cellular and humoral immunity, although oligoclonal intrathecal IgG production and oligoclonal elements of the T cell response occur. Markers of immune and phagocyte activation are nonspecifically elevated in both MS serum and CSF. Existing evidence does not provide support for a primary role of these changes in the initiation of MS. Individuals react against multiple pathogens and autoantigens, which likely represents sustained activation of the immune response.

Autoreactive cells and autoantibodies can be found more frequently in MS patients than in normal individuals, but the low frequencies of these cells and the low autoantibody titers argue against these elements having a primary pathogenic role. Indeed, some elements of this autoreactivity may be beneficial in limiting the primary immune response within the CNS or preventing a secondary immune response against cryptic antigens released by the demyelination or axonal loss. The understanding of cytokine actions is incomplete, although they have the potential to perturb myelination, neuronal physiology, and BBB function. Certain cytokines may be implicated in demyelination, but these molecules are also present at high concentrations in inflammatory diseases with no central demyelination. Alternatively, it is possible that some cytokines play a protective role in limiting the immunopathology or by enhancing remyelination. Therefore, the nature and timing of the inflammatory response do not provide evidence for an autonomous immunopathogenetic state in MS, even though the immune system appears to be the principal mediator of CNS injury and permanent neurological deficits.

References

1. AUSTIMS Research Group (1989) Interferon-α and transfer factor in the treatment of multiple sclerosis: a double-blind, placebo-controlled trial. J Neurol Neurosurg Psychiatry 52:566
2. Avrameas S, Ternynck T (1993) The natural autoantibodies system: between hypotheses and facts. Mol Immunol 30:1133
3. Back SA, Alhadeff JA (1992) Differential isoform profiles of α2-macroglobulin from plasma of patients with chronic-progressive or relapsing-remitting multiple sclerosis. Clin Chim Acta 211:27
4. Baig S, Olsson O, Olsson T, Love A, Jeansson S, Link H (1989) Cells producing antibody to measles and herpes simplex virus in cerebrospinal fluid and blood of patients with multiple sclerosis and controls. Clin Exp Immunol 78:390
5. Barnes D, Munro PM, Youl BD, Prineas JW, McDonald WI (1991) The longstanding MS lesion. A quantitative MRI and electron microscopic study. Brain 114:1271
6. Beck J, Rondot P, Jullien P, Wietzerbin J, Lawrence DA (1991) TGF-β-like activity produced during regression of exacerbations in multiple sclerosis. Acta Neurol Scand 84:452
7. Beck RW, Cleary PA, Trobe JD, Kaufman DI, Kupersmith MJ, Paty DW, Brown CH, Group ONT (1993) The effect of corticosteroids for acute optic neuritis on the subsequent development of multiple sclerosis. N Engl J Med 329:1764
8. Benvenuto R, Paroli M, Buttinelli C, Franco A, Barnaba V, Fieschi C, Balsano F (1991) Tumour necrosis factor-α synthesis by cerebrospinal fluid-derived T cell clones from patients with multiple sclerosis. Clin Exp Immunol 84:97
9. Booss J, Esiri MM, Tourtellotte WW, Mason DY (1983) Immunohistological analysis of T lymphocyte subsets in the central nervous system in chronic progressive multiple sclerosis. J Neurol Sci 62:219
10. Brueck W, Schmied M, Suchanek G, Brueck Y, Breitschopf H, Poser S, Piddleston S, Lassman H (1994) Oligodendrocytes in the early course of multiple sclerosis. Ann Neurol 35:65
11. Cacabelos R, Barquero M, Garcia P, Alvarez XA, Varela de SE (1991) Cerebrospinal fluid interleukin-1β (IL-1β) in Alzheimer's disease and neurological disorders. Methods Find Exp Clin Pharmacol 13:455
12. Chen IS, Haislip AM, Myers LW, Ellison GW, Merrill JE (1990) Failure to detect human T cell leukemia virus-related sequences in multiple sclerosis blood. Arch Neurol 47:1064
13. Chofflon M, Weiner HL, Morimoto C, Hafler DA (1988) Loss of functional suppression is linked to decreases in circulating suppressor inducer ($CD4^+$ $2H4^+$) T cells in multiple sclerosis. Ann Neurol 24:185
14. Chofflon M, Weiner HL, Morimoto C, Hafler DA (1989) Decrease of suppressor inducer ($CD4^+$ $2H4^+$) T cells in multiple sclerosis cerebrospinal fluid. Ann Neurol 25:494
15. Coyle PK (1985) CSF immune complexes in multiple sclerosis. Neurology 35:429
16. Dighiero G, Lymberi P, Guilbert B, Ternynck T, Avrameas S (1986) Natural autoantibodies constitute a substantial part of normal circulating immunoglobulins. Ann NY Acad Sci 475:135
17. Farrell MA, Kaufmann JC, Gilbert JJ, Noseworthy JH, Armstrong HA, Ebers GC (1985) Oligoclonal bands in multiple sclerosis: clinical-pathological correlation. Neurology 35:212
18. Fredrikson S, Karlsson-Parra A, Olsson T, Link H (1987) HLA-DR antigen expression on T cells from cerebrospinal fluid in multiple sclerosis and aseptic meningo-encephalitis. Clin Exp Immunol 68:298
19. Freedman MS, Muth KL, Trotter JL, Yoshizawa CN, Antel JP (1992) Prospective serial analysis of interleukin-2 and soluble interleukin-2 receptor in relapsing-remitting multiple sclerosis. Neurology 42:1596
20. Frei K, Fredrikson S, Fontana A, Link H (1991) Interleukin-6 is elevated in plasma in multiple sclerosis. J Neuroimmunol 31:147
21. Gallo P, Piccinno MG, Tavolato B, Siden A (1991) A longitudinal study on IL-2, sIL-2R, IL-4 and IFN-gamma in multiple sclerosis CSF and serum. J Neurol Sci 101:227
22. Ghatak NR, Leshner RT, Price AC, Felton III WL (1989) Remyelination in the human central nervous system. J Neuropathol Exp Neurol 48:507
23. Giang DW, Grow VM, Mooney C, Mushlin AI, Goodman AD, Mattson DH, Schiffer RB (1994) Clinical diagnosis of multiple sclerosis. The impact of magnetic resonance imaging and ancillary testing. Rochester-Toronto Magnetic Resonance Study Group. Arch Neurol 51:61

24. Gorny MK, Wroblewska Z, Pleasure D, Miller SL, Wajgt A, Koprowski H (1983) CSF antibodies to myelin basic protein and oligodendrocytes in multiple sclerosis and other neurological diseases. Acta Neurol Scand 67:338
25. Hafler DA, Hemler ME, Christenson L, Williams JM, Shapiro HM, Strom TB, Strominger JL, Weiner HL (1985) Investigation of in vivo activated T cells in multiple sclerosis and inflammatory central nervous system diseases. Clin Immunol Immunopathol 37:163
26. Halawa I, Lolli F, Link H (1989) Terminal component of complement C9 in CSF and plasma of patients with MS and aseptic meningitis. Acta Neurol Scand 80:130
27. Hammann KP, Hopf HC (1986) Determination of interferon-gamma in the peripheral blood of multiple sclerosis patients in remission, using a chemiluminescence technique. J Neuroimmunol. 13:9
28. Hartung HP, Reiners K, Schmidt B, Stoll G, Toyka KV (1991) Serum interleukin-2 concentrations in Guillain-Barre syndrome and chronic idiopathic demyelinating polyradiculoneuropathy: comparison with other neurological diseases of presumed immunopathogenesis. Ann Neurol 30:48
29. Hauser SL, Bhan AK, Gilles F, Kemp M, Kerr C, Weiner HL (1986) Immunohistochemical analysis of the cellular infiltrate in multiple sclerosis lesions. Ann Neurol 19:578
30. Hayano M, Sung JH, Mastri AR (1976) "Paramyxovirus-like" intranuclear inclusions occurring in the nervous system in diverse unrelated conditions. J Neuropathol Exp Neurol 35:287
31. Hedlund G, Sandberg-Wollheim M, Sjogren HO (1989) Increased proportion of $CD4^{+}CDw29^{+}$ $CD45R^{-}UCHL\text{-}1^{+}$ lymphocytes in the cerebrospinal fluid of both multiple sclerosis patients and healthy individuals. Cell Immunol 118:406
32. Henriksson A, Kam-Hansen S, Forsberg P, Grandien M (1986) Cerebrospinal fluid lymphocytes from patients with multiple sclerosis do not increase immunoglobulin or measles antibody production after stimulation with pokeweed mitogen. J Neuroimmunol 11:15
33. Hertzog PJ, Wright A, Harris G, Linnane AW, Mackay IR (1991) Intermittent interferonemia and interferon responses in multiple sclerosis. Clin Immunol Immunopathol 58:18
34. Huberman M, Shalit F, Roth-Deri I, Gutman B, Kott E, Sredni B (1993) Decreased IL-3 production by peripheral blood mononuclear cells in patients with multiple sclerosis. J Neurol Sci 118:79
35. Ichioka T, Uobe K, Stoskopf M, Kishimoto Y, Tennekoon G, Tourtellotte WW (1988) Anti-galactocerebroside antibodies in human cerebrospinal fluids determined by enzyme-linked immunosorbent assay (ELISA). Neurochem Res 13:203
36. Jingwu Z, Medaer R, Hashim GA, Chin Y, van den Berg-LE, Raus JC (1992) Myelin basic protein-specific T lymphocytes in multiple sclerosis and controls: precursor frequency, fine specificity, and cytotoxicity. Ann Neurol 32:330
37. Kerlero de RN, Milo R, Lees MB, Burger D, Bernard CC, Ben-Nun A (1993) Reactivity to myelin antigens in multiple sclerosis. Peripheral blood lymphocytes respond predominantly to myelin oligodendrocyte glycoprotein. J Clin Invest. 92:2602
38. Kolmel HW, Sudau C (1989) Cell count and ratio of helper/inducer to suppressor/cytotoxic T cells in the cerebrospinal fluid of patients with multiple sclerosis. J Neurol 236:424
39. Lee SC, Raine CS (1989) Multiple sclerosis: oligodendrocytes in active lesions do not express class II major histocompatibility complex molecules. J Neuroimmunol 25:261
40. Lee SC, Moore GR, Golenwsky G, Raine CS (1990) Multiple sclerosis: a role for astroglia in active demyelination suggested by class II MHC expression and ultrastructural study. J Neuropathol Exp Neurol 49:122
41. Lolli F, Halawa I, Link H (1989) Intrathecal synthesis of IgG, IgA, IgM and IgD in untreated multiple sclerosis and controls. Acta Neurol Scand 80:238
42. Losy J, Mehta PD, Wisniewski HM (1990) Identification of IgG subclasses' oligoclonal bands in multiple sclerosis CSF. Acta Neurol Scand 82:4
43. Lubetzki C, Lombrail P, Hauw JJ, Zalc B (1986) Multiple sclerosis: rat and human oligodendrocytes are not the target for CSF immunoglobulins. Neurology 36:524
44. Maimone D, Gregory S, Arnason BG, Reder AT (1991) Cytokine levels in the cerebrospinal fluid and serum of patients with multiple sclerosis. J Neuroimmunol 32:67
45. Martin JR, Holt RK, Webster HD (1988) Herpes-simplex-related antigen in human demyelinative disease and encephalitis. Acta Neuropathol 76:325
46. Martino G, Olsson T, Fredrikson S, Hojeberg B, Kostulas V, Grimaldi LM, Link H (1991) Cells producing antibodies specific for myelin basic protein region 70–89 are predominant in cerebrospinal fluid from patients with multiple sclerosis. Eur J Immunol 21:2971

47. Matsiota P, Blancher A, Doyon B, Guilbert B, Clanet M, Kouvelas ED, Avrameas S (1988) Comparative study of natural autoantibodies in the serum and cerebrospinal fluid of normal individuals and patients with multiple sclerosis and other neurological diseases. Ann Inst Pasteur Immunol 139:99
48. Matsuda M, Tsukada N, Miyagi K, Yanagisawa N (1991) Increased interleukin-1 production by peripheral blood mononuclear cells in patients with multiple sclerosis. J Neurol Sci 102:100
49. McCallum K, Esiri MM, Tourtellotte WW, Booss J (1987) T cell subsets in multiple sclerosis. Gradients at plaque borders and differences in nonplaque regions. Brain 110:1297
50. Meinl E, Weber F, Drexler K, Morelle C, Ott M, Saruhan-Direskeneli G, Goebels N, Ertl B, Jechart G, Giegerich G, Schonbeck S, Bannwarth W, Wekerle H, Hohlfeld R (1993) Myelin basic protein-specific T lymphocyte repertoire in multiple sclerosis. Complexity of the response and dominance of nested epitopes due to recruitment of multiple T cell clones. J Clin Invest 92:2633
51. Miller DJ, Sanborn KS, Katzmann JA, Rodriguez M (1994) Monoclonal autoantibodies promote central nervous system repair in an animal model of multiple sclerosis. J Neurosci 14:6230
52. Mix E, Olsson T, Correale J, Kostulas V, Link H (1990) $CD4^+$, $CD8^+$, and $CD4^-$ $CD8^-$ T cells in CSF and blood of patients with multiple sclerosis and tension headache. Scand J Immunol 31:493
53. Moller JR, Johnson D, Brady RO, Tourtellotte WW, Quarles RH (1989) Antibodies to myelin-associated glycoprotein (MAG) in the cerebrospinal fluid of multiple sclerosis patients. J Neuroimmunol 22:55
54. Muller FA, Hanny PE, Wichmann W, Fierz W, Dursteler MR (1989) Cerebrospinal fluid immunoglobulins and multiple sclerosis. Correspondence with magnetic resonance imaging and visually evoked potential changes. Arch Neurol 46:367
55. Nesbit GM, Forbes GS, Scheithauer BW, Okazaki H, Rodriguez M (1991) Multiple sclerosis: histopathological and MR and/or CT correlation in 37 cases at biopsy and three cases at autopsy. Radiology 180:467
56. Nobile-Orazio E, Spagnol G, Daverio R, Riboni L, Macchi R, Francomano E, Scarlato G (1987) Studies on anti-myelin antibodies in patients with multiple sclerosis. Riv Neurol 57:13
57. Panitch HS, Hirsch RL, Haley AS, Johnson KP (1987) Exacerbations of multiple sclerosis in patients treated with gamma interferon. Lancet I:893
58. Paty DW, Li DK (1993) Interferon-β-1b is effective in relapsing-remitting multiple sclerosis. II. MRI analysis results of a multicenter, randomized, double-blind, placebo-controlled trial. UBC MS/MRI Study Group and the IFNB Multiple Sclerosis Study Group. Neurology 43:662
59. Pedersen JS, Walker M, Toh BH, De Aizpurua HJ, Lolait SJ, Bernard CCA (1983) Flow microfluorometry detects IgM autoantibody to oligodendrocytes in multiple sclerosis. J Neuroimmunol 5:251
60. Perrella O, Carrieri PB, De MR, Buscaino GA (1993) Markers of activated T lymphocytes and T cell receptor γ/δ^+ in patients with multiple sclerosis. Eur Neurol 33:152
61. Polman CH, de GCJ, Koetsier JC, Sminia T, Veerman AJ (1987) Cerebrospinal fluid cells in multiple sclerosis and other neurological diseases: an immunocytochemical study. J Neurol 234:19
62. Prayoonwiwat N, Pease LR, Rodriguez M (1991) Human T cell lymphotropic virus type I sequences detected by nested polymerase chain reactions are not associated with multiple sclerosis. Mayo Clin Proc 66:665
63. Prineas JW, Connell F (1978) The fine structure of chronically active multiple sclerosis plaques. Neurology 28:68
64. Prineas JW, Connell F (1979) Remyelination in multiple sclerosis. Ann Neurol 5:22
65. Prineas JW, Wright RG (1978) Macrophages, lymphocytes, and plasma cells in the perivascular compartment in chronic multiple sclerosis. Lab Invest 38:409
66. Prineas JW, Kwon EE, Goldenberg PZ, Ilyas AA, Quarles RH, Benjamins JA, Sprinkle TJ (1989) Multiple sclerosis. Oligodendrocyte proliferation and differentiation in fresh lesions. Lab Invest 61:489
67. Prineas JW, Barnard RO, Kwon EE, Sharer LR, Cho ES (1993) Multiple sclerosis: remyelination of nascent lesions. Ann Neurol 33:137
68. Prineas JW, Barnard RO, Revesz T, Kwon EE, Sharer L, Cho ES (1993) Multiple sclerosis. Pathology of recurrent lesions. Brain 116:681
69. Procaccia S, Lanzanova D, Caputo D, et al. (1988) Circulating immune complexes in serum and in cerebrospinal fluid of patients with multiple sclerosis. Characterization and correlation with the clinical course. Acta Neurol Scand 77:373
70. Raine CS, Traugott U, Stone SH (1978) Chronic relapsing experimental allergic encephalomyelitis: CNS plaque development in unsuppressed and suppressed animals. Acta Neuropathol (Berl) 43:43

71. Rodriguez M, Lennon VA (1990) Immunoglobulins promote remyelination in the central nervous system. Ann Neurol 27:12
72. Rodriguez M, Sheithauer B (1994) Ultrastructure of multiple sclerosis. Ultrastruct Pathol 18:3
73. Rodriguez M, Wynn DR, Kimlinger TK, Katzmann JA (1990) Terminal component of complement (C9) in the cerebrospinal fluid of patients with multiple sclerosis and neurological controls. Neurology 40:855
74. Rodriguez M, Karnes WE, Bartleson JD, Pineda AA (1993) Plasmapheresis in acute episodes of fulminant CNS inflammatory demyelination. Neurology 43:1100
75. Rodriguez M, Scheithauer BW, Forbes G, Kelly PJ (1993) Oligodendrocyte injury is an early event in lesions of multiple sclerosis. Mayo Clin Proc 68:627
76. Rodriguez M, Siva A, Ward J, Stolp-Smith K, O'Brien P, Kurland L (1994) Impairment, disability, and handicap in multiple sclerosis: a population-based study in Olmsted County, Minnesota. Neurology 44:28
77. Ruijs TC, Freedman MS, Grenier YG, Olivier A, Antel JP (1990) Human oligodendrocytes are susceptible to cytolysis by major histocompatibility complex class I-restricted lymphocytes. J Neuroimmunol 27:89
78. Salmi AA, Hyypia T, Ilonen J, Reunanen M, Remes M (1989) Production of viral antibodies in vitro by CSF cells from mumps meningitis and multiple sclerosis patients. J Neurol Sci 90:315
79. Sanders ME, Koski CL, Robbins D, Shin ML, Frank MM, Joiner KA (1986) Activated terminal complement in cerebrospinal fluid in Guillain-Barre syndrome and multiple sclerosis. J Immunol 136:4456
80. Scolding NJ, Morgan BP, Houston WA, Linington C, Campbell AK, Compston DA (1989) Vesicular removal by oligodendrocytes of membrane attack complexes formed by activated complement. Nature 339:620
81. Selmaj K, Nowak Z, Tchorzewski H (1988) Multiple sclerosis: effect of myelin basic protein on interleukin 1, interleukin 2 production and interleukin 2 receptor expression in vitro. Clin Exp Immunol 72:428
82. Selmaj K, Brosnan CF, Raine CS (1991) Colocalization of lymphocytes bearing $\gamma\delta$ T cell receptor and heat shock protein hsp65$^+$ oligodendrocytes in multiple sclerosis. Proc Natl Acad Sci USA 88:6452
83. Selmaj K, Raine CS, Cannella B, Brosnan CF (1991) Identification of lymphotoxin and tumor necrosis factor in multiple sclerosis lesions. J Clin Invest 87:949
84. Sharief MK, Hentges R (1991) Association between tumor necrosis factor-α and disease progression in patients with multiple sclerosis. N Engl J Med 325:467
85. Sobel RA, Hafler DA, Castro EE, Morimoto C, Weiner HL (1988) The 2H4 (CD45R) antigen is selectively decreased in multiple sclerosis lesions. J Immunol 140:2210
86. Sobel RA, Mitchell ME, Fondren G (1990) Intercellular adhesion molecule-1 (ICAM-1) in cellular immune reactions in the human central nervous system. Am J Pathol 136:1309
87. Soderstrom M, Link H, Sun JB, Fredrikson S, Kostulas V, Hojeberg B, Li BL, Olsson T (1993) T cells recognizing multiple peptides of myelin basic protein are found in blood and enriched in cerebrospinal fluid in optic neuritis and multiple sclerosis. Scand J Immunol 37:355
88. Sun J, Link H, Olsson T, Xiao BG, Andersson G, Ekre HP, Linington C, Diener P (1991) T and B cell responses to myelin-oligodendrocyte glycoprotein in multiple sclerosis. J Immunol 146:1490
89. Sun JB, Olsson T, Wang WZ, Xiao BG, Kostulas V, Fredrikson S, Ekre HP, Link H (1991) Autoreactive T and B cells responding to myelin proteolipid protein in multiple sclerosis and controls. Eur J Immunol 21:1461
90. Tanaka R, Iwazaki Y, Koprowski H (1975) Ultrastructural studies of perivascular cuffing cells in multiple sclerosis brain. Am J Pathol 81:467
91. Tourtellotte WW, Baumhefner RW, Syndulko K, et al (1988) The long march of the cerebrospinal fluid profile indicative of clinical definite multiple sclerosis; and still marching. J Neuroimmunol 20:217
92. Traugott U (1987) Multiple sclerosis: relevance of class I and class II MHC-expressing cells to lesion development. J Neuroimmunol 16:283
93. Traugott U, Lebon P (1988) Interferon-γ and Ia antigen are present on astrocytes in active chronic multiple sclerosis lesions. J Neurol Sci 84:257
94. Trotter JL, Collins KG, van der Veen RC (1991) Serum cytokine levels in chronic progressive multiple sclerosis: interleukin-2 levels parallel tumor necrosis factor-α levels. J Neuroimmunol 33:29

95. Tsukada N, Behan WM, Behan PO (1985) Search for autoantibodies to endothelial and smooth muscle cells in patients with multiple sclerosis. Acta Neuropathol (Berl) 66:134
96. Van Engelen BGM, Hommes OR, Pinckers A, Cruysberg JRM, Barkhof F, Rodriguez M (1992) Improved vision after intravenous immunoglobulin in stable demyelinating optic neuritis. Ann Neurol 32:834
97. Vrethem M, Ernerudh J, Lindstrom F, Skogh T (1990) Immunoglobulins within the central nervous system in primary Sjogren's syndrome. J Neurol Sci 100:186
98. Warren KG, Catz I (1989) Cerebrospinal fluid autoantibodies to myelin basic protein in multiple sclerosis patients. Detection during first exacerbations and kinetics of acute relapses and subsequent convalescent phases. J Neurol Sci 91:143
99. Weber WE, Buurman WA, Vandermeeren MM, Medaer RH, Raus JC (1987) Fine analysis of cytolytic and natural killer T lymphocytes in the CSF in multiple sclerosis and other neurological diseases. Neurology 37:419
100. Weiner HL, Dau PC, Khatri BO, Petajan JH, Birnbaum G, McQuillen MP, Fosburg MT, Feldstein M, Orav EJ (1989) Double-blind study of true vs sham plasma exchange in patients treated with immunosuppression for acute attacks of multiple sclerosis. Neurology 39:1143
101. Woyciechowska JL, Dambrozia J, Leinikki P, Shekarchi C, Wallen W, Sever J, McFarland H, McFarlin D (1985) Viral antibodies in twins with multiple sclerosis. Neurology 35:1176
102. Wucherpfennig KW, Newcombe J, Li H, Keddy C, Cuzner ML, Hafler DA (1992) $\gamma\delta$ T-cell receptor repertoire in acute multiple sclerosis lesions. Proc Natl Acad Sci USA 89:4588
103. Wucherpfennig KW, Newcombe J, Li H, Keddy C, Cuzner ML, Hafler DA (1992) T cell receptor $V\alpha$-$V\beta$ repertoire and cytokine gene expression in active multiple sclerosis lesions. J Exp Med 175:993
104. Wynn DR, Rodriguez M, O'Fallon WM, Kurland LT (1990) A reappraisal of the epidemiology of multiple sclerosis in Olmsted County, Minnesota. Neurology 40:780
105. Zeman A, McLean B, Keir G, Luxton R, Sharief M, Thompson E (1993) The significance of serum oligoclonal bands in neurological diseases. J Neurol Neurosurg Psychiatry 56:32

Stress proteins: their role in the normal central nervous system and in disease states, especially multiple sclerosis

Gary Birnbaum

Department of Neurology, University of Minnesota, School of Medicine, Minneapolis, MN 55455, USA

Introduction

Heat shock proteins or stress proteins

The terms "heat-shock proteins" (hsp) and "stress proteins" will be used interchangeably in this chapter. While the more accurate term is stress proteins, hsp is used more commonly. Hsp are families of proteins present in almost all prokaryotic and eukaryotic cells. They are classified according to their molecular mass and range in size from 17 kDa, to more than 100 kDa. They are essential for normal cell functions and are expressed both constitutively and in increased amounts when cells are stressed in a variety of ways. One of the major functions of all hsp is the transport of peptides from one intracellular compartment to another. Such transport, often called "chaperoning", can occur during both synthetic and catabolic processes. Examples of stresses that increase expression of hsp are heat, viral infections, anoxia, and exposure to certain cytokines, such as tumor necrosis factor and interferon-gamma. There are excellent reviews on hsp [34, 45, 47].

As noted above, many hsp are expressed constitutively, and appear at different times during the ontogeny of both vertebrate and non-vertebrate nervous systems. Thus, it is reasonable to presume that hsp play a role in normal CNS development and function. Data in support of this presumption are presented here; however, emphasis will be placed on the role of hsp in abnormal states of the CNS, especially in autoimmune diseases, such as multiple sclerosis (MS). The chapter is, therefore, divided into two sections: (1) dealing with hsp expressed in the central nervous system (CNS) under normal conditions; and (2) dealing with hsp present, or increased, in abnormal CNS.

Several facts about hsp support their potential importance in the development of autoimmune diseases [32]. First, hsp are phylogenetically conserved. There is greater than 50% sequence homology between certain prokaryotic hsp and those of mammalian cells [32, 35, 57, 67]. Second, hsp are the immunodominant antigens for many infectious agents, including bacteria. mycobacteria, and parasites [34, 45, 47].

The resulting strong immune responses to the hsp of these agents have the potential to cross-react with either host hsp or the normal tissue components of the host.

Immune responses to hsp are implicated in the development of a number of human autoimmune diseases, including several inflammatory arthritides and type I diabetes mellitus [17, 32, 51, 57]. The possible roles of immune responses to hsp in autoimmune diseases of the CNS will be discussed here.

Defining the roles of stress proteins in normal and abnormal CNS is complicated by the large numbers of different cell types present in this organ. Neurons, astrocytes, oligodendrocytes, microglial cells, endothelial cells and choroidal cells are sufficiently different from one another, both functionally and metabolically, to preclude extrapolation of results from one cell type to the other. Superimposed on this complexity are the significant regional differences that exist within the CNS, as they relate to differences in function and differences in the patterns of constitutive stress protein expression. With these caveats in mind, I will proceed.

Stress proteins in normal nervous system

Most cells within the normal CNS express constitutive forms of stress proteins. This was demonstrated by Martin et al. [49] using a single monoclonal antibody (mAb) to a bacterial hsp 60-kDa protein, and by Brown and Rush [11] using mRNA probes for the 70-kDa hsp. In our laboratory we studied expression of hsp using a panel of 20 mAb with specificities either for the 65-kDa stress protein of mycobacteria, the human 70/72-kDa stress protein, or the human 60-kDa stress protein. A rabbit polyvalent antibody to the *E. coli* stress proteins was also used. These antibodies were used to immunocytochemically stain normal CNS [5]. Patterns of staining varied with the particular antibody. Most mAb to the 65-kDa mycobacterial stress protein stained neuronal, glial, and microglial cell bodies. Similar staining patterns were seen with the polyvalent rabbit antiserum. Little, if any, staining of oligodendrocytes was noted with any of the mAb tested. Three mAb, IH9, II-13, and N27F3-4, bound to normal CNS and peripheral nervous system myelin. IH9 is specific for an epitope on the *Mycobacterium leprae* 65-kDa stress protein, II-13 is specific for an epitope on the human 60-kDa stress protein, and N27F3-4 is specific for the both the constitutive and inducible form of the human 70/72-kDa stress protein. We could not determine, using our immunocytochemical method, whether these mAb bound to myelin epitopes that cross-reacted with stress proteins or to the stress proteins per se within the myelin sheaths.

Stress proteins during development of the nervous system

Increasing evidence suggests that stress proteins may play an important role during nervous system development. This was demonstrated in both drosophila and mammalian systems [13, 44, 52, 71, 72]. In drosophila [13, 22, 44], mRNA for hsp23, hsp26, and small hsp were constitutively expressed during particular stages of larval development, mainly in neurocytes and thoracic ganglion cells. In rat embryos [71, 72], severe heat shock resulted in marked deformities in the forebrain and eyes. Less severe heat shock resulted in lengthening of neuroectoderm cell cycle times. Both these phenomena were associated with increased levels of mRNA of hsp 27,

71, 73, and 88. Carter and Ross [12], using northern blot and riboprobe analyses of granule cells from 6-day postnatal mouse dentate gyrus and cerebellum, detected expression of mRNA coding for a protein that had 94% sequence homology with the first 33 residues of human hsp63, a protein induced during cardiogenic shock. These investigators proposed that this protein plays a role in granule cell differentiation and maintaining the mature phenotype.

The patterned expression of hsp mRNA and proteins during the ontogeny of vertebrate and invertebrate nervous systems suggests that stress proteins have important functions in regulating cell cycles, cellular differentiation, and cell maintenance at critical stages of organ development.

Stress proteins as markers of nervous system injury

Stress protein expression has been used as a marker to define the extent of injury caused by particular experimental or natural disease processes. Several different families of stress proteins are seen in areas of abnormal tissue. These include hsp70 proteins, hsp27 proteins, and the small hsp, ubiquitin. Expression of stress proteins in several experimental systems were used to define regions of ischemia [10, 13, 18, 28, 70], hyperthermia [10, 11, 26, 28, 42], and trauma [10, 75]. This technique was also used to identify injured cells in patients with Alzheimer's disease [14, 24, 26, 50, 53] and other neurodegenerative diseases [14, 26, 39, 50]. Expression of stress proteins in affected cells frequently occurred in the absence of other anatomic changes, suggesting that stress protein expression may be a sensitive marker of cell injury. The functions of stress proteins in the injured nervous system are not clear but they may play a protective role. For example, Lowenstein et al. [40] noted that induction of hsp72 in cultured rat cerebellar granular cells protected them from the damaging effects of the excitotoxin glutamate. Heat shock also protected neuronal cells from death by apoptosis [41].

Stress proteins as targets of immune responses

Two characteristics of stress proteins greatly increase the likelihood that these proteins may become autoantigens, involved either in initiating or perpetuating autoimmune diseases. First is their great phylogenetic conservation across major species barriers, from prokaryotes to mammals. The second is their immunodominance, that is their ability to evoke strong immune responses during the course of infections with a large number of infectious agents (detailed elsewhere in this issue). Thus, exposure to an infectious agent, in a genetically susceptible host at the appropriate time during the host's development, could result in an immune response to the stress proteins of the infecting agent that cross-reacts either with normal host stress proteins or with organ-specific proteins, resulting in an autoimmune disease. Several laboratories tested the feasibility of this hypothesis in the human disease MS, and in its animal model, experimental autoimmune encephalomyelitis (EAE).

Multiple sclerosis

MS is chronic, often progressive, inflammatory demyelinating disease of the CNS. Large numbers of immunocompetent cells are present in brain and cerebrospinal fluid (CSF) and concentrations of immunoglobulins also are increased in these compartments. The target of the inflammatory response is presumed to be the oligodendrocyte-myelin complex. As a result, there is extensive multifocal demyelination, little in the way of remyelination, and frequent astrocytic proliferation. Greater detail on MS is provided in the chapter by Rodriguez and Hunter in this issue.

The evidence that MS is an autoimmune disease, or for that matter, an immunologically mediated disease, is circumstantial. Numerous laboratories tried to find immune responses to myelin and other antigens that were unique to MS patients, but none succeeded. Persons with MS do have increased immune responses to a variety of myelin antigens, including myelin basic protein (MBP), proteolipid protein, and myelin oligodendrocyte glycoprotein [1, 2, 23, 31, 37, 54], and several of these are potent autoantigens capable of inducing EAE [20, 68]. However, it is not clear whether the heightened immune responses to these antigens are primary or secondary to the disease process.

A well-documented clinical phenomenon in MS is the association between disease activity and infections ([65] and personal observations). The study by Sibley et al. [65] showed a clear association between antecedent infections, usually viral, and attacks or worsening of disease. While preexistent symptoms of MS will worsen during the acute phase of an infection, especially if there is an associated fever, true disease exacerbations occur 3–14 days after the acute infection, during recovery.

There are several mechanisms whereby infections could cause exacerbations of an inflammatory, possibly autoimmune, disease:

1. Infections result in increased concentrations of circulating cytokines. Some of these, such as tumor necrosis factor or interferon-gamma, in addition to having direct effects on oligodendrocytes, could alter the blood-brain barrier, allowing easier entry of myelin-specific immunocompetent cells and/or antibodies. Cytokines could also nonspecifically activate anti-myelin T cells resident within the CNS of MS patients. The possible role of cytokines in the pathogenesis of MS is discussed by Grau and Chofflon in this issue.
2. Infections result in immune responses to the infectious orgamism's stress proteins. These immune responses could cross-react with phylogenetically conserved stress proteins expressed by oligodendrocytes in areas of inflammation.
3. Immune responses to the stress proteins of an infectious agent could cross-react with myelin antigens, resulting in myelin destruction.
4. A combination of above effects may be important.

Different approaches were used to determine whether stress proteins, or immune responses to stress proteins, are involved in the pathogenesis of MS. These included studies on: (a) the expression of stress proteins in the brains of persons with MS; (b) the distribution of γ/δ T lymphocytes in MS brains; (c) immune responses to stress proteins in MS patients; and (d) studies on cross-reactivity between stress proteins and CNS myelin.

Expression of stress proteins in MS CNS. Several laboratories studied expression of stress proteins in areas of MS demyelination. Selmaj et al. [62, 63] demonstrated expression of hsp65 in immature oligodendrocytes at the edges of MS plaques. There also was constitutive expression of hsp72 in MS and non-MS brain, especially in astrocytes. Wucherpfenning et al. [74] demonstrated expressions of hsp60 in foamy macrophages at the edges of acute plaques and expression of hsp90 in reactive astrocytes. As noted above, we utilized a panel of 20 mAb and polyvalent antibodies to human and mycobacterial stress proteins to immuno-cytochemically stain fixed, frozen sections of normal and MS brains [5]. Patterns of staining varied amongst antibodies. Some antibodies stained cell bodies of neurons and astrocytes, some stained neurofilaments, and others stained normal myelin. No increased staining was noted at the edges of either acute or chronic plaques and demyelinated areas did not stain at all.

Using a different approach, a number of laboratories studied the expression of stress proteins by glial cells in vitro. Again results varied with the techniques used and the antibodies used for the assays. Selmaj et al. [63], Freedman et al. [19], and Satoh et al. [60] identified oligodendrocytes as cells expressing stress proteins. Selmaj et al. [63] detected constitutive hsp65 expression in oligodendrocytes but not astrocytes, yet noted some hsp70/72 expression in astrocytes. Satoh et al. [60] detected constitutive expression of hsp60 in cultured murine oligodendrocytes and 'marginally detectable' levels in most astrocytes. Expression of hsp60 was increased in oligodendrocytes following heat stress. Freedman et al. [19] detected constitutive expression of hsp60 and hsp70/72 expression in oligodendrocytes with increased expression of hsp70/72 following heat stress. No astrocytes expressed hsp70/72 and only a small percentage of astrocytes expressed hsp60 proteins. In contrast, Marini et al. [42], using tissue cultures of rat astrocytes and neurons, demonstrated heat-inducible expression of hsp70 stress proteins predominantly in astrocytes, but oligodendrocytes may have been absent from these cultures. Using purified cultures of rat astrocytes, Dwyer et al. [16] demonstrated that such cells synthesized stress proteins of 30-34, 68, 70, 89, and 97 kDa and that exposure to heat readily induced expression of hsp65.

Since MS is an inflammatory disease, concentrations of cytokines are increased in MS brain [29, 61]. De Souza et al. [15] studied the effects of cytokines on the expression of stress proteins in mixed cultures of human glial cells [40, 41]. A mixture of cytokines induced expression of hsp72, predominantly in oligodendrocytes. The specific cytokines involved in this induction were interleukin-1, interferon-gamma, and tumor necrosis factor-alpha. Birnbaum et al. (manuscript in preparation) prepared cultures of purified murine astrocytes and exposed them to heat shock with or without exposure to a mixture of cytokines. Cytokines alone induced small amounts hsp70/72. However, the combination of heat shock and cytokine exposure augmented expression of hsp70/72 2- to 11-fold. The particular cytokines involved in this were interferon-gamma and tumor necrosis factor-alpha.

It is apparent from the above data that results can vary widely, depending on the specificities of the antibodies used and the particular methods, Nevertheless, there is good evidence that some antibodies to stress proteins bind to normal myelin, while others bind to oligodendrocytes or astrocytes. In addition, the inflammatory milieu within MS brains increases the expression of stress proteins within glia, increasing the possibility that immune responses to the stress proteins of the infectious agents could cross-react with their human homologues.

The distribution of γ/δ T cells in MS brains. While the vast majority of T cells responding to stress proteins express antigen-specific receptors composed of α and β chains, their numbers comprise only a small fraction of the total α/β expressing T cell pool. This is in contrast to the population of T cells expressing antigen-specific receptors composed of γ and δ chains. These cells comprise only about 1-5% of the total T cell pool yet a majority of this cells respond to stress proteins [8, 9, 25, 33, 55]. Investigators, therefore, studied the association between areas of MS demyelination, expression of stress proteins, and the distribution of γ/δ T cells. Their assumption was that the presence of γ/δ T cells could indicate the presence of an immune response to stress proteins in that region.

Selmaj et al. [62] and Wucherpfenning et al. [74] found accumulations of γ/δ T cells in areas of MS demyelination. Selmaj et al. [62] found them in regions of chronic demyelination, in proximity to immature oligodendrocytes that expressed hsp65. Wucherpfenning et al. [74] found γ/δ T cells in regions of acute demyelination; 60- and 90-kDa stress proteins were found in regions of normal myelin but were "overexpressed" in acute MS plaques. Cells expressing these stress proteins were foamy macrophages and reactive astrocytes, respectively. On the basis of sequence analyses, Wucherpfenning et al. [74] postulated there were oligoclonal expansions of γ/δ T cells in MS brains but since the repertoire of T cell receptors in this population is already restricted, these data do not necessarily indicate reactivity to a single antigen. Hvas et al. [30] looked for the presence of γ/δ cells in MS brains using polymerase chain reaction amplification of T cell receptor mRNA. They found γ/δ messages in all of their 12 MS brains, in 1 of their 10 non-neurological disease controls, and in 2 of their 3 neurological controls. Sequence analyses of the MS brain-derived cDNA failed to show evidence of clonal expansions of particular populations of γ/δ cells. Two groups studied γ/δ T cells in MS and control CSF. Shimonkevitz et al. [64] found increased numbers of γ/δ T cells in the CSF of MS patients with recent onset of disease but not in patients with chronic MS or in disease controls. Sequence analyses of the junctional regions of the antigen receptors of these cells suggested an oligoclonal expansion of this population, perhaps in response to a particular antigen. Perella et al. [53] found equivalent numbers of γ/δ T cells in MS and other neurological disease (OND) CSF, and did not feel that γ/δ T cells had a unique role to play in the pathogenesis of MS.

Since γ/δ T cells are present in brains of MS and other OND brains and CSF, their presence per se is not disease specific. However, their presence does suggest an in situ immune response to stress proteins. There may be disease specificity in these responses since it is possible that the antigens stimulating these cells may be different in persons with MS than in persons with OND.

Immune responses to stress proteins in MS patients and controls. Work in this area can be divided into studies involving cellular immune responses and those involving humoral responses to stress proteins. Cellular immune responses to hsp in persons with MS and OND have been described by Salvetti et al. [59] and our laboratory [6]. Salvetti et al. [59] studied peripheral blood T cell proliferative responses to recombinant hsp65 and hsp70 from *M. bovis* in 31 persons with MS, 19 individuals with OND and 19 normal controls. Proliferative responses to hsp70 were significantly more frequent in persons with MS compared to OND and healthy controls. Responses to hsp65 were equivalent in the three groups. Lines of T cells were established from 10 MS patients and 12 healthy controls, using purified protein derivative (PPD) as the antigen. Again, hsp70 reactive lines were significantly more common in MS patients than in healthy controls. Interestingly, cytofluorometric analyses of PPD-responsive lines revealed that only a minority of responding cells expressed γ/δ T cell receptors. Our laboratory studied T cell proliferative responses to mycobacterial stress proteins, tetanus toxoid, and recombinant hsp65 from *M. leprae*. The T cells were concurrently collected from the peripheral bloods and CSF of 20 persons with MS and 9 persons with inflammatory OND. Cells were cultured in vitro and stimulated with the above antigens. Significantly increased CSF lymphocyte proliferative responses to mycobacterial sonicate, relative to responses from paired peripheral blood lymphocytes, were present in 14 of the 20 specimens from patients with MS compared to 2 of 9 specimens from patients with other neurological diseases ($P < 0.025$). CSF lymphocytes also responded to tetanus toxoid, but differences between blood and CSF were not statistically significant. Lymphocytes from 1 patient with MS responded only to recombinant hsp65. When MS patients were classified according to duration of disease, 9 out of 10 with a disease duration less than 2 years had CSF T cells responding to *M. tuberculosis* compared to 5 out of 10 with a disease duration of longer than 2 years ($P < 0.012$). These data supplement the observations of Shimonkevitz et al. [64] who described increased numbers of activated γ/δ T cells predominantly in CSF from persons with recent onset MS.

Additional data in support of the hypothesis that immune responses to stress proteins play a role in autoimmune demyelination are the observations of Mor et al. [46]. These investigators prepared T cells from the spinal cords, blood, spleen and lymph nodes of rats during the acute phase of EAE or during recovery from EAE. Using limiting dilution analyses, they determined the frequency of T cell responses to the myelin protein MBP, and to recombinant hsp65 and hsp70. As expected, responses to MBP were enriched in the spinal cords of rats during and after acute EAE. However, there was also enrichment of T cells responsive to hsp65. T cell lines established from spinal cord lymphocytes responded to MBP, hsp65 and hsp70. When EAE was induced with an anti-MBP-responsive T cell line, similar patterns of enrichment for MBP- and hsp65-reactive T cells were noted, indicating that responses to stress protein occurred in the absence of exposure to adjuvant mycobacteria.

Initial studies of humoral immune responses to stress proteins in MS are mainly the work in our laboratory [3, 4] and that of Freedman and his associates [56]. We used immunoblots to detect antibodies to native and recombinant mycobacterial stress proteins and to bacterial stress proteins in CSF and paired sera from persons with MS and OND. Antibodies to many stress proteins, including those of the 60-kDa and 70-kDa families, were present in CSF and sera from all patient groups. Patterns of antibodies varied between CSF and sera, and between patients, but no disease-distinctive pattern was seen. When anti-stress protein antibodies were analyzed for isotypes, patients with MS had higher concentrations of antistress protein IgA antibodies than did OND patients. This suggested an in situ synthesis of such antibodies within the CNS in persons with MS. Prabhakar et al. [56] studied antibody concentrations to recombinant hsp60 using an ELISA. Titers of antibodies in MS CSF were significantly higher than those seen in persons with OND, and the higher titers correlated with the presence of oligoclonal bands in MS CSF but not OND CSF. In another recent study Gao et al. [21] demonstrated the presence of antibodies to mycobacterial hsp65 and human hsp60 in CSF of persons with MS as well as those with OND. In this study, immune responses to hsp were not disease specific.

The presence of immune responses to hsp in CSF of persons with inflammatory CNS diseases is not unexpected, given the stress responses expected from exposure to cytokines and inflammatory cells. However, all of the above studies used large protein molecules as antigens. Thus, immune responses to particular hsp epitopes could not be detected. Patterns of immune responses to particular hsp peptides vary among individuals, depending on differences in their immune response genes and their different illnesses [49]. Such differences may be important in determining susceptibility to autoimmune diseases such as MS. In support of this hypothesis are our recent (unpublished) observations that there are antibodies in MS CSF that bind to a hsp peptide that has sequence homology with the myelin protein 2', 3'-cyclic neucleotide 3'-phosphodiesterase (CNP). Such antibodies are rarely present in CSF from persons with other inflammatory diseases. This peptide is described below in greater detail.

Studies on cross-reactivity between stress proteins and CNS myelin. As noted above, certain anti-hsp antibodies stain normal myelin. We studied this phenomenon using immunoblots [7]. A panel of 20 anti-stress protein antibodies was assayed for the ability to bind to myelin proteins separated on SDS/PAGE. Three mAb, specific for either mycobacterial or human stress proteins, stained bands of normal myelin proteins. One murine mAb to *M. leprae* hsp65, IIH9, stained a 44 to 46-kDa doublet. This doublet was the same size as the myelin protein CNP. To study this observation further, purified CNP was used in immunoblots. IIH9 bound to this protein in a pattern identical to that seen with whole myelin. A nona-peptide region of sequence homology was identified between the epitope of hsp65 recognized by IIH9 and CNP. This peptide was synthesized and used in immunoblots. IIH9 bound strongly to this peptide, proving that this region of sequence homology was responsible for the observed cross-reactivity.

To determine whether immune responses to this peptide had biological consequences, we immunized rats with this peptide, using either complete (CFA) or incomplete Freund's adjuvant. No clinical disease or histological changes in the CNS were observed. After 4 weeks, animals were challenged with guinea pig spinal cord in CFA. Those animals that originally received peptide in CFA developed accelerated and enhanced EAE. Those animals previously immunized with peptide in incomplete Freund's adjuvant either remained healthy (40%) or developed very mild EAE. Differences between the two groups were statistically significant. Thus, an immune response to a hsp peptide having sequence homology with a myelin protein modulated the course of EAE. These interesting observations are being investigated further.

In a more circumspect manner, data from several other laboratories suggest that cross-reactivity occurs between stress protein and myelin. In 1975 Wisniewski and Bloom [73] noted primary demyelination in the brains of guinea pigs sensitized to tuberculin who were then subsequently challenged with PPD intracranially. The investigators interpreted these data to indicate that demyelination was the result of a passive bystander effect, induced by the localized delayed hypersensitivity response. In retrospect, an alternative explanation is that demyelination occurred because of a specific cross-reactivity between myelin and PPD. Additional observations that support this alternative conclusion come from several different laboratories [27, 36, 48, 69], all of them demonstrating that exposure of EAE-susceptible animals to either *Bordetella pertussis* or *M. tuberculosis* rendered them highly resistant to subsequent development of EAE. Yet in the experiments described in the previous paragraph, we successfully induced EAE in animals exposed to *M. tuberculosis* in CFA. These disparities may be due to the differences in the time intervals between initial exposure to *M. tuberculosis* and re-challenge.

It, thus, appears that at both the humoral and cellular level there is cross-reactivity between the stress proteins of infectious agents and normal myelin components. Immune responses to cross-reactive epitopes can modulate an autoimmune disease of the CNS.

Summary

Stress proteins are constitutively expressed in normal CNS tissues, in a variety of cell types (oligodendrocytes, astrocytes and neurons). Their function is uncertain but they may be critical during nervous system development and may protect cells from various stresses, such as hypoxia, anoxia, and excessive excitatory stimulation. Increased amounts of stress proteins are expressed in various cells of the CNS during acute toxic-metabolic states and in more chronic degenerative diseases. Increased expression of stress proteins may constitute a sensitive marker of cell injury.

Antibodies to mycobacterial stress proteins bind to normal human myelin and to oligodendrocytes in regions of MS demyelination. Cellular immune responses to stress proteins occur with increased frequency and magnitude in persons with MS, especially those with recent onset of disease. In addition. there are populations of T cells expressing γ/δ T cells in the brains and CSF of persons with MS, suggesting an in situ immune response to hsp. Humoral immune responses to stress proteins are found in CSF, but no disease specificity has been documented. Some myelin proteins have sequence homology with particular stress proteins. One instance is the homology between a peptide of mycobacterial hsp65 and the myelin protein CNP. Preliminary observations suggest that immune responses to such cross-reactive epitopes modify the course of EAE.

All in all, these data support the hypothesis that an immune response to the stress proteins of an infectious agent could result in a cross-reactive immune response to CNS myelin, resulting in demyelination. This may be an especially important mechanism in MS.

References

1. Allegretta M, Nicklas JA, Sriram S, Albertini RJ (1990) T cells responsive to myelin basic protein in patients with multiple sclerosis. Science 247:718
2. Baxevanis CN, Reclos GJ, Servis C, Anastasopoulos E, Arsenis P, Katsiyiannis A, Matikas N, Lambris JD, Papamichail M (1989) Peptides of myelin basic protein stimulate T lymphocytes from patients with multiple sclerosis. J Neuroinimunol. 22:23
3. Birnbaum G, Kotilinek L (1993) Antibodies to 70-kDa heat shock protein are present in CSF and sera from patients with multiple sclerosis. Abstracts of the 45th Annual Meeting of the American Academy of Neurology, New York. Neurology 43 [Suppl 2]:A162
4. Birnbaum G, Schlievert P (1992) Antibodies to mycobacterial antigens in the spinal fluids and blood from patients with multiple sclerosis and other neurologic diseases. Neurology 42:247
5. Birnbaum G, Clark HB, Psihos D (1991) Expression of heat shock proteins in the brains of patients with multiple sclerosis and other neurological diseases. Ann Neurol 30:305
6. Birnbaum G, Kotilinek L, Albrecht L (1993) Spinal fluid lymphocytes from a subgroup of multiple sclerosis patients respond to mycobacterial antigens. Ann Neurol 34:18
7. Birnbaum G, Kotilinek L, Schlievert P, Clark HB, Trotter J, Horvath E, Gao E, Cox M, Braun PE (1994) Immunologic cross-reactivity between a myobacterial heat shock protein and myelin 2', 3' cyclic nucleotide 3' phosphodiesterase. Neurology 44 [Suppl 2]:A146
8. Bluestone JA, Cron RQ, Barrett TA, Houlden B, Sperling AI, Dent A, Hedrick S, Rellahan B, Matis LA (1994) Repertoire development and ligand specificity of murine TCR γ/δ cells. Immunol Rev 120:5
9. Born W, Happ MP, Dallas A, Reardon C, Kubo R, Shinnick T, Brennan P, OBrien R (1990) Recognition of heat shock proteins and gamma delta T cell function. Immunol Today 11:40
10. Brown IR (1990) Induction of heat shock (stress) genes in the mammalian brain by hyperthermia and other traumatic events: a current perspective. J Neurosci Res 27:247

11. Brown IR, Rush SJ (1990) Expression of heat shock genes (hsp70) in the mammalian brain: distinguishing constitutively expressed and hyperthermia-inducible mRNA species. J Neurosci Res 25:14
12. Carter ML, Ross ME (1994) MN22, a heat shock protein-like gene, is developmentally regulated during granule neuron differentiation. Soc Neurosci Abstr 20:244
13. Chopp M (1993) The roles of heat shock proteins and immediate early genes in central nervous system normal function and pathology. Curr Opin Neurol Neurosurg 6:6
14. Cisse S, Perry G, Lacoste RG, Cabana T, Gauvreau D (1993) Immunochemical identification of ubiquitin and heat-shock proteins in corpora amylacea from normal aged and Alzheimer's disease brains. Acta Neuropathol 85:233
15. De Souza SD, Antel JP, Freedman MS (1994) Cytokine induction of heat shock protein expression in human oligodendrocytes: an interleukin-1-mediated mechanism. J Neuroimmunol 50:17
16. Dwyer BE, Nishimura RN, de VJ, Clegg KB (1991) Regulation of heat shock protein synthesis in rat astrocytes. J Neurosci Res 28:352
17. Elias D, Cohen IR (1994) Peptide therapy for diabetes in NOD mice. Lancet 43:704
18. Ferriero DM, Soberano HQ, Simon RP, Sharp FR (1990) Hypoxia-ischemia induces heat shock protein-like (HSP72) immunoreactivity in neonatal rat brain. Brain Res Dev Brain Res:145
19. Freedman MS, Buu NN, Ruijs TC, Williams K, Antel JP (1992) Differential expression of heat shock proteins by human glial cells. J Neuroimmunol 41:231
20. Fritz RB, Chou CH, McFarlin DE (1983) Relapsing murine experimental allergic encephalomyelitis induced by myelin basic protein. J Immunol 130:1024
21. Gao YL, Raine CS, Brosnan CF (1994) Humoral response to hsp65 in multiple sclerosis and other neurologic conditions. Neurology 44:941
22. Haass C, Klein U, Kloetzel PM (1990) Developmental expression of *Drosophila melanogaster* small heat-shock proteins. J Cell Sci 96:413
23. Hafler DA, Benjamin DS, Burks J, Weiner HL (1987) Myelin basic protein and proteolipid protein reactivity of brain- and cerebrospinal fluid-derived T cell clones in multiple sclerosis and postinfectious encephalomyelitis. J Immunol 139:68
24. Hamos JE, Oblas B, Pulaski SD, Welch WJ, Bole DG, Drachman DA (1991) Expression of heat shock proteins in Alzheimer's disease. Neurology 41:345
25. Haregewoin A, Soman G, Hom RC, Finberg RW (1989) Human gamma delta$^+$ T cells respond to mycobacterial heat-shock protein. Nature 340:309
26. Harrison PJ, Procter AW, Exworthy T, Roberts GW, Najlerahim A, Barton AJ, Pearson RC (1993) Heat shock protein (hsx70) mRNA expression in human brain: effects of neurodegenerative disease and agonal state. Neuropathol Appl Neurobiol. 19:10
27. Hempel K, Freitag A, Freitag B, Endres B, Mai B, Liebaldt G (1985) Unresponsiveness to experimental allergic encephalomyelitis in Lewis rats pretreated with complete Freund's adjuvant. Int Arch Allergy Appl Immunol 76:193
28. Higashi T, Takechi H, Uemura Y, Kikuchi H, Nagata K (1994) Differential induction of messenger-RNA species encoding several classes of stress proteins following focal cerebral ischemia in rats. Brain Res 650:239
29. Hofman FM, Hinton DR, Johnson K, Merrill JE (1989) Tumor necrosis factor identified in multiple sclerosis brain. J Exp Med 170:607
30. Hvas J, Oksenberg JR, Fernando R, Steinman L, Bernard CC (1993) Gamma delta T cell receptor repertoire in brain lesions of patients with multiple sclerosis. J Neuroimmunol 46:225
31. Johnson D, Hafler DA, Fallis RJ, Lees MB, Brady RO, Quarles RH, Weiner HL (1986) Cell-mediated immunity to myelin-associated glycoprotein, proteolipid protein, and myelin basic protein in multiple sclerosis. J Neuroimmunol 13:99
32. Jones DB, Coulson AF, Duff GW (1993) Sequence homologies between hsp60 and autoantigens. Immunol Today 14:115
33. Kabelitz D, Bender A, Schondelmaier S, Schoel B, Kaufmann SH (1990) A large fraction of human peripheral blood gamma/delta$^+$ T cells is activated by *Mycobacterium tuberculosis* but not by its 65-kDa heat shock protein. J Exp Med 171:667
34. Kaufmann SHE (1990) Heat shock proteins and the immune response. Immunol Today 11:129
35. Lamb JR, Bal V, Mendez-Samperio P, Mehlert A, So A, Rothbard J, Jindal S, Young RA, Young DB (1989) Stress proteins may provide a link between the immune response to infection and autoimmunity. Int Immunol 1:191

36. Lehmann D, Ben Nun A (1992) Bacterial agents protect against autoimmune disease. I. Mice pre-exposed to *Bordetella pertussis* or *Mycobacterium tuberculosis* are highly refractory to induction of experimental autoimmune encephalomyelitis. J Autoimmun 5:675
37. Liblau R, Tournier LE, Maciazek J, Dumas G, Siffert O, Hashim G, Bach MA (1991) T cell response to myelin basic protein epitopes in multiple sclerosis patients and healthy subjects. Eur J Immunol 21:1391
38. Lisak RP, Zweiman B (1977) In vitro cell-mediated immunity of cerebrospinal fluid lymphocytes to myelin basic protein in primary demyelinating diseases. N Engl J Med 297:850
39. Lowe J, Mayer RJ, Landon M (1993) Ubiquitin in neurodegenerative diseases. Brain Pathol 3:55
40. Lowenstein DH, Chan PH, Miles MF (1991) The stress protein response in cultured neurons: characterization and evidence for a protective role in excitotoxicity. Neuron 7:1053
41. Mailhos C, Howard MK, Latchman DS (1993) Heat shock protects neuronal cells from programmed cell death by apoptosis. Neuroscience 55:621
42. Marini AM, Kozuka M, Lipsky RH, Nowak TSJ (1990) 70-kilodalton heat shock protein induction in cerebellar astrocytes and cerebellar granule cells in vitro: comparison with immunocytochemical localization after hyperthermia in vivo. J Neurochem 54:1509
43. Martin JE, Swash M, Mather K, Leigh PN (1993) Expression of the human groEL stress-protein homologue in the brain and spinal cord. J Neurol Sci 118:202
44. Marin R, Valet JP, Tanguay RM (1993) hsp23 and hsp26 exhibit distinct spatial and temporal patterns of constitutive expression in Drosophila adults. Dev Genet 14:69
45. Möller G (ed) (1991) Heat-shock proteins and the immune system. Immunol Rev 121
46. Mor F, Cohen IR (1992) T cells in the lesion of experimental autoimmune encephalomyelitis. Enrichment for reactivities to myelin basic protein an d to heat shock proteins. J Clin Invest 90:2447
47. Morimoto RI, Tissières A, Georgopoulos C (eds) (1990) Stress proteins in biology and medicine. Cold Spring Harbor laboratory Press, Cold Spring Harbor
48. Mostarica-Stojkovic M, Vukmanovic S, Petrovic M, Ramic Z, Lukic ML (1988) Dissection of the adjuvant and suppressive effects of mycobacteria in experimental allergic encephalomyelitis production. Int Arch Allergy Appl Immunol 85:82
49. Mutis T, Corneliss YE, Datema G, van den Elsen PJ, Ottenhoff THM, De Vires RRP (1994) Definition of a human suppressor T-cell epitope. Proc Natl Acad Sci USA 91:9456
50. Namba Y, Tomonaga M, Ohtsuka K, Oda M, Ikeda K (1991) HSP 70 is associated with abnormal cytoplasmic inclusions characteristic of neurodegenerative diseases. No To Shinkei A 43:57
51. Ottenhoff TH, Torres P, De las Aguas JT, Fernandez R, Van EW, De Vries RR, Stanford JL (1986) Evidence for an HLA-DR4-associated immune response gene for *Mycobacterium tuberculosis.* A clue to the pathogenesis of rheumatoid arthritis? Lancet II:310
52. Pauli D, Tonka CH, Tissieres A, Arrigo AP (1990) Tissue-specific expression of the heat shock protein HSP27 during *Drosophila melanogaster* development. Cell Biol 111:817
53. Perrella O, Carrieri PB, De MR, Buscaino GA (1993) Markers of activated T lymphocytes and T cell receptor gamma/delta$^+$ in patients with multiple sclerosis. Eur Neurol 33:152
54. Pette M, Fujita K, Kitze B, Whitaker JN, Albert E, Kappos L, Wekerle H (1990) Myelin basic protein-specific T lymphocyte lines from MS patients and healthy individuals. Neurology 40:1770
55. Pfeffer K, Schoel B, Gulle H, Kaufmann SH, Wagner H (1991) Human gamma/delta T cells responding to mycobacteria. Behring Inst Mitt 88:36
56. Prabhakar S, Kurien E, Gupta RS, Zielinski S, Freedman MS (1994) Heat shock protein immunoreactivity in CSF – Correlation with oligoclonal banding and demyelinating disease. Neurology 44:1644
57. Quayle AJ, Wilson KB, Li SG, Kjeldsen KJ, Oftung F, Shinnick T, Sioud M, Forre Q, Capra JD, Natvig JB (1993) Peptide recognition, T cell receptor usage and HLA restriction elements of human heat-shock protein (hsp) 60 and mycobacterial 65-kDa hsp-reactive T cell clones from rheumatoid synovial fluid. Eur J Immunol 22:1315
58. Renkawek K, Bosman GJ, Gaestel M (1993) Increased expression of heat-shock protein 27 kDa in Alzheimer disease: a preliminary study. Neuroreport 5:14
59. Salvetti M, Buttinelli C, Ristori G, Carbonari M, Cherchi M, Fiorelli M, Grasso MG, Toma L, Pozzilli C (1992) T-lymphocyte reactivity to the recombinant mycobacterial 65- and 70-kDa heat shock proteins in multiple sclerosis. J Autoimmun 5:691
60. Satoh J. Nomaguchi H, Tabira T (1994) Constitutive expression of 65-kDa heat shock protein (HSP65)-like immunoreactivity in cultured mouse oligodendrocytes. Brain Res 595:281

61. Selmaj K, Raine CS, Cannella B, Brosnan CF (1991) Identification of lymphotoxin and tumor necrosis factor in multiple sclerosis lesions. J Clin Invest 87:949
62. Selmaj K, Brosnan CF, Raine CS (1991) Colocalization of lymphocytes bearing gamma delta T-cell receptor and heat shock protein hsp65$^+$ oligodendrocytes in multiple sclerosis. Proc Natl Acad Sci USA 88:6452
63. Selmaj K, Brosnan CF, Raine CS (1992) Expression of heat shock protein-65 by oligodendrocytes in vivo and in vitro: implications for multiple sclerosis. Neurology 42:795
64. Shimonkevitz R, Colburn C, Burnham JA, Murray RS, Kotzin BL (1993) Clonal expansions of activated gamma/delta T cells in recent-onset multiple sclerosis. Proc Natl Acad Sci USA 90:923
65. Sibley WA, Bamford CR, Clark K (1985) Clinical viral infections and multiple sclerosis. Lancet I:1313
66. Sprang GK, Brown IR (1987) Selective induction of a heat shock gene in fibre tracts and cerebellar neurons of the rabbit brain detected by in situ hybridization. Brain Res 427:89
67. Steinhoff U, Schoel B, Kaufmann SH (1990) Lysis of interferon-gamma activated Schwann cell by cross-reactive CD8$^+$ alpha/beta T cells with specificity for the myobacterial 65 kDa heat shock protein. Int Immunol 2:279
68. Tuohy VK, Sobel RA, Lees MB (1988) Myelin proteolipid protein-induced experimental allergic encephalomyelitis. Variations of disease expression in different strains of mice. J Immunol 140:1868
69. Vandenbark AA, Burger DR, Vetto RM (1975) Cell-mediated immunity in experimental allergic encephalomyelitis: cross-reactivity between myelin basic protein and mycobacteria antigens. Proc Soc Exp Biol Med 148:1223
70. Vass K, Welch WJ, Nowak TSJ (1988) Localization of 70-kDa stress protein induction in gerbil brain after ischemia. Acta Neuropathol 77:128
71. Walsh DA, Li K, Speirs J, Crowther CE, Edwards MJ (1989) Regulation of the inducible heat shock 71 genes in early neural development of cultured rat embryos. Teratology 40:321
72. Walsh D, Li K, Wass J, Dolnikov A, Zeng F, Zhe L, Edwards M (1993) Heat-shock gene expression and cell cycle changes during mammalian embryonic development. Dev Genet 14:127
73. Wisniewski HM, Bloom BR (1975) Primary demyelination as a nonspecific consequence of a cell-mediated immune reaction. J Exp Med 141:346
74. Wucherpfennig KW, Newcombe J, Li H, Keddy C, Cuzner ML, Hafler DA (1992) Gamma delta T-cell receptor repertoire in acute multiple sclerosis lesions. Proc Natl Acad Sci USA 89:4588
75. Xue ZY, Grossfeld RM (1993) Stress protein synthesis and accumulation after traumatic injury of crayfish CNS. Neurochem Res 18:209

Experimental immunotherapies for multiple sclerosis

Roland Martin[1,2], **Henry McFarland**[1]

[1] Neuroimmunology Branch, National Institute of Neurological Disorders and Stroke, National Institutes of Health, Building 10, Room 5B-16, Bethesda, MD 20892, USA
[2] Department of Neurology, University of Tübingen Medical School, Hoppe-Seyler-Str. 3, 72076 Tübingen, Germany

Summary. Multiple sclerosis (MS) is a chronic demyelinating disease affecting the central nervous system (CNS) principally in young adults. Although its etiology is as yet unknown current evidence suggests that tissue damage is mediated by autoimmune T cells. The examination of an experimental animal model for MS, experimental allergic encephalomyelitis (EAE), has demonstrated that myelin basic protein (MBP)- or proteolipid protein (PLP)-specific T cells mediate the destruction of CNS myelin. In recent years, elegant studies in EAE have shown that encephalitogenic T cells recognize short peptides of MBP or PLP in the context of MHC/HLA-class II molecules, express a restricted number of T cell receptor (TCR) molecules and secrete interferon-γ and tumor necrosis factor-α/β. Understanding the pathogenetic steps in lesion development at the molecular level led to highly specific immunotherapies for EAE targeting each individual molecule. It has been the hope of many investigators that immunological events resembling those in EAE can be found in patients with MS and that the specific immunotherapies effective in EAE could also be applied to MS. However, to date, the evidence for a unique immunological abnormality in MS is not strong. Although MBP- and PLP-specific T cells with properties similar to those that are encephalitogenic in animals can be isolated from patients, they are not specific for MS and occur with similar frequency in controls. In addition, the variability in specificity and TCR usage has raised questions regarding the relevance of these cells in patients. The importance of the T cell responses to myelin antigens in MS may not be established until the effects of abrogating their activity through specific therapies targeting the trimolecular complex (TMC) have been demonstrated. Consequently, attention has begun to focus on modifying the biology of the MS lesion rather than targeting the initiating event at the level of the TMC, and the success of this approach is reflected by the effect of interferon-β on lesion development in MS. The recent approval for the use of interferon-β for the treatment of relapsing-remitting MS has raised great interest in examining novel strategies for immunotherapies in MS. The

Correspondence to: H. McFarland

basic concepts as well as the current candidates for such new immunotherapies will be outlined in this short review.

Introduction

The etiology of multiple sclerosis (MS) is not known, but it is currently believed that autoimmune T cells specific for myelin antigens are involved in its pathogenesis [77]. Although the evidence for an immunological mechanism is largely circumstantial, it is becoming increasingly convincing. The pathology of MS resembles that associated with immunologically mediated diseases [100], and, in some patients, a beneficial effect of immunosuppressive therapy is seen. Further, the similarities between MS and the experimental autoimmune demyelinating disease, experimental allergic encephalomyelitis (EAE) support an autoimmune process in MS. While it is clear that the different forms of EAE that can be induced in different susceptible animal strains do not reflect all the clinical, histopathological and immunological features of MS, EAE is nevertheless an excellent model that has led to a better understanding of the pathogenetic steps leading to demyelination. In the chronic-relapsing form of EAE in the SJL mouse, the disease exacerbates and remits as in MS [99], and the pathology, characterized by inflammation and demyelination with lesions of different ages, bears a strong resemblance to that seen in MS [108, 109]. It is crucial to remember, however, that, while studies using the EAE model may provide important clues regarding potential treatment of MS, experimental treatments in EAE are mostly initiated during the period of disease induction. Even treatments that ameliorate the subsequent course when used after the disease is established operate in conditions considerably different from those in MS, where therapy is usually begun long after the disease has started. Evidence derived from magnetic resonance imaging (MRI) studies indicates that even at the time of the initial clinical episodes, many patients show multiple lesions, indicating considerable previous central nervous system (CNS) inflammation [46]. Based on evidence from the EAE model, some generalizations regarding the cause and potential therapy of MS can be made. First, it is possible that MS, like EAE, is caused by T cells recognizing antigens found in the CNS. Although studies of EAE have focused on two encephalitogenic proteins of myelin, myelin basic protein (MBP) [36] and proteolipid protein (PLP) [140], several other antigens, including proteins that are present predominantly in astrocytes and not in myelin, have been found to cause encephalitis [5, 60] although sometimes without demyelination. In addition, the T cell response to viruses which can persist in the CNS such as Theiler's virus [74], Corona virus [147] or a rat-adapted measles virus [71], can also produce an EAE-like picture. In MS, it is uncertain whether the immune response to one antigen triggers the disease or whether the relevant antigen varies from patient to patient. Further, in EAE, an immune response to multiple myelin antigens may occur as disease progresses and myelin breakdown occurs [86]. Thus, in MS an autoimmune response directed at multiple CNS myelin antigens could contribute to disease.

Since the general concepts about the activation of myelin-specific T cells and lesion development have been derived from EAE and the study of MS pathology, these findings shall be briefly summarized. It has been known for a long time that outbred animals such as mice are highly resistant to the induction of EAE, whereas disease can easily be induced in certain inbred rodent strains [36]. The factors that are most closely associated with disease susceptibility are members of the major histocompatibility complex (MHC) class II gene family, i.e., certain IA- and IE-MHC molecules (e.g., IA^5 in the SJL mouse) [37, 87]. Further examination of the influence of these MHC class II molecules has demonstrated that particular peptides derived from myelin antigens are preferentially recognized and are encephalitogenic in the various animal strains [36]. The portion of a specific myelin protein (i.e., MBP) that is encephalitogenic differs between species and strains and is influenced by the MHC makeup of the animal. Another important finding in EAE has been that encephalitogenic T cells in Lewis rats and B10.PL and PL/J mice preferentially use a very limited set of T cell receptor (TCR) chains (Vβ8.2 [19, 25] and Vα4.3, PL/J mice [1] or Vα4.2 and −2.3 in B10.PL mice [141]) in the recognition of the complex formed by disease-associated MHC class II molecules and encephalitogenic peptide. This relationship between disease and the trimolecular complex (TMC) comprised of TCR, MHC molecule and autoantigenic peptide has led to innovative therapeutic approaches targeting each of the components specifically. Blocking the MHC molecule with a nonstimulatory modified encephalitogenic peptide [156] or an unrelated peptide with high binding affinity [68] as well as T cell vaccination [11] or TCR peptide vaccination [48, 144] have been successful in the EAE model. Whether similar approaches will be successful in MS is not certain since the relationship between a specific TCR and disease is not yet as clear in MS as it is in some of the EAE models.

Considering the events that result in the induction of an autoimmune reaction, it is most likely that myelin-specific T cells are activated in the periphery. This conclusion is supported by the fact that T cells with specificity for a number of myelin antigens have been demonstrated in animals before induction of EAE and also in healthy human individuals [20, 76, 98]. Furthermore, it has been shown that encephalitogenic T cell lines (TCL) can be generated by repeated stimulation with MBP from lymph node cells of healthy Lewis rats [118] and marmosets [40]. These findings not only demonstrate that autoreactive T cells are a component of the T cell repertoire in healthy individuals, but also that they have the potential of mediating a demyelinating disease and are usually either actively suppressed or do not encounter their target autoantigen in the proper co-stimulatory environment such as the sufficient expression of CD28 and B7 antigens. Although the natural stimuli that lead to activation of myelin-reactive T cells in MS are as yet poorly understood, either viruses or bacterial superantigens could result in the activation of autoantigen-specific T cells. Viruses could also contribute to tissue tropism by inducing or up-regulating the expression of necessary adhesion and co-stimulatory molecules in the target organ.

Following activation the first event in lesion development appears to be the adhesion of activated T cells to endothelial cells in postcapillary venules preferentially in the periventricular white matter [110]. The relative importance of interactions between cell adhesion molecules (CAM) and selectins expressed on endothelial cells and their ligands on encephalitogenic T cells is not yet completely understood. The majority of studies demonstrate an up-regulation of constitutive expression of CAM and selectins on endothelial cells during the early phases of lesion development [9, 21, 34, 93, 152]. The need for up-regulation of these CAM as a requirement for T cell adherence and transmigration is currently unresolved [110]. The ability of activated T cells to transmigrate through the endothelial cells into the brain parenchyma has been clearly demonstrated in the rat [149] and mouse [30, 31], indicating that whichever CAM are expressed constitutively they are sufficient for this process. A relationship between encephalitogenicity and expression of very late antigen (VLA)-4 has been identified on PLP- [62] and MBP-specific T cells [9] and antibody to

VLA-4 reduces the severity of EAE [158]. Expression of vascular CAM (VCAM)-1 is, however, not usually high in acute active lesions but is increased in chronic active lesions suggesting that it occurs during a later stage of lesion evolution. Intercellular adhesive molecule (ICAM)-1, the ligand for LFA-1 on lymphocytes, is expressed at low levels constitutively [128]. Efforts to block disease with antibody to ICAM-1 or LFA-1 have met with variable success [22, 151]. Some investigators have reported no effect, while others have observed worsening of disease [22, 150]. The effect of anti-ICAM-1 and anti-LFA-1 seems to be related to the dose (M. Racke, unpublished results) and timing when antibody was given. Most likely, interactions with selectins and ICAM-1 are sufficient for the T cell to allow initial adhesion to the endothelial cell. Whether the T cell can then transmigrate through the endothelial cell barrier or whether some very local up-regulation of additional cell CAM such as VCAM-1 occurs is not known. Both interferon (IFN)-γ and tumor necrosis factor (TNF)-α, which are secreted by activated encephalitogenic T cells, can up-regulate CAM and MHC class molecules [22, 151]. Endothelial cells can present MBP to MBP-specific T cells in vitro, which results in lysis of the endothelial cell after MHC expression on these cells is up-regulated by IFN-γ [80]. The importance of antigen presentation at the level of the endothelial cell in EAE or MS is unknown.

Cells that have encountered their target antigen remain in the CNS and initiate an inflammatory response. Adoptive transfer of MBP-sensitized T cells labeled with ^{14}C [30] indicate that these cells can be found in the lesion during the early stage of lesion development. The labeled cells tend to remain in the perivascular cuff suggesting that antigen presentation takes place in the perivascular area [30]. These results are consistent with histochemistry and polymerase chain reaction (PCR) amplification studies of TCR expressed by encephalitogenic T cells in affected tissue, which have demonstrated the presence of specific cells early during the course of inflammation [10, 18, 54]. As the lesion progresses, secondarily recruited cells far outnumber the encephalitogenic T cells [10, 54]. Thus, the T cells that enter the CNS initially and encounter antigen induce a cascade of proinflammatory cytokines, which leads to activation of the endothelium and recruitment of additional inflammatory cells including macrophages. Among the factors secreted in the tissue by autoantigen-specific T cells, TNF-α/β (lymphotoxin; TNF-α/LT) and IFN-γ are probably the most important ones. Both TNF-α and IFN-γ are involved in the differentiation of Th1 CD4 T cells, in the up-regulation of MHC antigens and adhesion molecules [111, 151] on endothelial cells and putative antigen-presenting cells (APC) in the CNS, and the recruitment and activation of other T cells and mononuclear phagocytes. Furthermore, TNF-α can mediate direct toxicity to myelin membranes and oligodendrocytes [120]. The actual effector mechanisms causing demyelination are uncertain, but could include the abovementioned effects of TNF-α and IFN-γ and reactive oxygen and nitrogen intermediates (O_2^- and NO) that are secreted by macrophages upon activation. Finally, myelin-specific autoantibodies that may enter the inflammatory site through an open blood-brain barrier (BBB) and be secreted locally by plasma cells may mediate the attachment of activated macrophages to myelin molecules such as myelin-associated glycoprotein (MAG) or myelin oligodendrocyte glycoprotein (MOG) that are integral parts of the myelin membrane. Some experimental evidence for each of these mechanisms exists [73].

The inflammatory response in the CNS parenchyma is probably terminated by apoptosis of infiltrating antigen-specific and nonspecific T cells in situ [119, 133].

Table 1. Pathogenetic steps and factors involved in inflammatory demyelinating diseases

Sensitization/activation of myelin-specific T cells
EAE
Encephalitogenic T cells specific for myelin and glial antigens or viruses persistently infecting the CNS
Disease susceptibility linked to MHC class II makeup
Disease mediated by $CD4^+$, MCH class II-restricted, Th1 T cells
Restricted TCR usage seen in some animal strains but encephalitogenic T cells not limited to those TCR even in those strains
MS
T cell specificity of disease-mediating T cells unknown
Limited evidence for restricted TCR usage
Disease associated with HLA-class II makeup, but no absolute linkage
Entry of T cells into the CNS
EAE
Up-regulation of cell adhesion molecules may be required
Entry of T cells nonspecific and related to level of T cell activation
MS
Role of adhesion molecules uncertain
Parenchymal effector phase (induction of inflammatory response)
EAE
Disease correlated with Th1-type T cells
Secretion of proinflammatory cytokines associated with acute lesions
Lesion development related to secondary recruitment of T cells and monocytes
Induction of Th2-type cytokines reverses or prevents lesion development
MS
Acute lesions associated with both Th1- and Th2-type cytokine production
Demyelination
EAE
Oligodendrocytes preserved in acute lesions
Ultrastructural evidence for macrophage-myelin interaction
Demyelination enhanced by transfer of antibody to MOG
Cytokines such as TNF/LT may directly damage oligodendrocytes
Reactive oxygen and nitrogen intermediates may contribute to tissue damage
MS
Ultrastructural evidence for macrophage-myelin interaction
Oligodendrocytes probably not targeted in acute lesions
Oligodendrocytes lost in chronic lesions
Remyelination
EAE
Incomplete remyelination found in some animal strains
Can be augmented in some animal models using antibodies
MS
Incomplete remyelination in acute lesions

EAE, Experimental allergic encephalomyelitis; MS, multiple sclerosis; CNS, central nervous system; TCR, T cell receptor; MOG, myelin, oligodendrocyte glycoprotein; TNF/LT, tumor-necrosis factor-α/lymphotoxin-α

Remyelination is usually incomplete and the myelin protein composition of remyelinated plaques changes considerably during the course of the disease [86].

Thus, several stages of lesion development can be targeted therapeutically and are listed in Table 1. The discussion of current directions in the treatment of MS will be examined with respect to the phase of lesion development targeted by the treatment.

Therapies employing general immunosuppression

Since the specificity of the immune response producing disease is not known, the use of general and, in most cases, drug-induced immunosuppression was and continues to be an important approach in the treatment of MS. A large number of general immunosuppressive treatments have or are being studied. These are listed in Table 2. The evidence for significant effectiveness of any of these agents is marginal and often controversial. Many of the trials of immunosuppressive drugs that have been reported as showing effectiveness have been questioned because of the methods applied. The problems in identifying the usefulness of immunosuppressive drugs may be more a question of difficulties in conducting clinical trials in MS than a real lack of effectiveness of these drugs. Most immunosuppressive drugs have been tested in trials involving patients with chronic progressive MS (CP-MS) since the relationship between the potential risks and possible benefits is thought to be clearer. However, outcome measures in patients who already suffer from considerable disability are difficult and progression of disease in this patient population may also be related to mechanisms such as axonal degeneration further complicating the assessment of the results. Difficulties such as these may be overcome by trials involving larger numbers of patients and long periods of treatment, both of which are usually impossible because of the expense. Of interest, the recent meta analysis of the various trials using azathioprine indicate that the drug has some effectiveness in reducing progression of disability [159].

Since the role of antigen-specific suppression is uncertain because of our inability to identify a causative agent or the specific myelin antigen that is pathogenetically relevant in an individual patient, generalized immunosuppression may remain an important part of therapy in MS. Hopefully, studies using MRI as an outcome measure will permit assessment of these drugs over shorter study periods, thus limiting the potential risks of the treatment so that their effectiveness in patients with milder relapsing remitting MS (RR-MS) can be examined [83].

Table 2. Unspecific anti-inflammatory and immunosuppressive treatments

Corticosteroids
Azathioprine
Cyclophosphamide
Mitoxantrone
Cyclosporine
Methotrexate
Deoxyspergualine
2-Chlorodeoxyadenosine
Total body irradiation

Therapies targeting disease-producing T cells and the TMC

Although our understanding of the antigens and cells involved in myelin-specific T cell responses has greatly increased in recent years, the antigen or antigens which are the target of the immunopathological process in a specific patient or group of MS patients are still poorly understood. Some experimental therapies have built on the probability that the proteins with strong encephalitogenic potential in experimental animals such as MBP or PLP are likely candidate antigens in MS and on the increased number of MBP- and PLP-specific T cell in the peripheral blood and cerebrospinal fluid of MS patients [91, 92, 94, 163]. Based on this assumption, therapies designed to block or eliminate this response have been undertaken or are being designed (Table 3).

Therapies targeting the TMC can lead to a number of potential outcomes including blocking activation, induction of anergy or tolerance rather than activation, induction of programmed cell death (apoptosis) or change in functional characteristics of the T cell, so that it either no longer mediates disease or possibly even targets suppressive cells to the lesion. Specific therapies can be directed at each of the components of the TMC, i.e., the antigenic peptide, the TCR or the MHC class II molecule that presents the antigenic peptide to the TCR. Attempts to inactivate MBP-specific T cells in EAE by stimulating a potentially existing regulatory or suppressive network date back to the early 80s when Cohen and coworkers administered inactivated encephalitogenic T line cells [11] as a T cell vaccination. It was later shown that T cell vaccination may either be directed against molecules specific for the encephalitogenic cells, most likely the TCR idiotype [48, 132, 144] or against activation molecules on encephalitogenic T cells (anti-ergotypic therapy; [75]). The presence and in vivo activity of $CD8^+$ T cells that directly lyse encephalitogenic T line cells and suppress disease was demonstrated later [64, 132]. Recent studies in MS patients demonstrated that the administration of inactivated MBP-specific T cell clones not only induce populations of $CD8^+$ idiotype-specific T cells, but also result in a reduction of precursor frequency numbers of MBP-specific T cells [162].

Shortly after the demonstration of regulatory T cells in EAE in vitro, it became clear that encephalitogenic T cells in Lewis rats as well as in B10.PL and PL/J mice, although recognizing different MBP peptides in the context of different restriction elements, express a restricted TCR repertoire with the predominant expression of a particular β chain, Vβ8.2, and also a limited number of α chains [1, 19, 25, 141]. This observation has not only led to the successful application of monoclonal antibodies directed against Vβ8.2 [1, 19, 141, 160], but also allowed the development of therapies that employed the administration of peptides derived from these TCR chains rather than the injection of inactivated whole T cell populations [48, 144]. Such TCR peptides have been either derived from a region of the variable chain of the Vβ8.2 molecule (complementarity determining region 2, CDR2; [144]) or from the junctional region (CDR3) of the TCR [48]. The rationale for using a CDR2 peptide was to immunize with a peptide chain that is shared by all V$\beta8.2^+$ cells; this bears the inherent problem that every immune response that employs Vβ8.2 would be affected. The second approach, using a CDR3 peptide, assumed that targeting the area of the TCR that is known to interact with the antigenic peptide (CDR3; [53]), rather than with the MHC molecule (CDR2 and CDR1), would be more suitable and tailored specifically for the cells that react with the known encepalitogenic peptide. Both strategies were effective in EAE and the therapy that is based on immunization with the CDR2 peptide of Vβ8.2 presumably acts through the induction of $CD8^+$ regulatory T cells and TCR-

Table 3. Therapies targeting the trimolecular complex

1. Therapies unspecifically modifying the activation and function of antigen-specific CD4 T cells:
– Copolymer-1; interferes with antigen-presentation
– CD4 analogue peptides; interfere with T cell – APC conjugation
– Antibodies against HLA-class II antigens or CD4; interfere with antigen-presentation to class II-restricted T cells
– MHC-blocking peptides; interfere with antigen presentation
2. Antigen-specific inactivation of T cell responses against the encephalitogenic portions of myelin proteins:
– Antibodies against complexes of disease-associated MHC class II antigen and encephalitogenic peptide; exact mechanism of action not known
– Immunization with peptides derived from the third hypervariable region of disease-associated MHC class II antigen; induction of antibodies against this portion of the disease-related MHC class II antigen
3. Induction of anergy or tolerance:
– T cell vaccination and TCR peptide vaccination; induction of regulatory T cells of $CD4^+$ or $CD8^+$ phenotype as well as antibodies against portions of the TCR
– Anergy induction with peptide/MHC complexes, with membrane fractions of APC that had been prepulsed with encephalitogenic antigen or with fixed APC coupled with myelin antigen
– Oral tolerance; dose-dependent induction of anergy and/or active suppressive mechanisms including IL-4- and TGF-β secreting T cells
– Administration of altered peptide ligands based on the encephalitogenic peptide; induction of anergy or functional antagonism
4. Induction of apoptosis:
– High-dose i.v. injection of myelin antigen leads to apoptotic death of preactivated encephalitogenic T cells
6. Treatments inducing functional changes in encephalitogenic T cells:
– Altered peptide ligands based on the the encephalitogenic or immunodominant peptide may include phenotypic switches in encephalitogenic T cells from Th1 to Th2 cells

APC, antigen-presenting cell; IL, interleukin; TGF-β, transforming growth factor-β

specific antibodies [39, 89]. In addition to the induction of $CD8^+$ T cells, the presence of $CD4^+$ TCR peptide-specific T cells involved in specific regulatory network has been reported [65]. Unfortunately, only a few studies have shown that T cells that had been induced by either T cell or TCR peptide vaccination specifically recognize either the immunizing cell or the cell expressing the region from which the CDR3 TCR peptide was derived [64, 132]. Although the effectiveness of these approaches has been debated [56], it is now generally agreed that TCR peptide vaccination is effective in EAE. A clinical trial in MS that is based on the observation that Vβ5.2 and Vβ6.1 are overrepresented in MBP-specific TCL in MS patients [61] is currently under way [14, 27] and has shown that the administration of CDR2 peptides is well tolerated and, in some patients, results in a reduction of precursors for MBP-specific T cells and in T cell immunity against the immunizing peptides [14, 27]. The approach of TCR peptide and T cell vaccination has fascinated immunologists because of its extreme specificity in inducing an immune response only against the pathogenetically relevant cells. It relies, however, on restricted TCR usage by autoantigen-specific T cells, and that the particular TCR usage is known. Based on the current data, it seems unlikely that human autoantigen-specific T cells use a restricted TCR repertoire [42,

78]. Since we do not know which antigen and which T cell population is most relevant in an individual MS patient, we can currently only speculate whether TCR peptide vaccination will ever be used in MS. If so, it might be necessary to tailor an individual therapy for each patient. There is, however, some evidence that both the specificities and TCR usage of, for example, MBP-specific T cells in subgroups of patients such as those that are DR15 Dw2$^+$, show similarities [90, 94, 157]. Furthermore, the number of epitopes recognized and the heterogeneity of TCR expressed is probably smaller during the early stages of disease [124, 143]. Finally, the findings of restricted TCR usage in the brains of MS patients [90] as well as the skewing in the TCR repertoire that was observed in identical twins discordant for MS [142] argue for the involvement of T cells in the pathogenesis of MS and suggest that specific immunotherapies targeting subsets of T cells might be feasible at certain stages of the disease and in well-defined subgroups of patients.

Even before the TCR usage of encephalitogenic T cells was studied, it became clear that susceptibility for developing EAE in various animal strains was largely related to their MHC class types [37, 87]. Furthermore, depending on the disease-associated MHC class II phenotype, different regions of the encephalitogenic protein, such as MBP, were encephalitogenic in the different strains (for example, amino acids, aa, Ac1-11 in B10.PL and PL/J mice, aa 89–101 in SJL/J mice and aa 68–86 in Lewis rats; [36]). Based on these observations, a number of elegant studies have shown that modifications of those encephalitogenic peptides that are capable of binding to the restriction element with high affinity, but which do not induce an encephalitogenic response themselves, can block EAE [127, 156]. This could also be achieved with high affinity, binding peptides completely unrelated to the encephalitogen [68]. However, this approach has a number of problems. Theoretically, blockade of all available MHC class II molecules, for example HLA-DR2, in a homozygote individual could result in general immunosuppression. Although this seemed not to be the case in experimental therapies with MHC-blocking peptides, it has been recently shown that the peptide concentrations that are necessary to achieve MHC blockade in vivo can not be maintained over long periods of time [49]. Recently, a large multicenter trial [137] examined the effect of copolymer-1 (COP-1), a polypeptide composed of the four aa, tyrosine, lysine, alanine and glutamic acid in random sequence. This trial in early RR-MS patients showed a small, but significant therapeutic effect. Earlier reports claimed that COP-1 blocks EAE via the induction of MBP-specific suppressor cells [136]. Recent data suggests that it blocks antigen-presentation by interfering with binding of antigenic peptide to disease-associated MHC-class II molecules in vitro [103], although additional evidence that this mechanism accounts for the effects on disease in vivo is needed. In addition, monoclonal antibodies against disease-related MHC class II antigens have successfully been applied [131], but the problems mentioned above and the induction of xenogeneic antibodies have prevented this approach in humans. A recent study in which animals were vaccinated with a peptide derived from the third hypervariable region of the disease-associated MHC class II antigen (IA^s β chain 58–78; [139]) was effective in treating EAE via the induction of autoantibodies specific for the IA^s β chain. Finally, a monoclonal antibody that was generated against the antigen-IA complex also blocked disease [3].

Other experimental therapies have also successfully targeted the TMC in EAE by either including inactivated carrier cells such as spleen cells that had been coupled with antigen [57, 58, 134, 135] or by soluble MHC-peptide complexes [126] or by the administration of membrane fragments of cells that have been prepulsed with protein

[13]. These methods have led to an anergic state since encephalitogenic T cells are not properly activated in the absence of co-stimulatory signals.

A state of transient or long-term unresponsiveness resulting in anergy could also be achieved by intrathymic injection of antigen either in neonatal [28] or adult [44] animals. Administration of bacterial superantigen can result in either deletion or activation of T cells using TCR that are recognized by this superantigen [15, 106, 113, 117]. Treatment with superantigen reduced disease when it was injected before EAE was initiated [106]. These strategies are obviously not applicable to humans. However, anergy can also be induced by intraperitoneal [38] administration of encephalitogenic peptide and results in an improved course of EAE after rechallenge with antigen.

Recently, it has been shown that EAE can be influenced therapeutically by the induction of programmed cell death in encephalitogenic T cells [29]. This therapeutic strategy is based on the finding that preactivated T cells that are vigorously proliferating in the presence of IL-2 undergo apoptosis when they are rechallenged with antigen. This observation provided the basis for treating animals after the induction of EAE with high doses of MBP [29]. A therapeutic effect, due to apoptotic death of preactivated encephalitogenic T cells, could be demonstrated in this situation, but timing and dose of the administration of the apoptotic signal proved to be critical for the outcome.

It is obvious from the various therapeutic strategies that have been outlined above that some are too specific in that they only target one specific T cell clone bearing one TCR, and that others are too nonspecific and result in general immunosuppression. The solution to some of these problems may be found in a therapeutic concept called TCR antagonism. About 2 years ago, Sette and coworkers [33] observed that modified antigenic peptides are able to partially block or completely antagonize the stimulatory signal that is mediated through the recognition of the native peptide presented in the context of an appropriate MHC. The previous view that T cells receive either a positive signal (on) once their TCR engages the peptide-MHC complex, or no signal (off) if the complex is not recognized, did not explain a number of findings that followed this first observation. In the meantime, several studies have confirmed these results and have shown that TCR antagonists or altered peptide ligands (APL) can either mediate no signal, if the major TCR contact sites of an antigenic peptide are modified or cause a partial response. In the latter situation, APL peptides were able to induce lymphokine release or cytolysis in Th2 or Th1 cells, respectively, without concomitant proliferation (for review see [35]). Certain substitutions even cause phenotypic switches, i.e., Th1 cells no longer secrete IFN-γ and TNF/LT, but do produce IL-4 or other Th2 cytokines instead ([153], Vergelli et al., in preparation). The great advantages that TCR antagonism offers over the blockade of disease-associated MHC antigens by high-affinity binders and over TCR vaccination, are its effects in molar ranges that are orders of magnitude below those used for MHC blockade and that it is not specifically targeting one T cell population, but a broad range of T cells [63]. The latter notion has already been confirmed by the study of Kuchroo et al. [63] in SJL mice that showed that TCR antagonist peptides that are based on the encephalitogenic PLP peptide 138–153 can block EAE induced by a number of encephalitogenic PLP 138–153-specific T cell hybridomas although these cells express different TCR molecules. TCR antagonism has also been achieved in Lewis rat EAE by injection of soluble APL peptides [55]. Although the biological effects that are caused by these peptides are not yet entirely understood, they could act through the induction of anergy, by antagonizing the stimulus by the native peptide and by inducing a phenotypic switch in the response to the encephalitogenic peptide.

Another approach that has recently been studied intensively originates from the observation that the route of administration such as oral feeding or inhalation of antigen results in an immune response that is drastically different from that after intravenous, intramuscular or subcutaneous immunization [12, 45, 84, 85]. Oral administration of MBP, for example, establishes a state of tolerance that, depending on the dose of protein that is fed, is due to T cell anergy at higher doses [45] or to the induction of regulatory T cells that secrete transforming growth factor-β (TGF-β), IL-4 and IL-10 at lower doses [24, 85]. In addition, oral administration of MBP was able to cross-protect against PLP-induced EAE in SJL mice [4, 24]. Oral tolerization is currently being tested for a number of experimental and human autoimmune diseases including uveitis, insulin-dependent diabetes, rheumatoid arthritis and MS. Preliminary results from a pilot trial of oral tolerance with bovine myelin in MS patients were reported as suggesting some efficacy in a group of $DR2^-$ male patients [148]. Since the duration of the trial was short and treatment efficacy was not controlled by MRI, it is too early to draw conclusions from this data. A large placebo-controlled, double-blind multicenter trial is currently under way to confirm these interesting experimental data.

Although not directly involved in the TMC. CD4 serves an important accessory function in the recognition of antigen by $CD4^+$ T cells. Therefore, some time ago, attempts were made to interfere with the $CD4^+$ encephalitogenic T cells using monoclonal antibodies against the CD4 molecule, which stabilizes the conjugation between $CD4^+$ T cells and the MHC molecule on APC. Monoclonal antibodies against CD4 were able to block EAE [17, 46], and a humanized monoclonal is currently being tested in MS patients in Europe. Initial results of these trials show that the antibody is well tolerated, that the formation of interfering antibodies is not a problem and that the antibody successfully depletes $CD4^+$ T cells from the peripheral blood of the patients [72]. A recently published study employed a CD4 analogue that was designed with D-amino acids [51]. This approach was also able to block EAE. It is, however, anticipated that any therapy that will interfere with a molecule involved in general helper T cell function will eventually result in a very broad immunosuppression.

Therapies targeting T cell homing, adhesion to and migration through the BBB

Evidence derived from studies of the EAE model and examination of the pathology of acute lesions as well as MRI studies of MS patients indicate that the initial event of the development of a new MS lesion is breakdown of the BBB. This event is most likely associated with the transmigration of lymphocytes that have been activated in the periphery into the brain or spinal cord parenchyma. The relative importance of the properties of T cells compared to the expression of adhesion molecules on endothelial cells for the regions where the inflammatory infiltrate preferentially develops (i.e., around small postcapillary venules in the periventricular white matter) is unresolved. Evidence from studies in the rat and mouse indicate that activated T cells can transmigrate into the brain regardless of their specificity [30, 47, 149]. T cells specific for myelin or glial antigens that encounter their target autoantigen in the CNS are retained while T cells with other specificities tend to leave. Sequential immunohistological and PCR studies that followed the expression of TCR chains in the CNS tissue after adoptive transfer of well-defined myelin-specific TCL have demonstrated that encephalitogenic T cells and their TCR are found first, before a massive recruitment of cells with other receptors and presumably other specificities occurs [10, 18, 54]. Cells expressing the TCR of the encephalitogenic TCL stay longer and either leave or die in situ by apoptosis [199, 133].

Activated T cells express increased levels of several adhesion molecules and selectins [9, 21, 34, 62, 152]. For example, a relationship was described between encephalitogenicity of T cells raised against encephalitogenic determinants of MBP and PLP and the expression of VLA-4, the ligand of VCAM-1 [9, 62]. Only T cells expressing high levels of VLA-4 were able to transfer EAE without other manipulations of the mouse. Presumably, these cells cross the BBB more easily. Further, antibody to $\alpha4\beta1$ integrin can diminish the clinical and pathological severity of EAE [9, 158]. If interaction between VLA-4 on T cells and VCAM-1 is important in transmigration and encephalitogenicity, the expression of VCAM-1 must be up-regulated since VCAM-1 is expressed at only low levels on endothelial cells. Although constitutively expressed at higher levels on CNS endothelial cells, ICAM-1 is also up-regulated during the initial stages of lesion development and could contribute to T cell adhesion and subsequent transmigration. However, treatment of EAE with antibodies to ICAM-1 and its ligand, LFA-1, have produced variable results [7, 22, 149]. Depending on the timing and dose of monoclonal antibodies against ICAM-1 and LFA-1, the outcome was either beneficial, unaltered or deleterious [7, 22, 150]. It is also important to note that these CAM may contribute to cell-cell interactions other than those with endothelial cells or that engagement of these structures may itself provide an activation signal to T cells, such as stimulation by purified LFA-3 and ICAM-1 [123]. Before such treatments are considered for patients, carefully designed studies in experimental systems should address these questions and one should also determine whether there are phenotypic or functional differences in the various CD4 T cell populations that have been cultivated, for example, after stimulation with MBP, from MS patients and controls. Our own studies that compared MBP-specific cytotoxic and proliferative TCL could only find a slightly higher expression of LFA-1, but no differences for VLA-4 and LFA-3 (Vergelli et al., unpublished observations). Although human MBP-specific TCL in our hands are largely Th1 cells [145], like those mediating EAE in the various EAE systems [6, 62], it is not clear whether a specific subpopulation is pathogenetically more important than another.

With respect to treatment trials in MS, the use of humanized antibodies against different adhesion molecules has now entered or will enter clinical testing (see Table 4). Since treatments of this type would be expected to affect the occurrence of new lesions, MRI should provide a valuable tool for monitoring the effectiveness of these therapies.

Table 4. Therapies targeting the adhesion of autoantigenic T cells to the brain endothelium and their entry into the parenchyma

Antibodies against ICAM-1
Antibodies against LFA-1
Antibodies against VLA-4
Antibodies against VCAM-1

Therapies targeting proinflammatory cytokines

There is now overwhelming evidence that EAE can be mediated solely by the transfer of a $CD4^+$ Th1 T cell [67, 161]. These cells that mediate delayed-type hypersensitivity responses and tissue inflammation in various infectious diseases are characterized by their lymphokine secretion profile, i.e., the secretion of IFN-γ, TNF-α and lymphotoxin-α (LT) and IL-12, as well as by their ability to lyze antigen-presenting target cells [6, 16, 62]. As mentioned before, the encephalitogenic T cells are Th1 cells [6, 16, 62], and most of the human MBP-specific T cells probably also have this phenotype [76, 112, 145]. Th1-derived lymphokines contribute to the amplification of the local inflammatory response and to tissue damage in a number of ways. It is well known that IFN-γ up-regulates MHC antigen and adhesion molecule expression on a number of cells including endothelial cells, glial cells and cells of the lymphoid lineages themselves [81, 154, 155]. In addition, IFN-γ drives the differentiation of effector $CD4^+$ and $CD8^+$ T cells as well as macrophages and stimulates the release of cytocidal reactive oxygen (O_2^- and H_2O_2) and nitrogen (NO) intermediates from phagocytic cells and astrocytes [69]. Besides IL-12, IFN-γ is the most important molecule in skewing a $CD4^+$ T cell response in the direction of a Th1 response. Another Th1 lymphokine family, consisting primarily of macrophage-derived TNF-α and T cell-derived LT, also contributes to the inflammatory process [116, 120, 121]. Besides mediating effects similar to those of IFN-γ, TNF/LT has been shown to damage endothelial cells, resulting in changes to the BBB and oligodendroglial cells leading to demyelination [120]. TNF/LT is produced locally in inflammatory plaques in the CNS in EAE and MS. With respect to plaque development, it is currently hypothesized that encephalitogenic T cells adhere to CNS endothelial cells, up-regulate class II antigen and adhesion molecule expression at the BBB, and thus facilitate transmigration. In addition, they may directly damage endothelial cells [80], allowing the influx of serum components that contribute to tissue damage and then further enhance the local inflammatory process via promoting recruitment and activation of other incoming and resident cells. As mentioned before, TNF/LT has been shown to damage oligodendroglial cells in vitro and thus may contribute to demyelination. The importance of TNF/LT has been underlined by the effect of monoclonal antibodies against TNF/LT that effectively blocked EAE [116, 121]. Other strategies aimed at inactivating Th1 lymphokines include the administration of soluble TNF receptor-immunoglobin fusion proteins [8] or monoclonal antibodies against IL-12 [70], or targeting a macrophage-derived proinflammatory cytokine, IL-1β that is up-regulated during a Th1 response by administering soluble IL-1 receptors [50]. Each of these approaches has been successful in treating EAE.

The importance of proinflammatory cytokines in MS stems from a number of observations in patients. Therapy with IFN-γ resulted in an increase of exacerbations [96], and levels of TNF/LT have been reported to be increased in the blood and CSF of MS patients during relapses [26, 125]. The latter observation is, however, still controversial, since some studies have not confirmed the findings. Furthermore, it is known that soluble TNF is very difficult to determine in body fluids and tissue samples, since secretion of TNF rapidly induces the shedding of TNF receptors and consequently the formation of TNF-TNF receptor complexes. In this situation it is hard to determine the lymphokine level reliably.

Since the administration of monoclonal antibodies is costly and causes a number of problems during long-term treatment, there has been considerable interest in identifying other approaches to interfere with the production of TNF/LT and IFN-γ. It has been shown that phosphodiesterase (PDE) inhibitors are able to block lymphocyte activation and the secretion of TNF-α and IFN-γ via the increase of intracellular cAMP levels [122, 129]. This group of agents can be separated into phosphodiesterase inhibitors with relatively nonspecific action such as pentoxifylline and those that specifically block certain PDE types. A member of the latter group, the PDE type IV inhibitor rolipram, is particularly attractive for a number of reasons. In contrast to the unspecific PDE inhibitors, the PDE type IV is almost exclusively expressed in the CNS and in lymphoid cells and thus has significantly less side effects. The drug was originally developed and tested as an antidepressant but also has a potent suppressive effect on TNF and, to a lesser extent, on IFN-γ secretion [41, 129, 130]. Pentoxifylline was effective in EAE [88, 114]. The PDE type IV inhibitor rolipram could block TNF- and IFN-γ secretion in human MBP-specific T cells in a stereospecific manner and also inhibited EAE in the Lewis rats [129], in SJL mice (Sommer et al., in preparation) and also in the marmoset EAE model [41]. Since the substance has few side effects and can be administered orally it provides an attractive alternative to the treatments with monoclonal antibodies or soluble TNF receptors.

Another interesting approach that interferes with TNF employs inhibitors of metalloproteases [43]. These enzymes, which are usually matrix bound, as for example gelatinase B, cleave membrane-bound TNF from the cell membrane and thus release it at the inflammatory site. Inhibitors of metalloproteases are capable of inhibiting EAE and represent an attractive alternative to the above-mentioned strategies that interfere with Th1 lympkokines [43].

IFN-β, which was recently approved for the treatment of RR-MS, also belongs to the group of treatments that interfere with proinflammatory cytokines or their action at different levels, including the expression of adhesion molecules and MHC antigens, the recruitment of cells into the lesion, and the process of antigen presentation at the induction and effector stages [97, 138]. These therapies are summarized in Table 5.

Table 5. Therapies that modify the effector function of autoantigen-specific CD4 T cells (i.e., up-regulation of adhesion and MHC class molecule expression; activation of macrophages and resident glial cells) or unspecifically interfere with their activation

Interferon-beta (IFN-β)
IL-4
Transforming growth factor-beta (TGF-β)
IL-13
Soluble TNF receptor immunoglobulin fusion proteins
Antibodies against TNF/LT
Phosphodiesterase inhibitors (rolipram; pentoxifylline)
Inhibitor of the inducible nitric oxide synthase (aminoguanidine)
Metalloprotease inhibitors

Therapies employing regulatory cytokines

A number of cytokines have been identified which appear to down-regulate inflammation. One of the most potent is TGF-β. TGF-β has been shown to have immunosuppressive effects on T cells, B cells and natural killer cells and decreases leukocyte transmigration from blood vessels at inflammatory sites. Treatment of animals immunized with myelin antigens with antibodies to TGF-β have demonstrated accentuated disease [102]. TGF-β has also been examined for its ability to directly modify the clinical and pathological course of EAE. When administered prior to the onset of clinical disease in mice adoptively immunized with T cells sensitized to MBP, TGF-β significantly reduces the severity of clinical disease and produces a marked reduction of inflammation and demyelination pathologically [52, 66, 101, 104]. Further, immunohistochemistry indicates a decrease in expression of MHC class II molecules and proinflammatory cytokines such as TNF-α [101]. Administration of TGF-β after the onset of disease also reduces the severity and number of subsequent relapses [101]. It has also been demonstrated that amelioration of disease through feeding antigen (oral tolerance) induces T cells that secrete TGF-β [24, 85]. Together, these findings indicate that TGF-β can have a profound immunoregulatory effect on T cell-mediated autoimmune diseases such as EAE. Despite the encouraging initial results using TGF-β in the EAE, extrapolation into human disease must be done with caution since TGF-β has been thought to augment some inflammatory conditions and tissue fibrosis and may up-regulate expression of some CAM. A clinical phase I trial in MS patients to assess the safety and potential efficacy of systemically administered TGF-β has just begun at the National Institutes of Health.

In addition to TGF-β, other cytokines produced by Th2 T cells such as IL-4 [105] and IL-10 [115] and IL-13 [23] appear to down-regulate and antagonize Th1-mediated inflammation. Administration of IL-4 in the adoptive transfer EAE model in SJL mice has been shown to reduce clinical disease severity and to reduce production of proinflammatory cytokines produced by Th1 T cells [105]. Diets enriched for retinoic acids also ameliorated EAE in the mouse [107] and rat [79], and recent evidence suggests that the mechanism may involve the up-regulation of Th2-type cytokines. At present, it is not clear if the mechanism of retinoic acids is entirely due to increased production of IL-4. Along these lines, IL-10 expression increases several times in the CNS of mice that recover from an exacerbation of EAE [59], indicating that endogenous secretion of this Th2 lymphokine is critical for the recovery from CNS inflammation.

The modulation of experimental autoimmune diseases by immunoregulatory cytokines reflects their possible usefulness in modifying human diseases such as MS in a manner that is rather immunomodulatory than generally immunosuppressive.

Therapies targeting the effector phase of demyelination

As outlined above, activated macrophages and toxic radicals (NO and O_2^-) secreted by these cells as well as cytokines produced by macrophages and lymphocytes, in particular TNF/LT, are responsible for demyelination. A number of the approaches that have been mentioned to interfere with Th1 lymphokines will also act at this level and decrease the activation of macrophages, and consequently, also the extent of demyelination. The administration of immunoglobulins intravenously may not only antagonize TNF [2], but also interfere with myelin-specific antibodies such as anti-MOG antibodies and the engulfment of myelin membranes into macrophages via coated pits. Aminoguanidine, an inhibitor of the inducible NO synthase suppresses the secretion of the toxic nitrogen intermediate NO and is effective in treating EAE [37].

Conclusions

Until recently most attempts to treat MS have focused on general immunosuppressive drugs since the disease was presumed to be immunologically mediated. Although some of these drugs may have effects on the disease process, their clinical trials were often inconclusive and discouraging. In the last years, attention has been directed at therapies which can modify the TMC in a very specific manner similar to that seen in the experimental model EAE. However, since the target antigen or antigens in MS are not known and since the restricted TCR usage often found in experimental animals is generally not found in MS, these approaches have been difficult to initiate for MS. Thus, more recent treatment approaches have begun to focus on modifying the biology of the MS lesion.

Despite our inability to identify the cause of MS, our understanding of events involved in the evolution of the MS lesion is increasing. This, in turn, has led to a number of innovative therapies of MS which target the processes involved in lesion development, rather than the precise cause of the disease or early steps important during the activation of autoreactive T cells. These approaches range from attempts to alter events at the BBB, down-regulation of proinflammatory cytokines, or enhancement of cytokines thought to down-regulate and delete the effector mechanisms in the lesion. Currently, studies examining the effect of blocking TNF-α/LT or administering the regulatory cytokine TGF-β are under way. Finally, despite the fact that a unique trimolecular interaction has not yet been demonstrated in the pathogenesis of MS, there is evidence that immunodominant peptides of autoantigens are recognized in the context of disease-associated HLA class II molecules. Particularly during the early phases of disease, the T cell response may be primarily directed against these MHC peptide complexes, and studies of the requirements of T cell activation indicate that modifications of amino acids important for TCR contact on encephalitogenic peptides can therapeutically modify the T cell response. This approach has the potential for targeting T cells, which may have regulatory or suppressive functions at the lesion, and down-regulating inflammatory activity into the lesion, regardless of the initiating antigen. A similar mechanism may be achieved by oral tolerance. Thus, despite continued uncertainty regarding the cause of MS, our improved understanding of the pathogenetic process during EAE and MS has resulted in innovative therapies that hold hope for effective treatment of the disease.

References

1. Acha-Orbea H, Mitchell L, Timmermann L, Wraith DC, Tausch GS, Waldor KM, Zamvil SS, McDevitt HO, Steinman L (1988) Limited heterogeneity of T cell receptors from lymphocytes mediating autoimmune encephalomyelitis allows specific immune intervention. Cell 54:263
2. Achiron A, Margalit R, Hershkoviz R, Markovits D, Reshef T, Melamed E, Cohen IR, Lider O (1994) Intravenous immunoglobulin treatment of experimental T cell-mediated autoimmune disease.

Upregulation of T cell proliferation and downregulation of tumor necrosis factor α secretion. J Clin Invest 93:600

3. Aharoni R, Teitelbaum D, Arnon R, Puri J (1991) Immunomodulation of experimental allergic encephalomyelitis by antibodies to the antigen-Ia complex. Nature 351:147
4. Al-Sabbagh A, Miller A, Santos LMB, Weiner HL (1994) Antigen-driven tissue-specific suppression following oral tolerance: orally admistered myelin basic protein suppresses proteolipid protein-induced experimental autoimmune encephalomyelitis in the SJL mouse. Eur J Immunol 24:2104
5. Amor S, Groome N, Linington C, Morris MM, Dornmair K, Gardinier MV, Matthieu J-M, Baker D (1994) Identification of epitopes of myelin oligodendrocyte glycoprotein for the induction of experimental allergic encephalomyelitis in SJL and Biozzi AB/H mice. J Immunol 153:4349
6. Ando DG, Clayton J, Kono D, Urban JL, Sercarz EE (1989) Encephalitogenic T cells in the B10.PL model of experimental allergic encephalomyelitis (EAE) are of the Th-1 lymphokine subtype. Cell Immunol 124:132
7. Archelos JJ, Jung S, Mäurer M, Schmied M, Lassmann H, Tamatami T, Miyasaka M, Toyka KV, Hartung H-P (1993) Inhibition of experimental autoimmune encephalomyelitis by an antibody to the intercellular adhesion molecule ICAM-1. Ann Neurol 34:145
8. Baker D, Butler D, Scallon BJ, O'Neill JK, Turk JL, Feldmann M (1994) Control of established experimental allergic encephalomyelitis by inhibition of tumor necrosis factor (TNF) activity within the central nervous system using monoclonal antibodies and TNF receptor-immunoglobulin fusion proteins. Eur J Immunol 24:2040
9. Baron JL, Madri JA, Ruddle NH, Hashim G, Janeway CA (1993) Surface expression of α4 integrin by CD4 T cells is required for their entry into brain parenchyma. J Exp Med 177:57
10. Bell RB, Lindsey JW, Sobel RA, Hodgkinson S, Steinman L (1993) Diverse T cell receptor Vβ gene usage in the central nervous system in experimental allergic encephalomyelitis. J Immunol 150:4085
11. Ben Nun A, Wekerle H, Cohen IR (1981) Vaccination against autoimmune encephalomyelitis with T-lymphocyte line cells reactive against myelin basic protein. Nature 293:60
12. Bitar DM, Whitacre CC (1988) Suppression of experimental autoimmune encephalomyelitis by the oral administration of myelin basic protein. Cell Immunol 112:364
13. Boggs JM, Chang NH, Goundalkar A, Hashim GA (1992) Stimulation or tolerization of an anti-myelin basic protein T lymphocyte line with membrane fragments from antigen presenting cells. Cell Immunol 143:23
14. Bourdette DN, Whitham RH, Chou YK, Morrison WJ, Atherton J, Kenny C, Liefeld D, Hashim GA, Offner H, Vandenbark AA (1994) Immunity to TCR peptides in multiple sclerosis. I. Successful immunization of patients with synthetic Vβ5.2 and Vβ6.1 CDR2 peptides. J Immunol 152:2510
15. Brocke S, Gaur, A, Piercy C, Gautam A, Gijbels K, Fathman CG, Steinman L. (1993) Induction of relapsing paralysis in experimental allergic encephalomyelitis by bacterial superantigen. Nature 365:642
16. Broome Powell M, Mitchell D, Lederman J, Buckmeier J, Zamvil SS, Graham M, Ruddle N, Steinman L (1990) Lymphotoxin and tumor necrosis factor-alpha production by myelin basic protein-specific T cell clones correlates with encephalitogenicity. Int Immunol 2:539
17. Brostoff SW, Mason DW (1984) Experimental allergic encephalomyelitis: Successful treatment in vivo with a monoclonal antibody that recognizes T helper cells. J Immunol 133:1938
18. Buenafe AC, Vainiene M, Celnik B, Vandenbark AA, Offner H (1994) Analysis of Vβ8-CDR3 sequences derived from central nervous system of Lewis rats with experimental autoimmune encephalomyelitis. J Immunol 153:386
19. Burns FR, Li X, Shen N, Offner H, Chou YK, A. V, Heber-Katz A (1989) Both rat and mouse T cell receptors specific for the encephalitogenic determinant of myelin basic protein use similar Vα and Vβ chain genes even though the major histocompatibility complex and encephalitogenic determinants being recognized are different. J Exp Med 169:27
20. Burns J, Rosenzweig A, Zweiman B, Lisak RP (1983) Isolation of myelin basic protein-reactive T-cell lines from normal human blood. Cell Immunol 81:435
21. Cannella B, Cross AH, Raine CS (1990) Upregulation and coexpression of adhesion molecules correlate with relapsing autoimmune demyelination in the central nervous system. J Exp Med 172:1521
22. Cannella B, Cross AH, Raine CS (1993) Anti-adhesion molecule therapy in experimental autoimmune encephalomyelitis. J Neuroimmunol 46:43
23. Cash E, Minty A, Ferrara P, Caput D, Fradelizi D, Rott O (1994) Macrophage inactivation IL-13 suppresses experimental autoimmune encephalomyelitis in rats. J Immunol 153:4258

24. Chen Y, Kuchroo VK, Inobe J-I, Hafler DA, Weiner HL (1994) Regulatory T cell clones induced by oral tolerance: suppression of autoimmune encephalomyelitis. Science 265:1237
25. Chluba J, Steeg C, Becker A, Wekerle H, Epplen JT (1989) T cell receptor β chain usage in myelin basic protein-specific rat T lymphocytes. Eur J Immunol 19:279
26. Chofflon M, Juillard C, Juillard P, Gauthier G, Grau GE (1992) Tumor necrosis factor alpha production as a possible predictor of relapse in patients with multiple sclerosis. Eur Cytokine Netw 3:523
27. Chou YK, Morrison WJ, Weinberg AD, Dedrick R, Whitham R, Bourdette DN, Hashim G, Offner H, Vandenbark AA (1994) Immunity to TCR peptides in multiple sclerosis. II. T cell recognition of Vβ5.2 and Vβ6.1 CDR2 peptides. J. Immunol 152:2520
28. Clayton JP, Gammon GM, Ando DG, Kono DH, Hood L, Sercarz EE (1989) Peptide-specific prevention of experimental allergic encephalomyelitis. Neonatal tolerance induced to the dominant T cell determinant of myelin basic protein. J Exp Med 169:1681
29. Critchfield JM, Racke MK, Zuniga-Pflücker JC, Cannella B, Raine CS, Goverman J, Lenardo MJ (1994) T cell deletion in high antigen dose therapy of autoimmune encephalomyelitis. Science 263:1139
30. Cross AH, Cannella B, Brosnan CF, Raine CS (1990) Homing to central nervous system vasculature by antigen specific lymphocytes. I. Localization of 14C-labeled cells during acute, chronic and relapsing experimental allergic encephalomyelitis. Lab Invest 63:162
31. Cross AH, O'Mara T, Raine CS (1993) Chronologic localization of myelin-reactive cells in the lesions of relapsing EAE: implications for the study of multiple sclerosis. Neurology 43:1028
32. Cross AH, Misko TP, Lin RF, Hickey WF, Trotter JL, Tilton RG (1994) Aminoguanidine, an inhibitor of inducible nitric oxide synthase, ameliorates experimental autoimmune encephalomyelitis in SJL mice. J Clin Invest 93:2684
33. De Magistris MT, Alexander J, Coggeshall M, Altman A, Gaeta FCA, Grey HM, Sette A (1992) Antigen analog-major histocompatibility complexes act as antagonists of the T cell receptor. Cell 68:625
34. Dopp JM, Breneman SM, Olschowska JA (1994) Expression of ICAM-1, VCAM-1, L-selectin, and leukosialin in the mouse central nervous system during the induction and remission of experimental allergic encephalomyelitis. J Neuroimmunol 54:129
35. Evavold BD, Sloan-Lancaster J, Allen PM (1993) Tickling the TCR: selective T cell functions stimulated by altered peptide ligands. Immunol Today 14:602
36. Fritz RB, McFarlin DE (1989) Encephalitogenic epitopes of myelin basic protein. Chem Immunol 46:101
37. Fritz RB, Skeen MJ, Jen-Chou CH, Garcia M, Egorov IK (1985) Major histocompatibility complex-linked control of the murine immune response to myelin basic protein. J Immunol 134:2328
38. Gaur A, Wiers B, Liu A, Rothbard J, Fathman CG (1992) Amelioration of autoimmune encephalomyelitis by myelin basic protein synthetic peptide-induced anergy. Science 258:1491
39. Gaur A, Haspel R, Mayer JP, Fathman CG (1993) Requirement for $CD8^+$ cells in T cell receptor peptide-induced clonal unresponsiveness. Science 259:91
40. Genain CP, Lee-Parritz D, Nguyen M-H, Massacesi L, Joshi N, Ferrante R, Hoffman K, Moseley M, Letvin NL, Hauser SL (1994) In healthy primates, circulating autoreactive T cells mediate autoimmune disease. J Clin Invest 94:1339
41. Genain CP, Nguyen MH, Faulds D, Uccelli A, Davis RL, Hedgpeth J, Hauser SL (1994) Prevention of EAE in non-human primates by a type IV phosphodiesterase inhibitor that suppresses tumor necrosis factor. J Neuroimmunol 54:163
42. Giegerich G, Pette M, Meinl E, Epplen JT, Wekerle H, Hinkkanen A (1992) Diversity of T cell receptor alpha and beta chain genes expressed by human T cells specific for similar myelin basic protein peptide/major histocompatibility complexes. Eur J Immunol 22:753
43. Gijbels K, Galardy RE, Steinman L (1994) Reversal of experimental autoimmune encephalomyelitis with a hydroxamate inhibitor of matrix metalloproteases. J Clin Invest 94:2177
44. Goss JA, Nakafusa Y, Roland CR, Hickey WF, Flye M (1994) Immunological tolerance to a defined myelin basic protein antigen administered intrathymically. J Immunol 153:3890
45. Gregerson DS, Obritsch WF, Donoso LA (1993) Oral tolerance in experimental autoimmune uveoretinitis. Distinct mechanisms of resistance are induced by low dose vs high dose feeding protocols. J Immunol 151:5751

46. Harris JO, Frank JO, Patronas N, McFarlin DE, McFarland HF (1991) Serial gadolinium-enhanced magnetic resonance imaging scans in patients with early, relapsing-remitting multiple sclerosis: Implication for clinical trials and natural history. Ann Neurol 29:548
47. Hickey WF, Hsu BL, Kimura H (1991) T-lymphocyte entry into the central nervous system. J Neurosci. Res 28:254
48. Howell MD, Winters ST, Olee T, Powell HC, Carlo DJ, Brostoff SW (1989) Vaccination against experimental allergic autoimmune encephalomyelitis with T cell receptor peptides. Science 246:668
49. Ishioka GY, Adorini L, Guery J-C, Gaeta FCA, LaFond R, Alexander J, Powell MF, Sette A, Grey HM (1994) Failure to demonstrate long-lived MHC saturation both in vitro and in vivo. J Immunol 152:4310
50. Jacobs CA, Baker PE, Roux ER, Picha KS, Toivola B, Waugh S, Kennedy MK (1991) Experimental autoimmune encephalomyelitis is exacerbated by IL-1α and suppressed by soluble IL-1 receptors. J Immunol 146:2983
51. Jameson BA, McDonnell JM, Marini JC, Korngold R (1994) A rationally designed CD4 analogue inhibits experimental allergic encephalomyelitis. Nature 368:744
52. Johns LD, Flanders KC, Ranges GE, Sriram S (1991) Successful treatment of experimental allergic encephalomyelitis with transforming growth factor-β1. J Immunol 147:1792
53. Jorgensen JL, Esser U, Fazekas de St. Groth B, Reay PA, Davis MM (1992) Mapping T-cell receptor-peptide contacts by variant peptide immunization of single-chain transgenics. Nature 355:224
54. Karin N, Szafer F, Mitchell D, Gold DP, Steinman L (1993) Selective and nonselective stages in homing of T lymphocytes to central nervous system during experimental allergic encephalomyelitis. J Immunol 150:4116
55. Karin N, Mitchell DJ, Brocke S, Ling N, Steinman L (1994) Reversal of experimental autoimmune encephalomyelitis by a soluble peptide variant of a myelin basic protein epitope: T cell receptor antagonism and reduction of interferon γ and tumor necrosis factor α production. J Exp Med 180:2227
56. Kawano Y-I, Sasamoto Y, Kotake S, Thurau SR, Wiggert B, Gery I (1991) Trials of vaccination against experimental autoimmune uveoretinitis with a T-cell receptor peptide. Curr Eye Res 10:789
57. Kennedy MK, L.-J T, Dal Canto MC, Tuohy VK, Lu Z, Trotter JL, Miller SD (1990) Inhibition of murine relapsing experimental autoimmune encephalomyelitis by immune tolerance to proteolipid protein and its encephalitogenic peptides. J Immunol 144:909
58. Kennedy MK, Tan L-J, Dal Canto MC, Miller SD (1990) Regulation of the effector stages of experimental autoimmune encephalomyelitis via neuroantigen-specific tolerance induction. J Immunol 145:117
59. Kennedey MK, Torrance DS, Picha KS, Mohler KM (1992) Analysis of cytokine mRNA expression in the central nervous system of mice with experimental autoimmune encephalomyelitis reveals that IL-10 mRNA expression correlates with recovery. J Immunol 149:2496
60. Kojima K, Berger T, Lassmann H, Hinze-Selch D, Zhang Y, Gehrmann J, Reske K, Wekerle H, Linington C (1994) Experimental autoimmune panencephalitis and uveoretinitis transferred to the Lewis rat by T lymphocytes specific for the S100β molecule, a calcium binding protein of astroglia. J Exp Med 180:817
61. Kotzin BL, Karuturi S, Chou YK, Lafferty J, Forrester M, Better M, Nedwin GE, Offner H, Vandenbark A (1991) Preferential T-cell receptor Vβ-chain variable gene use in myelin basic protein-reactive T-cell clones from patients with multiple sclerosis. Proc Natl Acad Sci USA 88:9161
62. Kuchroo VK, Martin CA, Greer JM, Ju S-T, Sobel RA, Dorf ME (1993) Cytokines and adhesion molecules contribute to the ability of myelin proteolipid protein-specific T cell clones to mediate experimental allergic encephalomyelitis. J Immunol 151:4371
63. Kuchroo VK, Greer JM, Kaul D, Ishioka G, Franco A, Sette A, Sobel RA, Lees MB (1994) A single TCR antagonist peptide inhibits experimental allergic encephalomyelitis mediated by a diverse T cell repertoire. J Immunol 153:3326
64. Kuhröber A, Schirmbeck R, Reimann J (1994) Vaccination with T cell receptor peptides primes anti-receptor cytotoxic T lymphocytes (CTL) and anergizes T cells specifically recognized by these CTL. Eur J Immunol 24:1172
65. Kumar V, Sercarz EE (1993) The involvement of T cell receptor peptide-specific regulatory $CD4^+$ T cells in recovery from antigen-induced autoimmune disease. J Exp Med 178:909
66. Kuruvilla AP, Shah R, Hochwald GM, Liggitt HD, Palladino MA, Thorbecke GJ (1991) Protective effect of transforming growth factor β1 on experimental autoimmune diseases in mice. Proc Natl Acad Sci USA 88:2918

67. Lafaille JJ, Nagashima K, Katsuki M, Tonegawa S (1994) High incidence of spontaneous autoimmune encephalomyelitis in immunodeficient anti-myelin basic protein T cell receptor transgenic mice. Cell 78:399
68. Lamont AG, Sette A, Fujinami R, Colon SM, Miles C, Grey HM (1990) Inhibition of experimental autoimmune encephalomyelitis induction in SJL/J mice by using a peptide with high affinity for IAs molecules. J Immunol 145:1687
69. Lee SC, Dickson DW, Liu W, Brosnan CF (1993) Induction of nitric oxide synthase activity in human astrocytes by interleukin-1β and interferon-γ. J Neuroimmunol 46:19
70. Leonard JP, Waldburger KE, Goldman SJ (1995) Prevention of experimental autoimmune encephalomyelitis by antibodies against interleukin 12. J Exp Med 181:381
71. Liebert UG, Hashim GA, ter Meulen V (1990) Characterization of measles virus-induced cellular autoimmune reactions against myelin basic protein in Lewis rats. J Neuroimmunol 29:139
72. Lindsey JW, Hodgkinson S, Mehta R, Mitchell D, Enzmann D, Steinman L (1994) Repeated treatment with chimeric anti-CD4 antibody in multiple sclerosis. Ann Neurol 36:183
73. Linington C, Bradl M, Lassmann H, Brunner C, Vass K (1988) Augmentation of demyelination in rat acute allergic encephalomyelitis by circulating mouse monoclonal antibodies directed against a myelin/oligodendrocyte glycoprotein. Am J Pathol 130:443
74. Lipton HL, Dal Canto MC (1976) Theiler's virus-induced demyelination – prevention by immunosuppression. Science 192:62
75. Lohse AW, Mor F, Karin N, Cohen IR (1989) Control of experimental autoimmune encephalomyelitis by T cells responding to activated T cells. Science 244:820
76. Martin R, Jaraquemada D, Flerlage M, Richert J, Whitaker J, Long EO, McFarlin DE, McFarland HF (1990) Fine specificity and HLA restriction of myelin basic protein-specific cytotoxic T cell lines from multiple sclerosis patients and healthy individuals. J Immunol 145:540
77. Martin R, McFarland HF, McFarlin DE (1992) Immunological aspects of demyelinating diseases. Annu Rev Immunol 10:153
78. Martin R, Utz U, Coligan JE, Richert JR, Flerlage M, Robinson E, Stone R, Biddison WE, McFarlin DE, McFarland HF (1992) Diversity in fine specificity and T cell receptor usage of the human $CD4^+$ cytotoxic T cell response specific for the immunodominant myelin basic protein peptide 87–106. J Immunol 148:1359
79. Massacesi L, Castigli E, Vergelli M, Olivotto J, Abbamondi AL, Sarlo F, Amaducci L (1991) Immunosuppressive activity of 13-cis-retinoic acid and prevention of experimental autoimmune encephalomyelitis in rats. J Clin Invest 88:1331
80. McCarron RM, Spatz M, Kempski O, Hogan RN, Muehl L, McFarlin DE (1986) Interaction between myelin basic protein-sensitized T lymphocytes and murine cerebral vascular endothelial cells. J Immunol 137:3428
81. McCarron RM, Tanaka M, Spatz M (1990) Class II major histocompatibility complex antigen expression in central nervous system: microglia, astrocytes and endothelial cells. In: Pathophysiology of the blood-brain barrier. Johansson BB, Owman CO, Widmer H (eds), Elsevier, Amsterdam, pp 467–484
82. McCarron RM, Wang L, Racke MK, McFarlin DE, Spatz M (1993) Cytokine-regulated adhesion between encephalitogenic T lymphocytes and cerebrovascular endothelial cells. J Neuroimmunol 43:23
83. McFarland HF, Frank JA, Albert PS, Smith ME, Martin R, Harris JO, Patronas N, Maloni H, McFarlin DE (1992) Using gadolinium-enhanced magnetic resonance imaging lesions to monitor disease activity in multiple sclerosis. Ann Neurol 32:758
84. Metzler B, Wraith DC (1993) Inhibition of experimental autoimmune encephalomyelitis by inhalation but not oral administration of the encephalitogenic peptide: influence of MHC binding affinity. Int Immunol 5:1159
85. Miller A, Lider O, Roberts AB, Sporn MB, Weiner HL (1992) Suppressor T cells generated by oral tolerization to myelin basic protein suppress both in vitro and in vivo immune responses by the release of transforming growth factor beta after antigen-specific triggering. Proc Natl Acad Sci USA 89:421
86. Moscarello MA, Wood DD, Ackerley C, Boulias C (1994) Myelin in multiple sclerosis is developmentally immature. J Clin Invest 94:146
87. Mustafa M, Vingsbo C, Olsson T, Ljungdahl Å, Höjeberg B, Holmdahl R (1993) The major histocompatibility complex influences myelin basic protein 63–88-induced T cell cytokine profile and experimental autoimmune encephalomyelitis. Eur J Immunol 23:3089

88. Nataf S, Louboutin JP, Chabannes D, Feve JR, Muller JY (1993) Pentoxifylline inhibits experimental allergic encephalomyelitis. Acta Neurol. Scand 88:97
89. Offner H, Hashim GA, Vandenbark AA (1991) T cell receptor peptide therapy triggers autoregulation of experimental encephalomyelitis. Science 251:430
90. Oksenberg JR, Panzara MA, Begovich AB, Mitchell D, Erlich HA, Murray RS, Shimonkevitz R, Sherritt M, Rothbard J, Bernard CCA, Steinman L (1993) Selection for T-cell receptor Vβ-Dβ-Jβ gene rearrangements with specificity for a myelin basic protein peptide in brain lesions of multiple sclerosis. Nature 362:68
91. Olsson T, Sun J, Hillert J, Hojeberg B, Ekre HP, Andersson G, Olerup O, Link H (1992) Increased numbers of T cells recognizing multiple myelin basic protein epitopes in multiple sclerosis. Eur J Immunol 22:1083
92. Olsson T, Wei Zhi W, H jeberg B, Kostulas V, Yu-Ping J, Anderson G, Ekre HP, Link H (1990) Autoreactive T lymphocytes in multiple sclerosis determined by antigen-induced secretion of interferon-γ. J Clin Invest 86:981
93. O'Neill JK, Butter C, Baker D, Gschmeissner SE, Kraal G, Butcher EC (1991) Expression of vascular addressins and ICAM-1 by endothelial cells in the spinal cord during chronic relapsing experimental allergic encephalomyelitis in the Biozzi AB/H mouse. Immunology 72:520
94. Ota K, Matsui M, Milford EL, Mackin GA, Weiner HL, Hafler DA (1990) T-cell recognition of an immunodominant myelin basic protein epitope in multiple sclerosis. Nature 346:183
95. Owhashi M, Heber-Katz E (1989) Protection from experimental allergic encephalomyelitis conferred by a monoclonal antibody directed against a shared idiotype on rat T cell receptors specific for myelin basic protein. J Exp Med 168:2153
96. Panitch HS, Hirsch RL, Schindler J, Johnson KP (1987) Treatment of multiple sclerosis with gamma interferon: exacerbations associated with activation of the immune system. Neurology 37:1097
97. Paty DW, Li DKB, the UBC MS/MRI Study Group and the JFNB MS Study Group (1993) Interferon beta-1b is effective in relapsing-remitting multiple sclerosis. II. MRI analysis results of a multicenter, randomized, double-blind, placebo-controlled trial. Neurology 43:662
98. Pette M, Fujita K, Kitze B, Whitaker JN, Albert E, Kappos L, Wekerle H (1990) Myelin basic protein-specific T lymphocyte lines from MS patients and healthy individuals. Neurology 40:1770
99. Pettinelli CB, McFarlin DE (1981) Adoptive transfer of experimental allergic encephalomyelitis in SJL/J mice after in vivo activation of lymph node cells by myelin basic protein: requirement for Lyt-1+2$^-$T lymphocytes. J Immunol 127:1420
100. Prineas JW (1985) The neuropathology of multiple sclerosis. Handb Clin Neurol 47:213
101. Racke MK, Dhib-Jalbut S, Cannella B, Albert PS, Raine CS, McFarlin DE (1991) Prevention and treatment of chronic relapsing experimental allergic encephalomyelitis by transforming growth factor-β1. J Immunol 146:3012
102. Racke MK, Cannella B, Albert P, Sporn M, Raine CS, McFarlin DE (1992a) Evidence of endogenous regulatory function of transforming growth factor-β1 in experimental allergic encephalomyelitis. Int Immunol 4:615
103. Racke MK, Martin R, McFarland HF, Fritz RB (1992) Copolymer-1-induced inhibition of antigen-specific T cell activation: interference with antigen presentation. J Neuroimmunol 37:75
104. Racke MK, Sriram S, Carlino J, Cannella B, Raine CS, McFarlin DE (1993) Long-term treatment of chronic relapsing experimental allergic encephalomyelitis by transforming growth factor-β2. J Neuroimmunol 46:175
105. Racke MK, Bonomo A, Scott DE, Cannella B, Levine A, Raine CS, Shevach EM, Röcken M (1994a) Cytokine-induced immune deviation as a therapy for inflammatory autoimmune disease. J Exp Med 180:1961
106. Racke MK, Quigley L, Cannella B, Raine CS, McFarlin DE, Scott DE (1994) Superantigen modulation of experimental allergic encephalomyelitis: activation or anergy determines outcome. J Immunol 152:2051
107. Racke MK, Burnett D, Pak S-H, Albert PS, Cannella B, Raine CS, McFarlin DE, Scott DE. (1995) Retinoid treatment of experimental allergic encephalomyelitis. IL-4 production correlates with improved disease course. J Immunol 154:450
108. Raine CS (1983) Multiple sclerosis and chronic relapsing EAE: comparative ultrastructural neuropathology. In: Hallpike JF, Adams CW, Tourtelotte WW (eds) Multiple sclerosis, Williams & Wilkins, Baltimore, pp 413–478

109. Raine CS, Scheinberg LC (1988) On the immunopathology of plaque development and repair in multiple sclerosis. J Neuroimmunol 20:189
110. Raine CS, Cannella B, Duijvestijn AM, Cross AH (1990) Homing to central nervous system vasculature by antigen-specific lymphocytes. II. Lymphocyte/endothelial cell adhesion during the initial stages of autoimmune demyelination. Lab Invest 63:476
111. Renno T, Krakowski M, Piccirillo C, Lin J, Owens T (1995) TNF-α expression by resident microglia and infiltrating leukocytes in the central nervous system of mice with experimental encephalomyelitis. J Immunol 154:944
112. Richert JR, Robinson ED, Deibler GE, Martenson RE, Dragovic LJ, Kies MW (1989) Human cytotoxic T-cell recognition of a synthetic peptide of myelin basic protein. Ann Neurol 26:342
113. Rott O, Wekerle H, Fleischer B (1992) Protection from experimental allergic encephalomyelitis by application of a bacterial superantigen. Int Immunol 4:347
114. Rott O, Cash E, Fleischer B (1993) Phosphodiesterase inhibitor pentoxifylline, a selective suppressor of T helper type1- but not type 2-associated lymphokine production, prevents induction of experimental autoimmune encephalomyelitis in Lewis rats. Eur J Immunol 23:1745
115. Rott O, Fleischer B, Cash E (1994) Interleukin-10 prevents experimental allergic encephalomyelitis in rats. Eur J Immunol 24:1434
116. Ruddle NH, Bergman CM, McGrath KM, Lingenheld EG, Grunnet ML, Padula SJ, Clark RB (1990) An antibody to lymphotoxin and tumor necrosis factor prevents transfer of experimental allergic encephalomyelitis. J Exp Med 172:1193
117. Schiffenbauer J, Johnson HM, Butfiloski EJ, Wegrzyn L, Soos JM (1993) Stapylococcal enterotoxins can reactivate experimental allergic encephalomyelitis. Proc Natl Acad Sci USA 90:8543
118. Schlüsener H, Wekerle H (1985) Autoaggressive T lymphocyte lines recognize the encephalitogenic region of myelin basic protein; in vitro selection from unprimed rat T lymphocyte populations. J Immunol 135:3128
119. Schmied M, Breitschopf H, Gold R, Zischler H, Rothe G, Wekerle H, Lassmann H (1993) Apoptosis of T lymphocytes in experimental autoimmune encephalomyelitis: Evidence for programmed cell death as a mechanism to control inflammation in the brain. Amer. J Pathology 143:446
120. Selmaj K, Raine CS (1988) Tumor necrosis factor mediates myelin and oligodendrocyte damage in vitro. Ann Neurol 23:339
121. Selmaj K, Raine CS, Cross AH (1991) Anti-tumor necrosis factor therapy abrogates autoimmune demyelination. Ann Neurol 30:694
122. Semmler J, Wachtel H, Endres S (1993) The specific type IV phosphodiesterase inhibitor rolipram suppresses tumor necrosis factor-alpha production by human mononuclear cells. Int J Immunopharmacol 15:409
123. Semnani RT, Nutman TB, Hochman P, Shaw S, van Seventer GA (1994) Costimulation by purified intercellular adhesion molecule 1 and lymphocyte function-associated antigen 3 induces distinct proliferation, cytokine and cell surface antigen profiles in human "naive" and "memory" $CD4^+$ T cells. J Exp Med 180:2125
124. Sercarz EE, Lehmann PV, Ametani A, Benichou G, Miller A, Moudgil K (1993) Dominance and crypticity of T cell antigenic determinants. Annu Rev Immunol 11:729
125. Sharief MK, Hentges R (1991) Association between tumor necrosis factor-α and disease progression in chronic progressive multiple sclerosis. New Engl. J Med 325:467
126. Sharma SD, Nag B, Su XM, Spack E, Clark BR, Sriram S (1991) Antigen-specific therapy of experimental allergic encephalomyelitis by soluble class II major histocompatibility complex-peptide complexes. Proc Natl Acad Sci USA 88:11465
127. Smilek DE, Wraith DC, Hodgkinson S, Dwevedy S, Steinman L, McDevitt HO (1991) A single amino acid change in a myelin basic protein peptide confers the capacity to prevent rather than induce experimental autoimmune encephalomyelitis. Proc Natl Acad Sci USA 88:9633
128. Sobel RA, Mitchell ME, Fondren G (1990) Intercellular adhesion molecule-1 (ICAM-1) in cellular immune reactions in the human central nervous system. Am J Path 136:1309
129. Sommer N, Löschmann P-A, Northoff GH, Weller M, Steinbrecher A, Steinbach JP, Lichtenfels R, Meyermann, R, Riethmüller A, Dichgans J, Martin R (1994) Tumor necrosis factor-directed therapy of autoimmune encephalomyelitis by the phosphodiesterase IV inhibitor rolipram. J Neuroimmunol 54:198

130. Sommer N, Löschmann P-A, Northoff GH, Weller M, Steinbrecher A, Steinbach JP, Lichtenfels R, Meyermann R, Riethmüller A, Dichgans J, Martin R (1995) Cytokine-directed suppression of autoimmune encephalomyelitis by the antidepressant rolipram. Nature Med 1:244
131. Steinman L, Rosenbaum J, Sriram S, McDevitt HO (1981) In vivo effects of antibodies to immune response gene products: Prevention of experimental allergic encephalomyelitis. Proc Natl Acad Sci USA 78:7111
132. Sun D, Qin Y, Chluba J, Epplen JT, Wekerle H (1988) Suppression of experimentally induced autoimmune encephalomyelitis by cytotoxic T-T cell interactions. Nature 332:843
133. Tabi Z, McCombe P, Pender MP (1994) Apoptotic elimination of Vβ8.2+ cells from the central nervous system during recovery from experimental autoimmune encephalomyelitis induced by the passive transfer of Vβ8.2+ encephalitogenic T cells. Eur J Immunol 24:2609
134. Tan LJ, Kennedy MK, Dal CM, Miller SD (1991) Successful treatment of paralytic relapses in adoptive experimental autoimmune encephalomyelitis via neuroantigen-specific tolerance. J Immunol 147:1797
135. Tan LJ, Kennedy MK, Miller SD (1992) Regulation of the effector stages of experimental autoimmune encephalomyelitis via neuroantigen-specific tolerance induction. II. Fine specificity of effector T cell inhibition. J Immunol 148:2748
136. Teitelbaum D, Aharoni R, Arnon R, Sela M (1988) Specific inhibition of the T-cell response to myelin basic protein by the synthetic copolymer Cop-1. Proc Natl Acad Sci USA 85:9724
137. Teitelbaum D, Meiner Z, Brenner T, Abramsky O, Kott E, Schechter D, Gutman B, Nisipeanu P, Korczyn AD, Klein K, Fletcher S, Arnon R, Sela M (1994) Immunological parameters in a multicenter clinical trial of COP-1 in multiple sclerosis (MS): a 2 year follow-up. Neurology 44 [Suppl. 2]: A358
138. The IFNB Multiple Sclerosis Study Group (1993) Interferon beta-1b is effective in relapsing-remitting multiple sclerosis. I. Clinical results of a multicenter, randomized, double-blind, placebo-controlled trial. Neurology 43:655
139. Topham DJ, Nag B, Arimilli S, Sriram S (1994) A synthetic peptide from the third hypervariable region of major histocompatibility complex class II β chain as a vaccine for treatment of experimental autoimmune encephalomyelitis. Proc Natl Acad Sci USA 91:8005
140. Tuohy VK, Lu Z, Sobel RA, Laursen RA, Lees MB (1988) A synthetic peptide from myelin proteolipid protein induces experimental allergic encephalomyelitis. J Immunol 141:1126
141. Urban JL, Kumar V, Kono DH, Gomez C, Horvath SJ, Clayton J, Ando DL, Sercarz EE, Hood L (1988) Restricted use of the T cells receptor V genes in murine autoimmune encephalomyelitis raises possibilities for antibody therapy. Cell 54:577
142. Utz U, Biddison WE, McFarland HF, McFarlin DE, Flerlage M, Martin R (1993) Skewed T cell receptor repertoire in genetically identical twins with multiple sclerosis correlates with disease. Nature 364:243
143. Utz U, Brooks JA, McFarland HF, Martin R, Biddison WE (1994) Heterogeneity of T-cell receptor α-chain complementarity-determining region 3 in myelin basic protein-specific T cells increases with severity of multiple sclerosis. Proc Natl Acad Sci USA 91:5567
144. Vandenbark AA, Hashim G, Offner H (1989) Immunization with a synthetic T-cell receptor V-region peptide against experimental autoimmune encephalomyelitis. Nature 341:541
145. Voskuhl RR, Martin R, Bergman C, Dalal M, Ruddle NH, McFarland HF (1993) T helper 1 (TH1) functional phenotype of human myelin basic protein-specific T lymphocytes. Autoimmunity 15:137
146. Waldor MK, Sriram S, Hardy R, Herzenberg LA, Herzenberg LA, Lanier L, Lim M, Steinman L (1985) Reversal of experimental allergic encephalomyelitis with a monoclonal antibody to a T cell subset marker (L3T4). Science 227:415
147. Watanabe R, Wege H, Meulen V ter (1983) Adoptive transfer of EAE-like lesions from rats with coronavirus-induced demyelinating encephalomyelitis. Nature 305:150
148. Weiner HL, Mackin GA, Matsui M, Orav EJ, Khoury SJ, Dawson DM, Hafler DA (1993) Double-blind pilot trial of oral tolerization with myelin antigens in multiple sclerosis. Science 259:1321
149. Wekerle H, Linington C, Lassmann H, Meyermann R (1986) Cellular immune reactivity within the CNS. Trends Neuro Sci 9:271
150. Welsh CT, Rose JW, Hill KE, Townsend JJ (1993) Augmentation of adoptively transferred experimental allergic encephalomyelitis by administration of a monoclonal antibody specific for LFA-1α. J Neuroimmunol 43:161

151. Welsh J, Sapatino B, Rosenbaum B, Smith R, Linthicum S (1993) Correlation between susceptibility to demyelination and interferon-γ induction of major histocompatibility complex class II antigens on murine cerebrovascular endothelial cells. J Neuroimmunol 48:91
152. Wilcox CE, Ward AMV, Evans A, Baker D, Rothlein R, Turk JL (1990) Endothelial cell expression of the intercellular adhesion molecule-1 (ICAM-1) in the central nervous system of guinea pigs during acute and chronic relapsing experimental allergic encephalomyelitis. J Neuroimmunol 30:43
153. Windhagen A, Scholz C, Fukaura H, Sette A, Hafler DA (1994) Switch of autoreactive human T cell clones from a Th2 to a TGFβ1 secreting phenotype with low affinity T-cell receptor stimulating peptide/MHC complex. J Neuroimmunol 54:206
154. Wong GHW, Bartlett PF, Clark-Lewis I, Battye F, Schrader JW (1984) Inducible expression of H-2 and Ia antigens on brain cells. Nature 310:688
155. Wong GHW, Clark-Lewis J, Harris AW, Schrader JW (1984) Effect of cloned interferon-γ on expression of H-2 and Ia antigen on cell lines of hemopoietic, lymphoid, epithelial, fibroblastic and neuronal origin. Eur J Immunol 14:52
156. Wraith DC, Smilek DE, Mitchell DJ, Steinman L, McDevitt HO (1989) Antigen recognition in autoimmune encephalomyelitis and the potential for peptide-mediated immunotherapy. Cell 59:247
157. Wucherpfennig KW, Ota K, Endo N, Seidman JG, Rosenzweig A, Weiner HL, Hafler DA (1990) Shared human T cell receptor V beta usage to immunodominant regions of myelin basic protein. Science 248:1016
158. Yednock TA, Cannon C, Fritz LC, Sanchez-Madrid F, Steinman L, Karin N (1992) Prevention of experimental autoimmune encephalomyelitis by antibodies against $\alpha 4\beta 1$ integrin. Nature 356:63
159. Yudkin PL, Ellison GW, Ghezzi A, Goodkin DE, Hughes RAC, McPherson K, Mertin J, Milanese C (1991) Overview of azathioprine in multiple sclerosis. Lancet 338:1051
160. Zaller DM, Osman G, Kanagawa O, Hood L (1990) Prevention and treatment of murine experimental allergic encephalomyelitis with T cell receptor V beta-specific antibodies. J Exp Med 171:1943
161. Zamvil SS, Steinman L (1990) The T lymphocyte in experimental allergic encephalomyelitis. Annu Rev Immunol 8:579
162. Zhang J, Medaer R, Stinissen P, Hafler DA, Raus J (1993) MHC-restricted depletion of human myelin basic protein-reactive T cells by T cell vaccination. Science 261:1451
163. Zhang J, Markovic-Plese S, Lacet B, Raus J, Weiner HL, Hafler DA (1994) Increased frequency of interleukin 2-responsive T cells specific for myelin basic protein in peripheral blood and cerebrospinal fluid of patients with multiple sclerosis. J Exp Med 179:973

The immune response to infectious diseases of the central nervous system: a tenuous balance

Andrew R. Pachner

Department of Neurology, Georgetown University Medical Center, 3800 Reservoir Road NW, Washington DC 20007-2197, USA

Introduction

Imagine that you are visiting a large computerized factory making rocket parts. Access to the factory is controlled by security guards at the entrance and the hustle, bustle, noise, heat, and grime of the main construction floor bombards your senses as soon as you walk in. You then are led through a narrow corridor to the computer control room. Entrance to this room is only allowed by special arrangements. You enter a quiet, cool room in which you see the large computers whirring and calm, white-coated computer specialists at the keyboards. In this room the turmoil of the main floor would be devastating to the function of the "nerve" center of the factory.

Similarly, in the human body, the daily functions of life, with their physical and chemical turmoils, are separated in space from the "computer control center", the brain. The body has gone to great lengths evolutionarily to protect the brain from the hustle and bustle of the rest of the body. In no aspect of the body's tumultuous operations is this more obvious than in the immune response. Cells of the central nervous system (CNS) do not generally express antigens of the major histocompatibility complex (MHC). Access to the CNS by immune cells and molecules within the blood is very limited by the blood-brain barrier (BBB). The potent forces of the immune system are kept out of the CNS just as the hustle, bustle, noise, and dirt of the assembly floor are kept out of the computer control room.

However, terrorists can sometimes infiltrate the factory's front gate and the security of the computer control room. When this happens, the goal of the factory's security is to rid the factory of the terrorists with the minimal damage to the control room. Similarly, infection and inflammation can disturb the CNS and the consequences can be dire; the application of potentially very powerful forces in the sensitive control center can cause damage to both the invader and the host. Thus, infection in the CNS results in two potentially conflicting demands on the immune response: to eradicate the infection and to limit the resultant damage to CNS tissue to a minimum. This is the "tenuous balance" which the host must achieve attempt to in all CNS infections.

The immune response has a variable level of success in reconciling these conflicting demands. There are clearly host factors that determine the balance. The most dramatic clinical example of this can be seen in leprosy, in which lepromatous and tuberculous forms of the infection represent two extremes in the immune respone to the same organism. The balance is also determined by the nature of the infecting agent. In some infections, such as acute viral encephalitis, as in Western equine or acute herpes simplex encephalitis (HSE), an inadequate immune protective response results in damage from rapid spread of a cytolytic virus. In other infections, e.g, Theiler's virus in mice, damage to the CNS is caused by an excessive immune response; CNS lesions can be avoided by immunosuppression. Another variable is that of duration of infection: some infectious agents are rapidly destructive once they enter the CNS (i.e., Western equine encephalitis virus), while others are chronic infections which may remain latent for years (i.e., *Treponema pallidum*).

It should be stressed here that most of the strong anti-infectious weaponry of the body is present outside the CNS, and thus, compared to the high incidence of infections in extraneural tissues, it is rare for an invading infectious agent to penetrate into the CNS, and there cause significant disease. Thus, the reticuloendothelial system and the cellular and humoral immune system are potent forces which usually slow down and eventually clear any invaders. In addition, the blood-brain barrier also effectively limits penetration of infectious organisms into the brain. Thus, the vast majority of infections, even with highly neurotropic infectious agents, do not result in CNS infection; the invader is eliminated before he can even get to the control room.

Acute encephalitis

Sometimes invasion of the brain by an infection is so aggressive that the immune response is overwhelmed. In such a situation havoc is wrought by the rapid replication and spread of the virus.

The most commonly implicated agent of encephalitis in humans is herpes simplex virus (HSV). Invasion can be rapid and the immune system frequently has little time to respond. In a recently developed model of HSE in SJL mice [25], mononuclear cells infiltrating the infected temporal lobes were predominantly T cells, $CD4^+$ and $CD8^+$ in equal numbers. Many of them were cytotoxic. Cytotoxic T cells in contrast could not be isolated from spleens or lymph nodes. These data demonstrate that the immune response to acute infection may be highly compartmentalized. In another model of HSV type 1 (HSV-1) infection of the brain in BALB/c mice, in situ hybridization studies documented the progressive spread through the trigeminal ganglion, perirhinal cortex, amygdala, and hippocampus. The route of inoculation appears to be critical in this model since inoculation into cerebral ventricles, oral cavity, and peritoneal cavity does not mimic the spread of infection in humans, while inoculation of virus into tooth pulp does [9].

The relative importance of cellular versus humoral immunity in this infection is not known. Anti-HSV antibodies develop in infected animals and humans. Although their role in the immune response to HSV appears minor relative to cellular immunity, their presence in the cerebrospinal fluid (CSF) of infected humans can be helpful diagnostically [6, 7].

HSV also is present in the brain in a latent form, generally in the trigeminal ganglion. In mice recovering from acute encephalitis the virus was detectable in brain parenchyma as well as the trigeminal ganglion [31], raising the question of whether parenchymal latency of the virus exists in humans. The factors, immune and otherwise, that determine latency versus active replication in HSV infection, as in other infections in the CNS, remain unknown. A similar situation applies to varicella zoster virus (VZV) which can become latent in dorsal root ganglion after the primary infection, and re-activate many years later to cause shingles. It is possible that reactivation may be due to deficiencies in immune surveillance; particularly aggressive forms of VZV infection of the CNS in patients with AIDS have recently been reviewed [20].

Viral meningitis

Viral meningitis, the most common infection of the CNS, is usually caused by enteroviruses, adenoviruses, togaviruses, mumps, or lymphocyte choriomeningitis virus. Virus is almost always limited to the leptomeninges, and a strong, inflammatory response there limits the spread of the organism. Generally, in less than 2 weeks, the virus is nearly eradicated, and clinical improvement is rapid, without significant clinical sequelae. Rarely, severe encephalomyelitis may occur.

The benignity of viral meningitis is to a great extent determined by the fact that the CNS is not the primary site of multiplication of the causative viruses. As in models of HSE mentioned above, the route of infection is critical. The skin, pharynx, or alimentary tract are the portals of entry at which point the virus initially multiplies. During this period the immune response responds with selection and expansion of virus-specific lymphocytes and generation of virus-specific antibody; local immunity, such as IgA antibody and reticuloendothelial structures, can also operate during this time. The virus eventually disseminates in the blood stream, reaching the leptomeninges, but by that time the immune response is generally fairly well primed, and the inflammation in the meninges effectively results in clearance of the virus.

Infections targeting specific cell types in the nervous system

The most dramatic examples of infections targeting specific cell types within the CNS are progressive multifocal leukoencephalopathy (PML), polio and its chronic counterpart, post-polio syndrome. The clinical manifestations of these infections are directly related to their cell tropism within the CNS. The interplay between the immune system and the virus is poorly understood in each of these infections. For both infections "adequate" immunity prevents the spread of the virus to the nervous system; once established in the CNS, however, immune mechanisms are less potent because of the ability of these viruses to penetrate cells and infect chronically.

PML, first described in 1958 [4] is caused by the JC virus (JCV), which targets oligodendroglia. The predilection of JCV for oligodendroglia may be due to oligo-specific transcription factors which are permissive for expression of the JCV genome [2]; these nuclear factors may also be present within B cells, a common site for JCV infection outside the CNS [5]. Patients with PML develop, over the course of a few months, focal abnormalities including weakness and ataxia, confusion, aphasia, problems with coordination, and sensory loss; these findings are associated with multiple areas of demyelination on a magnetic resonance imaging (MRI) scan. The disease is seen predominantly in immunosuppressed individuals, and has increased greatly in prevalence because of the AIDS epidemic. PML is particularly aggressive in AIDS patients, with 50% of patients dying within 4 months of diagnosis [18]. JCV is a ubiquitous organism, and antibodies to the virus are found in approximately two thirds of healthy adults [43]. PML is felt to occur because of activation of latent infection and subsequent lack of immune-mediated suppression of increasing JCV numbers and virulence. Patients with PML have not only diffuse immunosuppression, caused by AIDS, Hodgkin's disease, or CLL but also have selective cellular deficiencies in the response to the JCV [54]. Immunosuppression seems to be a necessary, but not sufficient, condition for PML. The theory that PML is due to reactivation of latent infection is supported by the findings that JC viruria can occur in immunosuppressed individuals without PML, and the JC virus is present in peripheral blood lymphocytes in 38% of patients who are HIV-1 seropositive but in none of the immunocompetent patients with Parkinson's disease [51]. Another item of data in support of the importance of constant immune surveillance of this infection is that the immune response to the infection in the brain is predominantly macrophage infiltration; when lymphocytes are present the course of the disease tends to be less malignant [32].

Latent infection is also an important consideration in poliovirus infection, because of the increasing numbers of patients with post-polio syndrome. These patients have had symptomatic polio when younger leading to some impairment of motor neuron function. After an asymptomatic hiatus of many years, their neuromuscular problems re-activate, i.e., they develop new weakness and atrophy and acute findings on electromyography in previously quiescent muscles [14]. The etiology of post-polio syndrome is unknown, but may be due to reactivation of chronic persistent poliovirus infection [33]. In acute poliovirus infections, cell tropism is due to the presence of a poliovirus receptor (PVR), a member of the immunoglobulin superfamily, on anterior horn cells. Poliovirus and its interaction with PVR are currently a focus of interest in a number of laboratories.

Because of the clinical similarities between post-polio syndrome and amyotrophic lateral sclerosis (ALS), investigators have searched for virus sequences in the spinal cords from patients with ALS. Enteroviral sequences closely related to coxsackie B virus have recently been detected in the spinal cords from ALS patients, raising the possibility that ALS is related to a chronic persistent infection of motor neurons [55].

Other examples of persistent but latent infection within the nervous system abound. *Toxoplasma gondii* and cytomegalovirus are both neurotropic pathogens, which are probably latent infections and which activate during immunosuppression to cause CNS disease. In all of these infections, it is primarily the cellular immune response, dominated by the T lymphocyte, which is responsible for surveillance and control. Thus, these are infections which commonly affect the CNS in patients with AIDS, in whom T cell function is poor.

Chronic infection within the CNS-spirochetal disease

CNS syphilis and Lyme neuroborreliosis represent infections of the CNS in which the immune mechanisms are not potent enough to completely clear the infection in a substantial percentage of patients. Fortunately, in both of these diseases, serious CNS damage associated with late neurological infection is not a predictable outcome of initial infection [1, 13, 22, 23, 30, 38, 47]. However, a substantial percentage of patients develop long-term complications either because of persistent inflammation or even in its apparent absence [50]. The factors which dictate CNS latency with inflammation and CNS disease in one patient, versus CNS latency without damage in another, or ultimate CNS clearence in a third are unknown.

Both infections begin with skin manifestations following initial exposure of skin or mucous membrane to infectious agent, and subsequent spirochetemia with seeding of target organs. More extensive animal experimentation has been performed in Lyme borreliosis than in syphilis because *Borrelia burgdorferi* can be readily cultivated and expanded in vitro, while *Treponema pallidum* cannot. In murine Lyme borreliosis, dissemination to the brain follows the relatively slow development of antibody in peripheral blood [35, 36]. Culture positives in the brain are rare, but polymerase chain reaction (PCR) positivity is common, and a significant percentage of infected mice develop poor maze performance after infection (Pachner, preliminary observations).

The best animal model of Lyme neuroborreliosis is in Rhesus macaques [41, 42]. Within the 1st week after inoculation of infectious spirochetes intradermally, specific antibody appears in the peripheral blood. Erythema migrans follows shortly thereafter and can last for weeks. Within the first few weeks, spirochetemia develops despite increasing amounts of anti-*B. burgdorferi* antibody directed at multiple bacterial proteins. Reproducibly, at 4–5 weeks after inoculation, the CNS is invaded, as demonstrated by culture and PCR positivity (for *B. burgdorferi* DNA) and pleocytosis of the CSF. Once the spirochetes enter the CSF and inflammation begins, CSF at subsequent days and weeks are no longer culture positive. However, CSF PCR positivity persists as does CSF pleocytosis, and in postmortem analyses *B. burgdorferi* can be found throughout the neuraxis by PCR [40]. Despite the extensive inflammation which persists for months, and the continued presence of spirochete in the brain, immunocompetent monkeys develop no overt signs of chronic CNS infection. Careful neuropsychological studies of infected animals have not been done, and it is possible that higher cognitive functions, such as memory or learning, may be impaired. Mild encephalopathy is a major sequela of Lyme borreliosis in humans [13, 21, 23, 28, 30, 37] and can be a persistent symptom even after treatment [50].

The immune response to *B. burgdorferi* in animal models is highly dependent on route of inoculation [36]. When an infective strain is injected intradermally in mice, systemic infection occurs despite the development of a strong antibody response [35]. With low to moderate inocula, antibody to the major protein of the spirochete, the outer surface protein A (OspA), does not develop, while injection of heat-killed spirochetes or spirochetal sonicate results primarily in an anti-OspA response. This situation is similar to human infection, where an anti-OspA response is absent until many months to years after initial infection. It appears likely that OspA gene expression is down-regulated early in the infection in the skin; a major change in the OspA gene seems unlikely since the gene sequenced from infected brains in the rabbit model has an identical sequence to the original infecting strain [39]. This ability to down-regulate major surface immunogenic proteins, along with the ability of the organism to penetrate endothelial cells and wrap around collagen fibers, accounts to some extent for its ability to persist in tissue, including the CNS [42], despite the presence of a strong immune response by the host.

Post viral inflammatory syndromes and the immune response as a "loose cannon"

The generally mild course of viral infections of the CNS can be contrasted to the frequently devastating effects of postinfectious encephalomyelitis [27, 46]. Its relatively common clinical occurrence has been decreased by the discontinuation of smallpox vaccines and the widespread use of attenuated live measles vaccine. Currently postinfectious encephalomyelitis is seen most frequently after viral upper respiratory infections without identification of the causative virus. It is generally believed that it is caused by virally triggered initiation of a cell-mediated autoimmune response against CNS tissue.

Important analogs to human postinfectious encephalomyelitis are viral models of chronic demyelination in mice, particularly those due to Theiler's murine encephalomyelitis virus (TMEV) or mouse hepatitis virus (MHV). In these infections, it is the immune response and not the direct effects of the infectious agent which cause most damage. These murine models of postinfectious encephalomyelitis require a critical combination of virus phenotype (DA strain of TMEV) and host genetic immune response (SJL/J mice). Unlike the GDVII strain of TMEV, which causes an acute encephalitis, and is subsequently cleared, the DA strain causes persistent infection and subsequent demyelination; critical areas in the capsid protein seem to be responsible for the phenotype of persistence [26]. Demyelination appears to result from a nonspecific bystander response in a virus-specific delayed-type hypersensitivity (DTH) reaction within the CNS [45]. The inflammatory demyelination caused by TMEV in SJL mice is dependent on $CD4^+$, DTH-mediating, TMEV-specific T cells specific for a virus capsid protein [16, 17]. This mechanism is consistent with data that implicates the H-2 gene as critical for viral persistence and subsequent demyelination [8]. Interestingly, T cells are also necessary for clearance of virus from the brain [29]. Susceptibility to TMEV has been correlated with the ability of interferon-γ (IFN-γ) to induce expression of class II MHC antigens on cerebrovascular endothelial cells [52]. Demyelination in TMEV encephalitis is correlated with decreased proteolipid gene expression [48]. Thus, in human postinfectious encephalomyelitis, inflammation and CNS damage may be similarly dependent on the action of virus-specific T-cells within the brain and spinal cord. Not surprisingly, immunosuppressive therapy, such as high-dose corticosteroids, is usually successful.

Infection outside the CNS

The CNS can be affected not only by local infection, but also by neurotoxic factors released by infection and inflammation outside the CNS. Clinically revelant examples of this abound. Probably the most dramatic and commonly encountered clinical example of this is encephalopathy (i.e., confusion, agitation, disorientation sometimes with lethargy) due to sepsis. In septic encephalopathy, direct infection of the brain is absent, but as-yet-uncharacterized soluble products of sepsis, which cross the blood-brain barrier, mediate the neurotoxic effect. Another well-characterized clinical situation is that of fever in which a systemically circulating immune product, the cytokine interleukin-1 (IL-1), produced primarily by macrophages, results in a neuroendocrine result. "Toxic encephalopathies" clear when the inciting process outside the CNS is resolved. We currently know very little about the molecular nature of neurotoxic "humors" but the rapidly increasing clinical importance of AIDS-dementia complex (ADC) has brought this phenomenon into the limelight.

The CNS is considered by many to be a major reservoir for HIV, and the virus productively replicates in brain macrophages and microglia. The most obvious potential pathogenesis of ADC, direct HIV infection of neural cells, is supported by some studies [53] but not by others [10]. Direct disruption in neuronal infection by HIV itself may not be the primary cause of encephalopathy and many investigators postulate that toxins of some sort are the primary effectors of ADC. Toxins may have a number of effects including causing damage to myelin, affecting neuronal function, or disrupting neurotransmitter release, or a combination of the above. Tumor necrosis factor-α (TNF-α) has also been implicated in ADC. Using reverse transcriptase PCR to detect mRNA for TNF-α in the brains of patients with ADC, Glass et al. [19] found elevated levels of TNF-α mRNA. Many investigators in this field believe that TNF or other cytokines, possibly in conjunction with other neurotoxins and permeability of the BBB, cause ADC [11, 15].

The role of cytokines in CNS infections

There is considerable evidence that normally quiescent CNS cells can express on their surface or secrete immunologically relevant molecules when stimulated by a local immune response. The nature of the bidirectional effects of neural cells and molecules interacting with immunological cells and molecules has generated an increasing amount of research in the past decade.

Since the first characterization of IL-1 in 1980, cytokines have increasingly been searched for and found within the brain [34]. The most commonly implicated and most studied cytokines in neuropathology are IL-1, IL-2, IL-6, TNF-α, lymphotoxin (TNF-β) IFN-γ, and transforming growth factor-β. Since cytokines are molecules which generally function locally, and can have inhibitors, they may not be easy to detect in body fluids. Thus, measurement of cytokines in the CSF and serum in multiple sclerosis (MS) have not been particularly helpful [12, 14], although some elevations have been found [24]. TNF-α has attracted the most interest from neuroscientists, since this cytokine has been shown to be cytotoxic for oligodendrocytes and responsible for myelin degradation in vitro. Using immunohistochemistry, TNF-α has been found to be associated with astrocytes and macrophages in the center of MS lesions and in endothelial cells at the lesion edge [49]. On the other hand, based on in vitro studies it has been postulated that $CD4^+$ T cells do not need to secrete TNF to lyse oligodendrocytes [3]. The role of cytokines in the CNS, both those produced locally and by infiltrating immune cells, will be a major area of investigation in the coming decade.

Summary

The outcome of CNS infection is dependent on both the organism and host. Acute infections are usually cleared with minimal residua. However, chronic infections of the CNS, such as HIV, PML, HSV, polio, Lyme neuroborreliosis, and neurosyphilis, are becoming increasingly recognized as a cause of severe neurological morbidity, and are poorly understood. We will need to learn more about the CNS as an immune compartment to increase our knowledge of these infections.

References

1. Ackermann RE, Gollmer E, Rehse-Kupper B (1985) Progressive Borrelien-Enzephalomyelitis. Chronische Manifestation der Erythema-chronicum-migrans Krankeit am Nervensystem. Dtsch Med Wochenschr 110:1039
2. Amemiya K, Traub R, Durham L, Major EO (1992) Adjacent nuclear factor-1 and activator protein binding sites in the enhancer of the neurotropic JC virus. J Biol Chem 267:14204
3. Antel JP, Williams K, Blain M, McRea E, McLaurin J (1994) Oligodendrocyte lysis by $CD4^+$ T cells is independent of TNF. Ann Neurol 35:341
4. Astrom KE, Mancall EL, Richardson EP (1958) Progressive multifocal leukoencephalopathy. Brain 81:93
5. Atwood WJ, Amemiya K, Traub R, Harms J. Major E (1992) Interaction of the human polyoma virus, JCV, with human B lymphocytes. Virology 190:716
6. Aurelius E (1993) Herpes simplex encephalitis. Early diagnosis and immune activation in the acute stage. Scand J Infect Dis [Suppl] 89:3
7. Aurelius E, Johansson B, Skoldenberg B, Forsgren M (1993) Encephalitis in immunocompetent patients due to herpes simplex virus type 1 or 2 as determined by type-specific polymerase chain reaction and antibody assays of cerebrospinal fluid. J Med Virol 39:179
8. Azoulay A, Brahic M, Bureau JF (1994) FVB mice transgenic for the H-2D gene become resistant to persistent infection by Theiler's virus. J Virol 68:4049
9. Barnett EM, Jacobsen G, Evans G, Cassell M, Perlman S (1994) Herpes simplex encephalitis in the temporal cortex and limbic system after trigeminal nerve inoculation. J Infect Dis 169:782
10. Bell JE, Busutil A, Ironside JW, et al (1993) Human immunodeficiency virus and the brain: investigation of virus load and neuropathological changes in pre-AIDS subjects. J Infect Dis 168:818
11. Brew B (1994) The clinical spectrum and pathogenesis of HIV encephalopathy, myelopathy, and peripheral neuropathy. Curr Opin Neurol 7:209
12. Carrieri P (1994) The role of cytokines in the pathogenesis of MS. Int MS J 1:53
13. Cintron R, Pachner AR (1994) Spirochetal diseases of the nervous system. Curr Opin Neurol 7:217
14. Dalakas M, Illa I (1991) Post-polio syndrome: concepts in clinical diagnosis, pathogenesis, and etiology. Adv Neurol 56:495
15. Genis P, Jett M, Bernton EW, et al (1992) Cytokines and arachidonic acid metabolites produced during human immunodeficiency virus (HIV)-infected macrophage-astroglia interactions: implications for the neuropathogenesis of HIV disease. J Exp Med 176:1703
16. Gerety SJ, Karpus WJ, Cubbon AR, et al (1994) Class II-restricted T cell responses in Theiler's murine encephalomyelitis virus-induced demyelinating disease. V. Mapping of a dominant immunopathologic VP2 T cell epitope in susceptible SJL/J mice. J Immunol 152:908
17. Gerety SJ, Rundell MK, DalCanto MC, Miller SD (1994) Class II-restricted T cell responses in Theiler's murine encephalomyelitis virus-induced demyelinating disease. VI. Potentiation of demyelination with and characterization of an immunopathologic $CD4^+$ T cell line specific for an immunodominant VP2 epitope. J Immunol 152:919
18. Gillespie SM, Chang Y, Lemp G, et al (1991) Progressive multifocal leukoencephalopathy in persons infected with human immunodeficiency virus, San Francisco, 1981–1989. Ann Neurol 30:597
19. Glass JD, Wesselingh SL, Selnes OA, McArthur JC (1993) Clinical-neuropathological correlation in HIV-associated dementia. Neurology 43:2230
20. Gray F, Belec L, Chretien F, et al (1994) Varicella zoster virus infection of the central nervous system in the acquired immune deficiency syndrome. Brain 117:987
21. Halperin JJ, Heyes MP (1992) Neuroactive kynurenines in Lyme borreliosis. Neurology 42:43
22. Halperin J, Luft BJ, Anand AK, et al (1989) Lyme neuroborreliosis: central nervous system manifestations. Neurology 39:753
23. Halperin J. Krupp LB, Golightly MG, Volkman DJ (1990) Lyme borreliosis-associated encephalopathy. Neurology 40:1340
24. Hauser SL, Doolittle TH, Lincoln R, et al (1990) Cytokine accumulations in CSF of MS patients: frequent detection of IL-1 and TNF but not IL-6. Neurology 40:1735

25. Hudson SJ, Streilein JW (1994) Functional cytotoxic T cells are associated with focal lesions in the brains of SJL mice with experimental herpes simplex encephalitis. J Immunol 152:5540
26. Jarousse N, Grant RA, Hogle JM, et al (1994) A single amino acid change determines persistence of a chimeric Theiler's virus. J Virol 68:3364
27. Johnson RT, Griffin DE (1987) Post-infectious encephalomyelitis. In: Kennedy PGE, Johnson RT (eds) Infections of the nervous system. London, Butterworths, pp 209–226
28. Kaplan RF, Meadows ME, Vincent LC, Logigian EL, Steere AC (1992) Memory impairment and depression in patients with Lyme encephalopathy. Neurology 42:1263
29. Kohanawa M, Nakane A, Asano M, Minagawa T (1994) Theiler's virus is eliminated by a gamma-interferon-independent mechanism in the brain. J Neuroimmunol 52:79
30. Logigian EL, Kaplan RF, Steere AC (1990) Chronic neurological manifestations of Lyme disease. N Engl J Med 323:1438
31. Lynas C, Hill TJ, Maitland NJ, Love S (1993) Latent infection with the MS strain of herpes simplex virus type 2 in the mouse following intracerebral inoculation. J Neurol Sci 120:107
32. Major EO, Amemiya K, Tomatore CS, Hour SA, Berger JR (1992) The pathogenesis and molecular biology of progressive multifocal leukoencephalopathy (PML): JC virus induced demyelination disease of the human brain. Clin Microbiol Rev 5:49
33. Miller JR (1981) Prolonged intracerebral infection with poliovirus in asymptomatic mice. Ann Neurol 9:590
34. Morganti-Kossman M, Kossman T, Wahl SM (1992) Cytokines and neuropathology. Trends Pharmacol Sci 13:286
35. Pachner AR, Delaney E (1993) PCR compared to culture and serology in murine experimental Lyme borreliosis. J Clin Microbiol 31:208
36. Pachner AR, Delaney E (1993) Murine Lyme borreliosis: route of inoculation determines immune response and infectivity. Reg Immunol 4:345
37. Pachner AR, Steere AC (1986) CNS manifestations of third stage Lyme disease. Zentralbl Bakteriol Hyg [A]1986; 263:301
38. Pachner AR, Duray P, Steere AC (1989) Central nervous system manifestations of Lyme disease. Arch Neurol 46:790
39. Pachner AR, Braswefl ST, Delaney E, Amemiya K, Major E (1994) A rabbit model of Lyme neuroborreliosis: characterization by PCR, serology, and sequencing of the OspA gene from brain spirochetes. Neurology 44:1938
40. Pachner AR, Delaney E, O'Neill T (1994) Lyme neuroborreliosis is a persistent infection in the central nervous system in the Rhesus model. Ann Neurol 36:299
41. Pachner AR, O'Neill TP, Delaney E, Major E (1994) inoculation of non-human primates with the N40 strain of *Borrelia burgdorferi* induces a model of Lyme neuroborreliosis faithful to the human disease. Neurology 1994; 44:1355
42. Pachner AR, Delaney E, O'Neill T, Major E (1995) Inoculation of non-human primates with the N40 strain of *Borrelia burgdorferi* leads to a model of Lyme neuroborreliosis faithful to the human disease. Neurology 45:165
43. Padgett BL, Walker DL (1973) Prevalence of antibodies inhuman sera against JC virus, an isolate from a case of progressive multifocal leukoencephalopathy. J Infect Dis 127:467
44. Peter GB, Boctor FN, Tourtellotte WW (1991) Serum and CSF levels of IL-2, sIL-2r, TNF-α, and IL-1β in chronic progressive multiple sclerosis: expected lack of clinical utility. Neurology 41:121
45. Peterson JD, Karpus WJ, Clatch RJ, Miller SD (1993) Split tolerance of Th1 and Th2 cells in tolerance to Theiler's murine encephalomyelitis virus. Eur J Immunol 23:46
46. Reik L (1982) Immune-mediated central nervous system disorders in childhood viral infections. Semin Neurol 2:106
47. Reik L (1987) Spirochetal infections of the nervous system. In: Kennedy Peter GE, Johnson RT, eds. Infections of the nervous system. London, Butterworths, pp43–75
48. Rodriguez M, Prayoonwiwat N, Howe C, Sanborn K (1994) Proteolipid protein gene expression in demyelination and remyelination of the central nervous system: a model for multiple sclerosis. J Neuropathol Exp Neurol 53:136
49. Selmaj KW, Raine CS, Cannella B, Brosnan CF (1991) Identification of lymphotoxin and tumor necrosis factor in MS lesions. J Clin Invest 87:949
50. Shadick NA, Phillips CB, Logigian EL, et al (1994) The long-term clinical outcomes of Lyme disease. Ann Intern Med 121:560

51. Tornatore C, Berger JR, Houff SA, et al (1992) Detection of JC virus DNA in peripheral lymphocytes from patients with and without progressive multifocal leukoencephalopathy. Ann Neurol 31:454
52. Welsh J. Sapatino B, Rosenbaum B, Smith R, Linthicum S(1993) Correlation between susceptibility to demyelination and interferon-gamma induction of major histocompatibility complex class II antigens on murine cerebrovascular endothelial cells. J Neuroimmunol 48:91
53. Wiley AA, Achim C (1994) Human immunodeficiency virus encephalitis is the pathologic correlate of dementia in acquired immunodeficiency syndrome. Ann Neurol 36:673
54. Willoughby ER, Price RW, Padgett BL, Walker DL, Dupont B (1980) Progressive multifocal leukoencephalopathy (PML): in vitro cell-mediated immune responses to mitogens and JC virus. Neurology 30:256
55. Woodall CJ, Riding MH, Graham DI, Clements GB (1994) Sequences specific for enterovirus detected in spinal cord from patients with motor neuron disease. BMJ 308:1541

Immune control of the brain

Monika Bradl

Max-Planck-Institut für Psychiatrie, Abteilung Neuroimmunologie, Am Klopferspitz 18a,
D-82152 Martinsried, Germany

Introduction

Immunosurveillance of the brain is the result of elaborate cellular interactions. As the first step in the process of immune surveillance, T lymphocytes of the immune system control the integrity of the central nervous system (CNS) parenchyma. They enter the brain in search of pathogens, and recruit CNS cells as antigen-presenting cells. Upon contact with "their" specific antigens, they initiate local immune responses. Then, in a second step, cells of the local glia and of the immune system communicate bidirectionally with each other and interact to jointly control the immune reactivity within the CNS.

An important tool for studying these interactions is represented by autoaggressive T cells, which recognize well-defined proteins of the nervous system and specifically migrate into the CNS upon transfer into naive animals. Here, we first describe how CNS-specific, autoaggressive T cells arise, and using the animal model of experimental allergic encephalomyelitis (EAE) show how these T lymphocytes gain access to the CNS, how they initiate the formation of inflammatory lesions, and how they cooperate in the immune response with glia cells of the CNS. Finally, we speculate on the mechanisms that result in a self limitation of immune responses within the CNS.

Autoaggressive T cells

The vast majority of T lymphocytes mature and differentiate in the thymus. There, different gene elements of the T cell receptor (TCR) rearrange randomly, giving rise to T cells with specificity for non-self and self proteins. Strong intrathymic selection processes ensure, that most of the self reactive T cells die within the thymus, possibly following the recognition of the self peptides presented by medullary stroma cells [65]. Some autoreactive T cells, however, escape from this negative selection and reach the peripheral immune system. Here, some of them are rendered unreactive (anergic), and others are kept in a safe state of rest. It is this latter population of autoreactive

T cells that is of special importance for autoimmunity against CNS antigens, since they contain the myelin-specific T cells present in the immune repertoire of healthy rodents [69] and men [59]. These autoreactive cells are usually not harmful to the organism. Due to the blood-brain barrier, they do not easily gain access to the CNS.

The blood-brain barrier

The blood-brain barrier was identified by Paul Ehrlich at the beginning of this century [20]. He administered certain dyes intravenously into experimental animals, and noticed that all organs but the brain of an experimental animal were stained by the dye. Based on these observations he proposed a barrier between brain and bloodstream that differs from barriers between other organs and blood. From ultrastructural examinations several decades later [7, 64], we know now that the blood-brain barrier is formed by the vascular interface between the blood stream and the CNS parenchyma. This interface is comprised of endothelial cells surrounded by a basal lamina, pericytes, and tightly apposed astrocyte endfeet [8]. Endothelial cells of cerebral blood vessels form a network of tight junctions [7], thus differing fundamentally from endothelia of other organs [18]. Due to this peculiarity of cerebral blood vessels, a physical barrier exists controlling the exchange of macromolecules and cells between the blood and the CNS.

How can T lymphocytes pass the blood-brain barrier and initiate immune reactions within the brain parenchyma?

Migration of T lymphocytes through the blood-brain barrier

Studies of CNS inflammation transferred by autoaggressive, myelin-specific T cells in different states of activation revealed that the blood-brain barrier effectively excludes *resting* T lymphocytes from entering the CNS parenchyma [53]. This observation fits well with the notion that autoreactive cells in a safe state of rest are not harmful to the organism. The same studies, however, convincingly demonstrated that *activated* myelin-specific T cells are able to migrate through the blood-brain barrier, to initiate CNS lesions and to mediate clinically overt autoimmune disease [53]. It has further been shown that activated T cells enter the CNS parenchyma irrespective of their antigen specificity [35, 80]. Isotope labeled T cells specific for a CNS antigen (myelin basic protein, MBP) or a non-CNS antigen (ovalbumin) were used. These cells were activated in vitro with their specific antigen and subsequently injected intraveneously into naive Lewis rats. The cellular migration pattern to the CNS was monitored by autoradiography of histological sections. Interestingly, MBP as well as ovalbumin-specific activated T cell blasts could be found in the CNS within a few hours of injection [80]. Thus, it has been unequivocally shown that:

1. Activated, but not resting T cells are able to migrate through the blood-brain barrier [80]

2. Activated T cells can enter the CNS parenchyma, irrespective of their antigen specificity.

Other experiments have convincingly demonstrated that activated T cells are even able to enter the CNS parenchyma when the brain endothelial cells are mis-matched with the antigen presentation requirements of the T cells examined [36]. Thus, it is probably not a specific antigen recognition on the luminal surface of CNS endothelial cells that is responsible for the entrance of activated T cells into the brain parenchyma. Rather, the presence or absence of certain cell adhesion molecules (CAM) on the CNS vascular endothelial cells seems to account for the observed differences in the migratory behavior of activated and resting T lymphocytes [11, 16]. Current concepts favor the presence of brain-endothelial cell-specific CAM profiles, although the exact nature of these profiles is still elusive [79].

It is now quite obvious that the blood-brain barrier is, after all, not such a dense barrier. However, it does radically limit the number of immune cells patrolling the CNS by effectively excluding resting T cells from the CNS parenchyma, and restricts the entry to a minority of circulating T cells, which are activated and represent the spectrum of current immune reactions in the organism. This has consequences for immunosurveillance *and* the development of autoimmune diseases of the CNS.

Development of autoimmune diseases of the CNS

As stated previously, the healthy immune repertoire contains T cells with specificity for many different myelin or glia autoantigens. These T cells are in a state of rest, and they cannot, therefore, overcome the blood-brain barrier. In addition, under normal conditions, they cannot be activated by their respective antigens because most of these proteins are hidden behind the blood-brain barrier.

However, spontaneous autoimmune diseases of the CNS occur, albeit at a rather low frequency. In these cases, autoreactive T cells have been activated and their autoaggressive potential turned against the body's own tissues. Therefore, in the absence of the antigen proper, other mechanisms must exist resulting in an activation of autoreactive T lymphocytes. The most obvious candidates for this are molecular mimicry reactions or T cell activation by bacterial superantigens. The term molecular mimicry describes a situation, in which some microbial agents share determinants with host proteins. If the host mounts an immune response against a specific microbial determinant, T cells specific for the infectious agent may cross-react with the mimicked host sequence. Autoimmunity could be the consequence [56], and there is much evidence in favor of this hypothesis. A number of viral proteins contain sequence homologies to the encephalitogenic epitope of MBP. Immunization of animals with these viral peptides induces inflammatory infiltrates around blood vessels of the CNS that are hard to distinguish from infiltrates invoked by immunization with the corresponding peptide of MBP [28]. In contrast to the rather specific T cell activation by cross-reactive epitopes is the unspecific activation by bacterial superantigens. These superantigens activate all T cells bearing certain TCR isotypes, irrespective of antigen specificity [33], and thus also certain autoreactive T cells [66]. Staphylococcal superantigen, for example, is able to activate proliferation and cytotoxic function of human T lymphocytes reactive with different myelin autoantigens such as MBP and proteolipid protein (PLP) [66].

Molecular mimicry and activation by bacterial superantigens could result in *spontaneous* autoimmunity. *Experimental* autoimmunity can be induced by immunization of animals with self proteins.

Experimental autoimmune encephalomyelitis

One example for such an organ-specific autoimmune disease is EAE, which is widely used to study the initial, inflammatory stages of multiple sclerosis (MS). EAE is characterized by massive infiltrations of mononuclear cells, especially T lymphocytes, into the CNS, resulting in an activation of local glia cells [79]. EAE can be induced actively by immunization with CNS autoantigens. In this case, autoreactive T lymphocytes are activated outside the CNS target tissue near the site of antigen inoculation. These T cells belong to the $CD4^+$ (T helper) subset of lymphocytes, use the $\alpha\beta$ TCR for antigen recognition, recognize their antigen in combination with class II proteins of the major histocompatibility complex (MHC), and synthesize interleukin-2 (IL-2) and interferon-γ (IFN-γ) following activation. They can be isolated from the animal, and can be expanded in vitro by alternating antigen-specific T cell activation with IL-2-dependent propagation. Moreover, they can be reinjected into naive, syngeneic rodents and transfer the organ-specific autoimmune disease. Thus, EAE can also be induced passively by transfer of CNS-specific T cells into naive animals.

All T cells that are able to induce EAE are specific for myelin or glia proteins of the nervous system. Historically, MBP was the antigen of choice to elicit CNS-specific immune responses [57]. The reasons for that are obvious: MBP is a very abundant protein of the CNS. It is the main protein component of the myelin sheath, is easy to purify, and due to its solubility in water extremely easy to handle. We now know, however, that other protein components of the CNS or peripheral nervous system (PNS) are also able to induce autoimmune responses. This was first shown for myelin PLP [78] and myelin oligodendrocyte glycoprotein (MOG) [49], which were both components of the CNS, and myelin P2 and P0 proteins [48], found in the PNS. Even nonmyelin proteins with properties strikingly different from MBP, PLP, or MOG are targets for an autoimmune attack, for example the calcium-binding protein S100β [41], which is mostly expressed by astrocytes. Probably every protein found in the nervous system could play an important role in CNS or PNS autoimmunity, since all proteins tested so far were potent autoantigens to induce EAE and evoked severe inflammatory responses in the nervous system, albeit with subtle differences in neurological dysfunction and lesion distribution [41, 49, 57, 78].

The inflammatory lesion

As shown in the animal model of EAE, activated T lymphocytes pass the blood-brain barrier, irrespective of their antigen specificity, due to interactions with CAM on the naive blood-brain barrier endothelium. Once within the brain parenchyma, activated T cells produce IFN-γ, no matter whether they are specific for CNS or non-CNS proteins. As a result, MHC class II gene products are induced on a variety of parenchymal cells. If these parenchymal cells are then unable to present the appropriate antigen (which would be the case for non-CNS-specific T cells), the number of T cells will eventually decrease. However, CNS-specific T cells do encounter "their" antigens presented by MHC-expressing parenchymal cells and initiate an autoimmune response. They produce a wealth of cytokines and inflammatory factors. Some of these factors, for example IL-1β released early in the immune response, serve as powerful chemoattractants. They guide additional immune cells to inflammatory lesions and amplify local inflammations [27, 44]. Other factors and signals induce an activation of the local blood-brain barrier endothelium, which becomes leaky and permits an uncontrolled entry of plasma molecules along with blood-borne cells into the CNS parenchyma. As a consequence, focal edema develops, and massive cellular infiltration takes place [54]. In EAE, these inflammatory infiltrates are mainly concentrated around postcapillary microvessels of the CNS white matter and within the CNS parenchyma [79], and are dominated by T lymphocytes (Fig. 1).

Are all T cells found in inflammatory lesions specific for CNS antigens, or is the vast majority recruited as "innocent" bystander cells? This question was addressed in a recent experiment taking advantage of the restricted TCR usage found in MBP-specific T cell lines [15]. Briefly, MBP-specific T cells of the Lewis rat preferentially use the Vβ8.2 gene element of the TCR receptor, while such a particular TCR preference is not noted in the general T cell population of these animals. For this reason, the presence of Vβ8.2 TCR elements could be used as a cellular marker to trace the migration of antigen-specific T cells. MBP-specific, V$\beta 8.2^+$ T cells were activated in vitro and injected into naive Lewis rats to induce T cell line-transferred EAE. Then, the TCR usage of inflammatory infiltrates was determined at the peak of clinical disease.

The results were striking: in general, the TCR repertoire used by infiltrating lymphocytes appeared to be highly diverse. Pronounced differences in individual infiltrate compartments were, however, noted. V$\beta 8.2^+$ T cells representing the transferred antigen-specific cells preferentially accumulated within the CNS parenchyma, while T cells with random usage of Vβ elements as seen in the naive Lewis rat dominated the perivascular infiltrates. Two important conclusions can be drawn from this study:

1. The CNS infiltrate of Lewis rat EAE is composed of at least two different compartments, the perivascular and the parenchymal infiltrate compartment.

2. These compartments differ fundamentally in their origin and function: most of the perivascular infiltrate T cells in the acute transfer EAE lesion are host derived and seem to be passively recruited from the recirculating T cell pool. Contrarily, most of the parenchymal immune cells are the progeny of the transferred T cell lines [45].

As seen in this and other experiments, T cells are the disease initiating components in EAE. However, in the course of an ongoing autoimmune response within the CNS, the rupture of the blood-brain barrier paves the way for other cellular infiltrations. Thus, it is not surprising that B cells and macrophages are found in inflammatory lesions of EAE [79] and MS [60, 61]. In fact, intrathecal immunoglobulin synthesis even serves as a diagnostic hallmark of MS [73], although little is known about the target specificity of these antibodies. Evidence does exist, however, showing that antibodies may play an important role in myelin destruction [47, 82], along with cytokine [10] or T cell-mediated [40] killing of oligodendrocytes. This evidence was provided by studies of EAE induced by transfer of myelin-specific T cells with or without antibodies directed against an outer surface protein of the myelin sheath, MOG. Transfer of myelin-specific T cells alone into naive Lewis rats resulted in inflammatory lesions of the CNS without noticeable demyelination. However, when myelin-specific T cell lines were transferred together with the anti-MOG antibodies, widespread demyelination occurred [46, 47].

Ongoing demyelination can be further enhanced by macrophages. Activated macrophages remove degenerating myelin, and secrete cytotoxic compounds, such as reactive oxygen intermediates, nitric oxide, and proteases [71]. The role of macrophages in disease exacerbations could be convincingly demonstrated in vivo. For these experiments, EAE was induced in rats. Prior to the onset of overt disease,

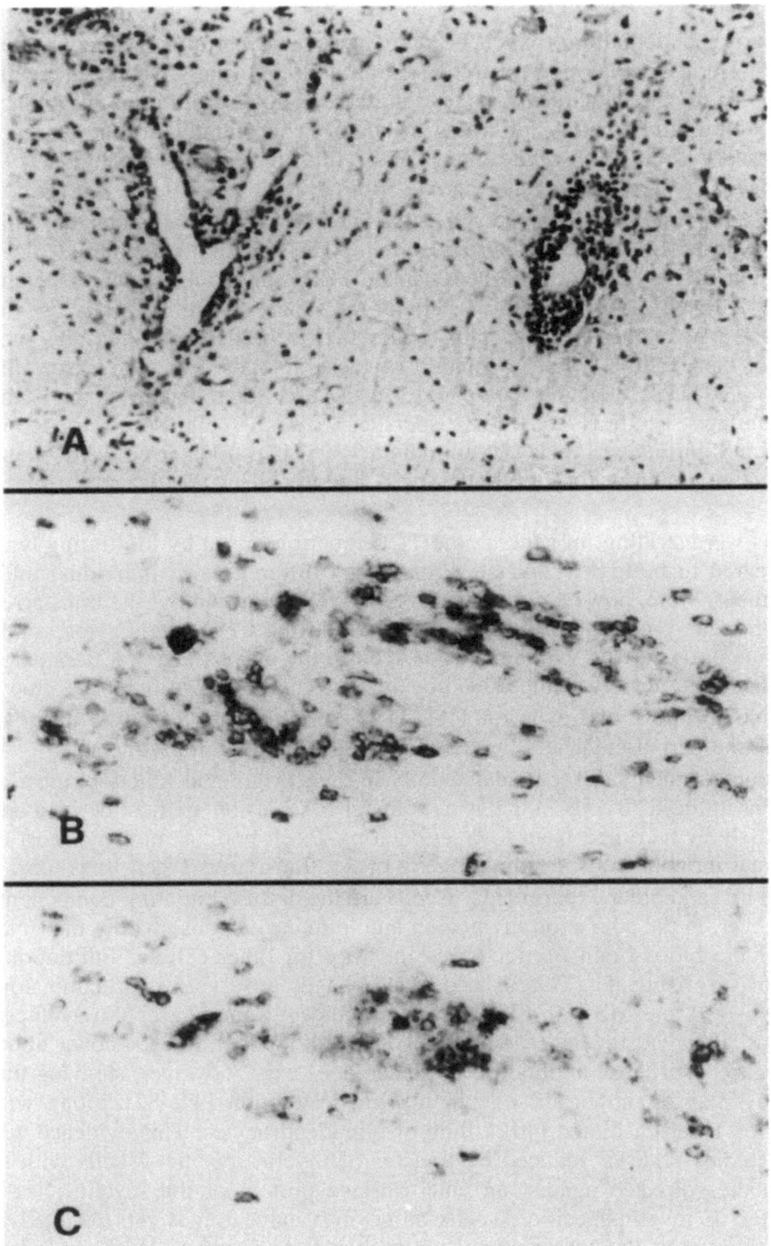

Fig. 1. A Perivascular lesions are a pathological hallmark of T cell-mediated experimental allergic encephalomyelitis (EAE) of the Lewis rat. Hematoxylin-eosin stained section. **B** They are composed in part of $\alpha\beta$TCR$^+$ T lymphocytes. **C** A subpopulation of these T lymphocytes expresses the Vβ8.2 T cell receptor chain

the animals were depleted of macrophages either by silica or by liposome treatment. While untreated animals developed severe EAE, the depleted rats displayed significantly milder clinical and histopathological symptoms [9, 37].

Thus, autoantibody production by B cells, and uptake of myelin by phagocytic macrophages might be crucial for the development of large demyelinated lesions, which are routinely observed in CNS lesions of patients with MS.

So far, we have given an overview on the development of immune responses within the CNS, describing how autoreactive T cells enter the CNS parenchyma, and how they initiate immune responses there. We will now describe in more detail how immune cells interact with glia cells of the CNS, and how together they orchestrate the immune reactivity within the CNS, starting with the induction of MHC expression on brain parenchymal cells, which is *the* crucial event for CNS immunity.

MHC expression within the CNS

Under normal circumstances, i.e., in the intact, non-diseased nervous system of rodents and men, expression of MHC products within the CNS is rare [19, 30, 32]: MHC class I expression is confined to blood vessels, and MHC class II molecules are found on macrophages of the meninges and the choroid plexus, and on perivascular cells (Fig. 2A–D) [75]. This low or undetectable expression of MHC products was in part responsible for the classical view of the nervous system as an immunologically privileged site [80]. During inflammation, however, MHC expression can be observed in areas with T cell infiltration. In EAE, the majority of these brain infiltrating encephalitogenic T cells belong to the T helper 1 (Th1) subtype of $CD4^+$ lymphocytes [1], and in MS the $CD4^+$ infiltrating T cells resemble the T helper 0 (Th0) subtype [17]. Interestingly, both of these subtypes produce IFN-γ upon activation. IFN-γ induces MHC expression on a variety of cell types, among them glia cells of the CNS. However, not all parenchymal cell types of the CNS respond equally well with MHC expression to IFN-γ stimuli. In fact, there is a hierarchic sequence of MHC class II induction in the brain. This was elegantly shown by the work of Vass and Lassmann [75]. The authors administered interferon-γ intrathecally to Lewis rats and subsequently monitored MHC class II expression in the brain. Initially after interferon-γ treatment, perivascular and microglia cells were induced to express MHC class II products, followed at later stages by other glial components such as astrocytes and ependymal cells [21, 75].

The MHC expression in the brain has important consequences since it is a prerequisite for a variety of cell types to present antigens to adjacent T lymphocytes. MHC-expressing microglia and perivascular cells count among the most efficient antigen-presenting cells of the brain parenchyma [26]. They both are bone marrow derived. While perivascular cells undergo constant replacement by bone marrow precursor cells in rats [34] and man [74], microglia form a stable pool of resident CNS cells displaying minimal turnover with the bone marrow [29]. Both cell types play an important role as a rapidly activated immune defense system within the CNS [31]. This importance is underlined by two different observations.

First, microglia cells become rapidly activated in response to even mild types of injury. If, for example, remote areas of the facial or optic nerves are damaged, MHC class II products are expressed in CNS target regions of the damaged nerves (Fig. 2E–H). Such mechanisms are of special interest since they may serve to link

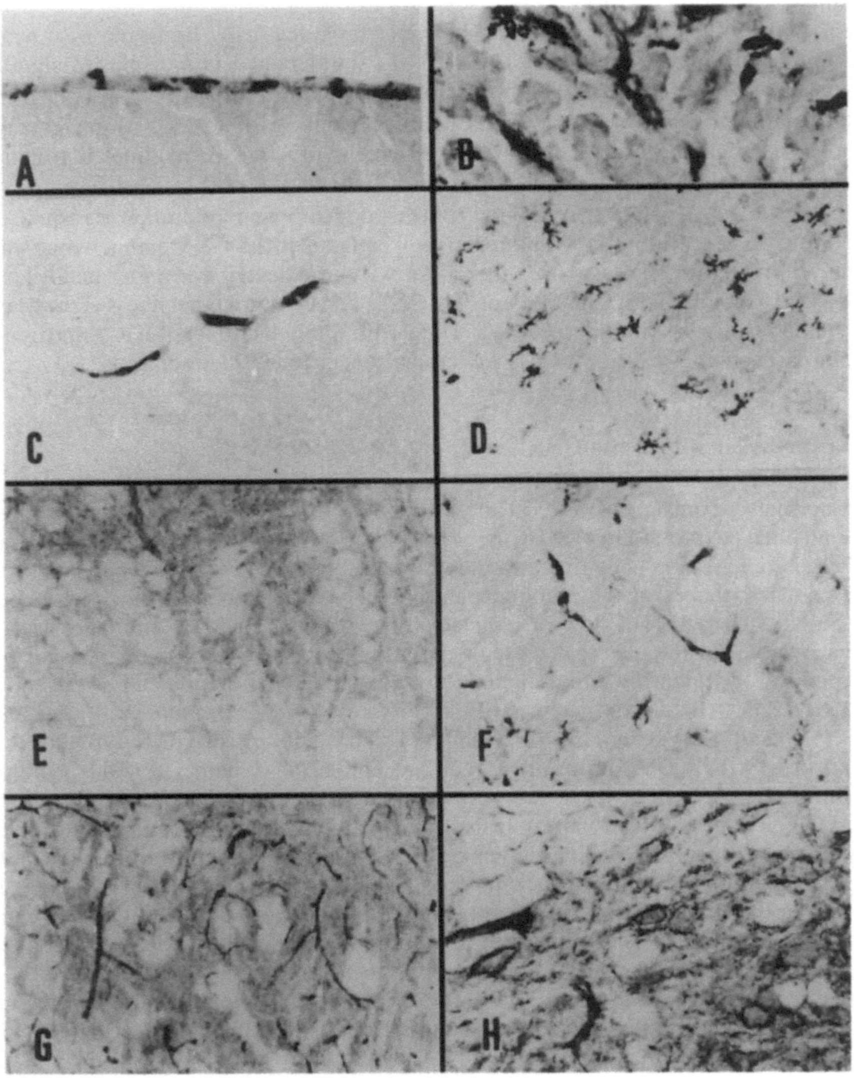

Fig. 2. Major histocompatibility complex (MHC) class I and II molecule expression in the normal (**A–D**) and in the injured rat central nervous system (CNS) following facial nerve axotomy, a remote type of injury (**E–H**). In the normal CNS constitutive MHC class II expression is observed on meningeal macrophages (**A**), choroid plexus macrophages (**B**), perivascular cells (**C**) and occasionally on resting microglia for example in the cerebellar white matter (**D**). At 7 days after facial nerve axotomy MHC class II molecule expression is induced on microglia in addition to perivascular cells (**E** control nucleus; **F** axotomized nucleus). At the same time MHC class I molecules, which are constitutively found only on vascular surfaces (**G**), are also induced on activated microglia in the axotomized facial nucleus (**H**). **A, B** ×480; **C–H** ×240

trauma to the establishment of inflammation in the brain. Severe inflammation occurs within these MHC-expressing areas after transfer of encephalitogenic T cells, even if these areas were otherwise not preferential targets of CNS inflammation [42, 43, 50].

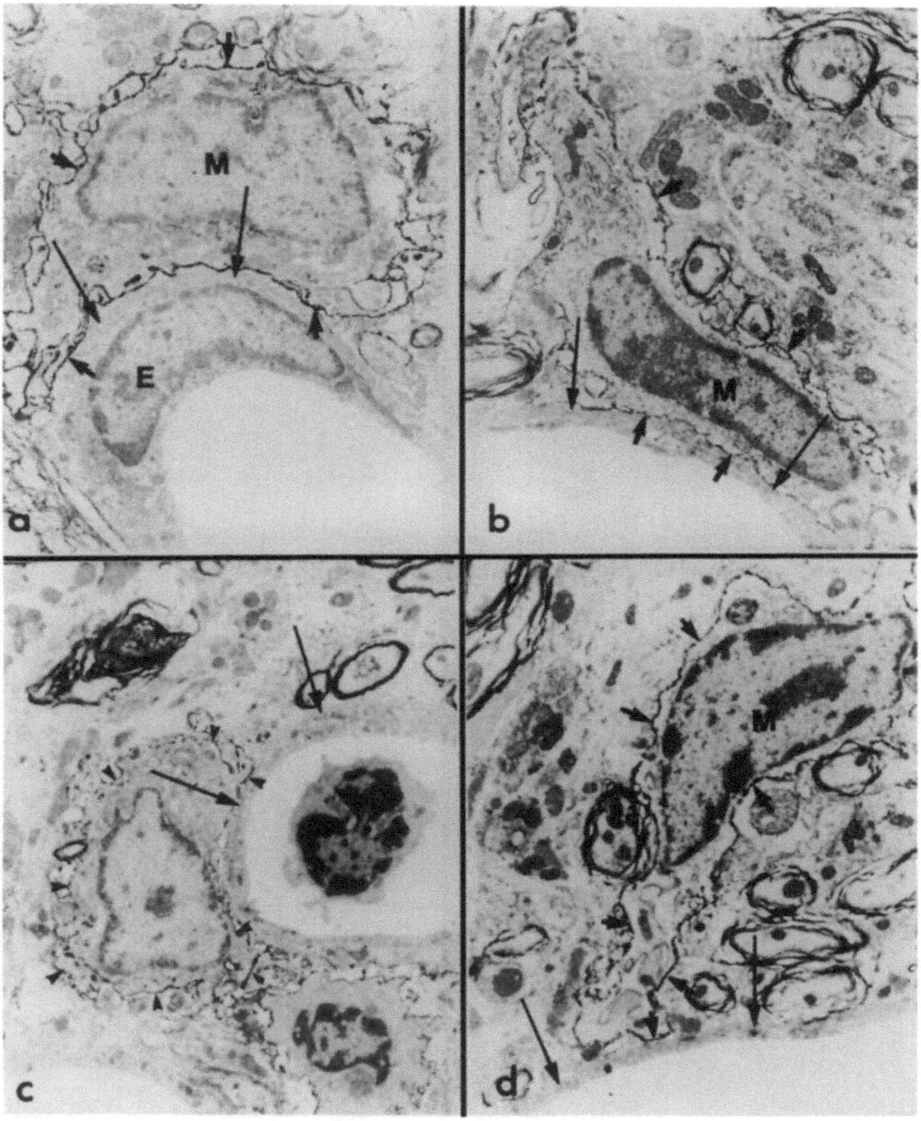

Fig. 3. Ultrastructure of juxtavascular MHC class II-immunoreactive microglia (*arrows* indicate the sites of diaminobenzidine reaction product) during T cell-mediated EAE of the Lewis rat. These cells are directly incorporated within the perivascular glia limitans and are only separated from adjoining endothelial cells (*E*) by basal lamina (long *arrows*). They thus appear to be targets of infiltrating T lymphocytes (*c*). **A, B, D** ×10 500; **C** ×6300

Second, in EAE (Fig. 3) and MS lesions (Fig. 4), activated microglia cells express high levels of MHC class II molecules [6, 76].

It is not only the bone marrow-derived cells of the CNS that can be recruited by IFN-γ to function as antigen-presenting cells. Many other glia cells can express MHC gene products and present antigen, at least in vitro and probably also in vivo,

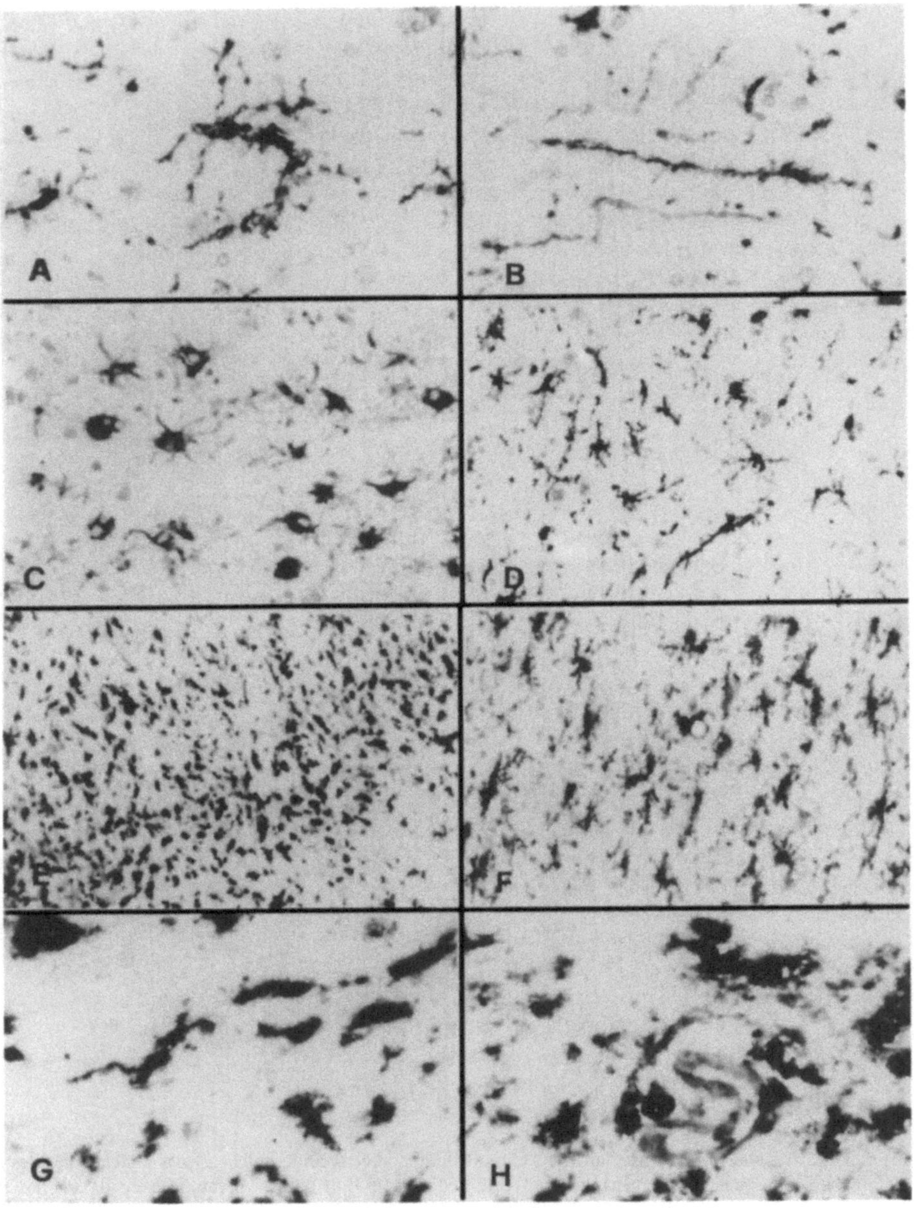

Fig. 4A–H. Glial reactions in multiple sclerosis. **A, B** Typical ramified (**A**) or rod-cell morphology; CD68-immunoreactivity. **D–E** At the border zone of chronic-inactive lesions activated microglia are present in varying numbers but consistently express large amounts of HLA-DR (MHC class II) molecules. **C** In addition, numerous glial fibrillary acidic protein-immunoreactive hypertrophic astrocytes are observed. **G, H** In the vicinity of active, demyelinating lesions strongly hypertrophic, HLA-DR-immunoreactive microglia occur, which are frequently seen in a close perivascular position (**H**). **A, B, G, H** ×640; **C, D, F** ×280; **E** ×140

e.g. astrocytes [22, 51], which provide structural, metabolic, and trophic support for neurons and play an important role in maintaining neurotransmitter metabolism and intercellular signaling within neuronal networks [2, 23]. These cells are stimulated to proliferate by products released by activated T cells [4, 5] and to express MHC class II gene products. IFN-γ and tumor necrosis factor (TNF)-α often cooperate in the induction of MHC class II expression [12, 77], and antibodies against TNF-α block the development of EAE [67]. Detailed studies in vitro revealed that MHC-expressing astrocytes are able to activate T lymphocytes immunospecifically [22, 24]. Whether they actually do so as well in vivo is still a matter of debate. In addition, differences in MHC class II inducibility exist within different species [52]. Even in vitro, not all astrocytes are fully competent antigen-presenting cells: some display severe suppressive effects on T cell activation, which may reflect their ability to down-modulate immune reactions [81].

Self limitation of immune responses within the CNS

In spite of induction of MHC class II products on local glia cells, immune reactivity in the brain is self limited. Powerful counterregulatory mechanisms exist.

These can probably be ascribed to the action of the transforming growth factors (TGF) type β, produced by activated inflammatory cells such as T cells and macrophages. TGF-β antagonizes the induction of MHC class II expression by IFN-γ and TFN-α [68] and by this means successfully suppresses the presentation of autoantigens by astrocytes. Thus, it diminishes lymphocyte function in an inflammatory site [25]. The important role of TGF-β in immunomodulation is also revealed in experiments, where it is administered during ongoing EAE: consistent improvements in the clinical course and marked reductions in CNS damage are observed [62]. Counterregulatory mechanisms cannot only be ascribed to the communication between cells of the immune system and glia of the CNS. They probably also result from interactions of immune cells within the target organ.

$CD4^+$ T cells are the main disease-initiating components [3] of EAE, while $CD8^+$ T cells become more numerous at later stages of the disease. Little is known as yet about their antigen specificity or mode of action. Thus, different scenarios can be envisaged: $CD8^+$, antigen-specific T cells could directly kill oligodendrocytes [38] or, alternatively, could eliminate $CD4^+$ effector T cells through direct cytotoxicity [72]. They might also have dramatic effects on the immunoregulation of the EAE course. When animals recover from a first episode of EAE, they are usually resistant to a second induction of this disease. This resistance, however, is abolished when the animals are experimentally depleted of $CD8^+$ T cells by antibody-mediated clearance [39]. When mice depleted of $CD8^+$ T cells by genetic ablation due to homologous recombination and control mice were challenged with EAE, both types of mice displayed a similar onset and severity of disease. However, mice lacking $CD8^+$ T cells had a much higher frequency of relapse [40]. Since in both experimental situations spontaneous recovery of EAE was noted, one may conclude that $CD8^+$ T cells are not required in down-regulatory events within the CNS parenchyma. What other mechanisms then exist to terminate immune reactions locally within the brain?

Death of infiltrating T cells

Many of the T cells entering the CNS do not persist there for prolonged periods of time [55]. With spontaneous clinical recovery from EAE, they are cleared from the nervous tissue. In situ nick translation on EAE lesions revealed that T cells [70] along with some oligodendrocytes [58] undergo apoptosis. This programmed cell death, obviously providing a non-inflammatory means of removing T cells within CNS lesions, is most developed when the animals have recovered from EAE. Thus, apoptosis could be of general importance in terminating immune reactions within the CNS.

Conclusions

In contrast to traditional views, cells of the lymphohematopoietic system can and do enter the CNS. Some of them, for example perivascular cells and cells of the microglia, enter the CNS during development. Others, such as T lymphocytes previously engaged in immune responses, gain access to the CNS parenchyma in the process of immune surveillance, or during inflammatory reactions. Elaborate control systems then ensure that the immune reactivity within the CNS system is limited to an absolute minimum.

References

1. Ando DG, Clayton J, Kono D, Urban JL, Sercarz EE (1989) Encephalitogenic T cells in the B10.PL model of experimental allergic, encephalomyelitis (EAE) are of the Th1 lymphokine subtype. Cell Immunol 124:132
2. Barres BA, Koroshetz WJ, Chun LLY, Corey DP (1990) Ion channel expression by white matter glia: the type-1 astrocyte. Neuron 5:527
3. Ben-Nun A, Wekerle H, Cohen IR (1981) The rapid isolation of clonable antigen-specific T lymphocyte lines capable of mediating autoimmune encephalomyelitis. Eur J Immunol 11:195
4. Benveniste EN, Merrill JE, Kaufman SE, Golde DW, Gasson JC (1985) Purification and characterization of a human T-lymphocyte-derived glial growth-promoting factor. Proc Natl Acad Sci USA 85:3930
5. Benveniste EN, Sparacio SM, Bethea JR (1989) Tumor necrosis factor-α enhances interferon-mediated class II antigen expression on astrocytes. J Neuroimmunol 25:209
6. Boyle EA, McGeer PL (1990) Cellular immune response in multiple sclerosis plaques. Am J Pathol 137:575
7. Brightman MW, Reese TS (1969) Junctions between intimately apposed cell membranes in the vertebrate brain. J Cell Biol 40:648
8. Brightman MW, Zis K, Anders J (1983) Morphology of cerebral endothelium and astrocytes as determinants of the neuronal microenvironment. Acta Neuropathol (Berl) [Suppl]VIII:21
9. Brosnan CF, Bernstein MB, Bloom BR (1981) The effects of macrophage depletion on the clinical and pathologic expression of experimental allergic encephalomyelitis. J Immunol 126:614
10. Brosnan CF, Selmaj K, Raine CS (1988) Hypothesis: a role for tumor necrosis factor in immune-mediated demyelination and its relevance to multiple sclerosis. J Neuroimmunol 18:87
11. Butcher EC (1990) Cellular and molecular mechanisms that direct leukocyte traffic. Am J Pathol 136:3
12. Cannella B, Raine CS (1989) Cytokines upregulate Ia expression in organotypic cultures of central nervous tissue. J Neuroimmunol 24:239
13. Reference deleted
14. Reference deleted
15. Chluba J, Steeg C, Becker A, Wekerle H, Epplen JT (1989) T cell receptor β chain usage in myelin basic protein-specific rat T lymphocytes. Eur J Immunol 19:279

16. Dailey MO, Gallatin WM, Weissman IL (1985) The in vivo behavior of T cell clones: altered migration due to loss of the lymphocyte homing receptor. J Mol Cell Immunol 2:27
17. Del Prete G, Maggi E, Romagnani S (1994) Biology of disease. Human Th1 and Th2 cells: functional properties, mechanisms of regulation, and role in disease. Lab Invest 70:299
18. Dermietzel R, Krause D (1991) Molecular anatomy of the blood-brain barrier as defined by immunocytochemistry. Int Rev Cytol 127:57
19. Dickson DW, Mattiace LA, Kure K, Hutchins K, Lyman WD, Brosnan CF (1991) Microglia in human disease, with an emphasis on acquired immune deficiency syndrome. Lab Invest 64:135
20. Ehrlich P (1904) über die Beziehungen chemischer Constitution, Verheilung, und pharmakologischer Wirkung. In: Gesammelte Arbeiten zur Immunitätsforschung, Berlin
21. Fabry Z, Waldschmidt MM, Moore SA, Hart MN (1990) Antigen presentation by brain microvessel smooth muscle and endothelium. J Neuroimmunol 28:63
22. Fierz W, Endler B, Reske K, Wekerle H, Fontana A (1985) Astrocytes as antigen presenting cells. I. Induction of Ia antigen expression on astrocytes by T cells via immune interferon and its effect on antigen presentation. J Immunol 134:3785
23. Finkbeiner S (1992) Calcium waves in astrocytes – filling in the gaps. Neuron 8:1101
24. Fontana A, Fierz W, Wekerle H (1984) Astrocytes present myelin basic protein to encephalitogenic T cell lines. Nature 307:273
25. Frater-Schröder M, Müller G, Birchmeier W, Bohlen P (1986) Transforming growth factor-β inhibits endothelial cell proliferation. Biochem Biophys Res Commun 137:295
26. Frei K, Siepl C, Groscurth P, Bodmer S, Schwerdel C, Fontana A (1987) Antigen presentation and tumor cytotoxicity by interferon-γ treated microglial cells. Eur J Immunol 17:1271
27. Freidin M, Bennett MVL, Kessler JA (1992) Cultured sympathetic neurons synthesize and release the cytokine interleukin 1β. Proc Natl Acad Sci USA 89:10 440
28. Fujinami RS, Oldstone MBA (1985) Amino acid homology between the encephalitogenic site of myelin basic protein (MBP) and virus: mechanism for autoimmunity. Science 230:1043
29. Gehrmann J, Kreutzberg GW (1996) Microglia in experimental neurophathology. In: Kettenmann H, Ransom B (eds) Neuroglial cells. Oxford University Press, Oxford (in press)
30. Gehrmann J, Banati RB, Kreutzberg GW (1993) Microglia in the immune surveillance of the brain: human microglia constitutively express HLA-DR molecules. J Neuroimmunol 48:189
31. Gehrmann J, Gold R, Linington C, Lannes-Vieira J, Wekerle H, Kreutzberg GW (1993) Microglial involvement in experimental autoimmune inflammation of the central and peripheral nervous system. Glia 7:50
32. Hayes GM, Woodroofe MN, Cuzner ML (1987) Microglia are the major cell type expressing MHC class II in human white matter. J Neurol Sci 80:25
33. Herman A, Kappler JW, Marrack P, Pullen AM (1991) superantigens: mechanisms of T cell stimulation and role in immune responses. Annu Rev Immunol 9:745
34. Hickey WF, Kimura H (1988) Perivascular microglial cells of the CNS are bone-marrow derived and present antigen in vivo. Science 239:290
35. Hickey WF, Hsu BL, Kimura H (1991) T lymphocyte entry into the central nervous system. J Neurosci Res 28:254
36. Hinrichs DJ, Wegmann KW, Dietsch GN (1987) Transfer of experimental allergic encephalomyelitis to bone marrow chimeras: endothelial cells are not the restricting element. J Exp Med 166:1906
37. Huitinga I, Van Rooijen N, De Groot CJA, Uitdehaag BMJ, Dijkstra CD (1990) Suppression of experimental allergic encephalomyelitis in Lewis rats after elimination of macrophages. J Exp Med 172:1025
38. Jewtoukoff V, Lebar R, Bach MA (1989) Oligodendrocyte-specific autoreactive T cells using an α/β T-cell receptor kill their target without self restriction. Proc Natl Acad Sci USA 86:2824
39. Jiang H, Zhang SL, Pernis B (1992) Role of $CD8^+$ T cells in murine experimental allergic encephalomyelitis. Science 256:1213
40. Koh DR, Fung-Leung WP, Ho A, Gray D, Acha-Orbea H, Mak TW (1992) Less mortality but more relapses in experimental allergic encephalomyelitis in $CD8^{-/-}$ mice. Science 256:1210
41. Kojima K, Berger T, Lassmann H, Hinze-Selch D, Zhang Y, Gehrmann J, Reske K, Wekerle H, Linington C (1994) Experimental autoimmune panencephalitis and uveoretinitis in the Lewis rat transferred by T lymphocytes specific for the S100β molecule, a calcium binding protein of astroglia. J Exp Med 180:817

42. Konno H, Yamamoto T, Iwasaki Y, Suzuki H, Saito T, Terunama H (1989) Wallerian degeneration induces Ia-expression in the rat brain. J Neuroimmunol 25:151
43. Konno H, Yamamoto T, Suzuki H (1990) Targeting of adoptively transferred experimental allergic encephalomyelitis lesion at the site of Wallerian degeneration. Acta Neuropathol 80:521
44. Kovacs EJ (1991) Fibrogenic cytokines: the role of immune mediators in the development of scar tissue. Immunol Today 12:17
45. Lannes-Vieira J, Gehrmann J, Kreutzberg GW, Weckerle H (1994) The inflammatory lesion of T cell line transferred experimental autoimmune encephalomyelitis of the Lewis rat: distinct nature of parenchymal and perivascular infiltrates. Acta Neuropathol 87:435
46. Linington C, Lassmann H (1987) Antibody responses in chronic relapsing experimental allergic encephalomyelitis: correlation of serum demyelinating activity with antibody titer to myelin/oligodendrocyte glycoprotein (MOG). J Neuroimmunol 17:61
47. Linington C, Bradl M, Lassmann H, Brunner C, Vass K (1988) Augmentation of demyelination in rat acute allergic encephalomyelitis by circulating mouse monoclonal antibodies directed against a myelin/oligodendrocyte glycoprotein. Am J Pathol 130:443
48. Linington C, Lassmann H, Ozawa K, Kosin S, Mongan L (1992) Cell adhesion molecules of the immunoglobulin supergene family as tissue-specific autoantigens: induction of experimental allergic neuritis (EAN) by P0 protein-specific T cell lines. Eur J Immunol 22:1813
49. Linington C, Berger T, Perry L, et al (1993) T cell specific for the myelin oligodendrocyte glycoprotein (MOG) mediate an unusual autoimmune inflammatory response in the central nervous system. Eur J Immunol 23:1364
50. Maehlen J, Olsson T, Zachau-A, Klareskog L, Kristenssen K (1989) Local enhancement of major histocompatibility complex (MHC) class I and class II expression and cell infiltration in experimental allergic encephalomyelitis around axotomized motor neurons. J Neuroimmunol 23:125
51. Reference deleted
52. Massa PT, Ter Meulen V, Fontana A (1987) Hyperinducibility of Ia antigen on astrocytes correlated with strain specific susceptibility to experimental autoimmune encephalomyelitis. Proc Natl Acad Sci USA 84:4219
53. Naparstek Y, Ben-Nun A, Holoshitz J, et al (1983) T lymphocyte line producing or vaccinating against autoimmune encephalomyelitis (EAE). Functional activation induces peanut agglutinin receptors and accumulation in the brain and thymus of line cells. Eur J Immunol 13:418
54. Newcombe J, Hawkins CP, Henderson CL, et al (1991) Histopathology of multiple sclerosis lesions detected by magnetic resonance imaging in unfixed postmortem central nervous system tissue. Brain 114:1013
55. Ohmori K, Hong Y, Fujiwara M, Matsumoto Y (1992) In situ demonstration of proliferating cells in the rat central nervous system during experimental autoimmune encephalomyelitis. Evidence suggesting that most infiltrating T cells do not proliferate in the target organ. Lab Invest 66:54
56. Oldstone MBA (1987) Molecular mimicry and autoimmune disease. Cell 50:819
57. Paterson PY (1960) Transfer of allergic encephalomyelitis in rats by means of lymph node cells. J Exp Med 111:119
58. Pender MP, Nguyen KB, McCombe PA, Kerr JFR (1991) Apoptosis in the nervous system in experimental allergic encephalomyelitis. J Neurol Sci 104:81
59. Pette M, Fujita K, Wilkinson D, et al (1990) Myelin autoreactivity in multiple sclerosis: recognition of myelin basic protein in the context of HLA-DR2 products by T lymphocytes of multiple sclerosis patients and healthy donors. Proc Natl Acad Sci USA 87:7968
60. Prineas JW (1975) Pathology of the early lesion in multiple sclerosis. Hum Pathol 6:531
61. Prineas JW, Wright RG (1978) Macrophages, lymphocytes and plasma cells in the perivascular compartment in chronic multiple sclerosis. Lab Invest 38:409
62. Racke MK, Dhib-Jalbut S, Cannella B, Albert PS, Raine CS, McFarlin DE (1991) Prevention and treatment of chronic relapsing experimental allergic encephalomyelitis by transforming growth factor-β1. J Immunol 146:3012
63. Reference deleted
64. Reese TS, Karnovsky MJ (1967) Fine structural localization of a blood-brain barrier to exogenous peroxidase. J Cell Biol 4:207
65. Robey E, Fowlkes BJ (1994) Selective events in T cell development. Annu Rev Immunol 12:675
66. Rott O, Wekerle H, Fleischer B (1992) Protection from experimental allergic encephalomyelitis by application of a bacterial superantigen. Int Immunol 4:347

67. Ruddle NH, Bergman CM, McGrath KM, et al (1990) An antibody to lymphotoxin and tumor necrosis factor prevents transfer of experimental allergic encephalomyelitis. J Exp Med 172:1193
68. Schluesener HJ (1990) Transforming growth factors $\beta 1$ and $\beta 2$ suppress rat astrocyte autoantigen presentation and antagonize hyperinduction of class II major histocompatibility complex antigen expression by interferon-γ and tumor necrosis factor-α. J Neuroimmunol 27:41
69. Schluesener HJ, Wekerle H (1985) Autoaggressive T lymphocyte lines recognizing the encephalitogenic region of myelin basic protein: in vitro selection from unprimed rat T lymphocyte populations. J Immunol 135:3128
70. Schmied M, Breitschopf H, Gold R, et al (1993) Apoptosis of T lymphocytes A mechanism to control inflammation in the brain. Am J Pathol 143:446
71. Smith ME (1993) Phagocytosis of myelin by microglia in vitro. J Neurosci Res 35:480
72. Sun D, Qin Y, Chluba J, Epplen JT, Wekerle H (1988) Suppression of experimentally induced autoimmune encephalomyelitis by cytolytic T-T cell interactions. Nature 332:843
73. Tourtelotte WW, Walsh MJ (1984) Cerebral spinal fluid profile in multiple sclerosis. In: Poser CM, Paty DW, Scheinberg L, McDonald WI, Eberts GC (eds) The diagnosis of multiple sclerosis. Thieme-Stratton, New York
74. Unger ER, Sung JH, Manivel JC, Chenggis ML, Blazar BR (1993) Male donor-derived cells in the brains of female sex-mismatched bone marrow transplant recipients: a Y chromosome specific in situ hybridization study. J Neuropathol Exp Neurol 52:460
75. Vass K, Lassmann H (1990) Intrathecal application of interferon gamma: progressive appearance of MHC antigens within the rat nervous system. Am J Pathol 137:789
76. Vass K, Lassmann H, Wekerle H, Wisniewski HM (1986) The distribution of Ia antigen in the lesions of rat acute experimental allergic encephalomyelitis. Acta Neuropathol (Berl) 70:149
77. Vidovic M, Sparacio SM, Elovitz M, Benveniste EN (1990) Induction and regulation of class II major histocompatibility complex mRNA expression in astrocytes by interferon-γ and tumor necrosis factor-α. J Neuroimmunol 30:189
78. Waksman BH, Porter H, Lees MB, Adams RD, Folch JA (1954) A study of the chemical nature of components of bovine white matter effective in producing allergic encephalomyelitis in the rabbit. J Exp Med 100:451
79. Wekerle H (1992) Myelin specific, autoaggressive T cell clones in the normal immune repertoire: their nature and their regulation. Int Rev Immunol 9:231
80. Wekerle H, Linington C, Lassmann H, Meyermann R (1986) Cellular immune reactivity within the CNS. Trends Neurosci 9:271
81. Wekerle H, Sun D, Oropeza-Wekerle RL, Meyermann R (1987) Immune reactivity in the nervous system: Modulation of T-lymphocyte activation by glial cells. J Exp Biol 132:43
82. Zajicek JP, Wing M, Scolding NJ, Compston DAS (1992) Interactions between oligodendrocytes and microglia. A major role for complement and tumor necrosis factor in oligodendrocyte adherence and killing. Brain 115:1611

Superantigens in demyelinating disease

Stefan Brocke[1,2], Christopher Piercy[1], Lawrence Steinman[1]

[1] Department of Neurology and Neurological Sciences, Beckman Center for Molecular and Genetic Medicine, B002, Stanford University Mediacal Center, Stanford, CA 94305-5429, USA
[2] National Institutes of Health, National Institute of Neurological Disorders and Stroke, Building 10, Room 5B-16, 10 Center Dr MSC 1400, Bethesda, MD 20892-1400, USA

The role of infection in multiple sclerosis

Autoreactive T cells are part of the normal lymphocyte repertoire of healthy individuals [4]. In the immune system of healthy individuals, potent regulatory mechanisms control T and B cell tolerance to self antigens. While genetic factors influence susceptibility to autoimmunity in humans and experimental animals, environmental factors contribute to disease penetration. For autoimmune disease to occur, autoreactive lymphocytes have to be activated in vivo or in vitro. This process leads to the clonal expansion of autoreactive cells. In addition, during the initiation phase of demyelinating disease, only activated autoreactive lymphocytes migrate through the endothelial cell layer of the intact blood-brain barrier (BBB).

The central nervous system (CNS) has long been considered to be an immunologically privileged site, protected from immune surveillance by the BBB. The BBB is located in the endothelial cells of the capillaries in the brain. These endothelial cells differ from those of systemic capillaries. They are rich in mitochondria, rarely have pinocytotic vesicles and are fused by tight junctions of high electrical resistance. In addition, they make frequent contact on their abluminal (brain) side with projections from astrocytes. Through these contacts, astrocytes are thought to regulate capillary permeability. The concept of a BBB was developed early in the century when several investigators noted that intravenous injection of cationic dyes in experimental animals rapidly diffuse from capillaries and stained tissues in most organs but not in the brain. These observations promoted the notion that the two major CNS components, the brain and the cerebrospinal fluid, are separated or excluded from the blood-lymphatic compartment. It is now clear, however, that the BBB is a two-way regulatory interface between blood and nervous system rather than an impermeable fence. The endothelial cells selectively transport nutrients into the brain, such as sugars and amino acids, along their concentration gradients. Further, they protect brain cells from extraneous fluctuations in extracellular ion concentrations, neurotransmitters and

Correspondence to: S. Brocke (address see [2] above)

growth factors. The vascular tight junctions which normally prevent leukocytes from entering the CNS, may become permeable in a number of degenerative, infectious, epileptic, leukemic and autoimmune diseases. For example, the inflammation during bacterial and aseptic meningitis is characterized by leukocytic infiltration and increased vascular permeability to water-soluble drugs such as penicillin.

The dynamics of the autoimmune T cell infiltration into and out of the CNS parenchyma have proven difficult to investigate in humans with multiple sclerosis (MS). A more complete picture is emerging from studies in animal models such as experimental autoimmune encephalomyelitis (EAE). The establishment of inflammatory EAE lesions and clinical disease is a multistep event. It has been proposed that the first step requires that activated T cells cross the BBB. Activated lymphocytes bear a memory cell-like phenotype (characterized by the expression of the surface marker CD44 in the absence of the marker Mel 14), suggesting previous activation by antigen, and also express the adhesion molecule integrin $\beta 1\alpha 4$. The activated T cells become attached to appropriate receptors on endothelial cells at parajunctional areas, adjacent to the endothelial tight junctions, and then proceed to pass directly into the interstitial matrix. This process has been shown to be dependent on the activation state of the T cells, but independent of their antigen specificity. Thus, myelin basic protein (MBP)-reactive cells cross the BBB with no greater efficiency than do activated cells that do not recognize a CNS antigen. Next, CNS antigen-specific $CD4^+$ T cells are reactivated in situ by fragments of antigens such as myelin, presented in the framework of major histocompatibility complex (MHC) class II molecules on the surface of antigen-presenting cells (macrophages, microglia, and perhaps astrocytes). This leads to a second wave of inflammatory recruitment and clinical EAE. Proinflammatory cytokines such as tumor necrosis factor-α (TNF-α) are probably key mediators of the full-blown inflammatory response. Encephalitogenic myelin-specific T cells may not be capable of mediating EAE in the absence of this secondary leukocyte recruitment. In addition, the adherence and extravasation of T cells increase endothelial permeability to pathogenic demyelinating antibodies, indicating that T cells and antibodies may act synergistically to induce demyelination.

Infectious pathogens are known to be the most potent activators of cells of the immune system. Bacteria and viruses have long been regarded as relevant exogenous factors in autoimmune diseases. Various infectious agents, including many viruses and bacteria, have been implicated in the pathogenesis of demyelinating diseases [25]. Several mechanisms associated with infection could account for the activation of autoreactive lymphocytes leading to disease. One such mechanism is molecular mimicry: viruses and bacteria have determinants that are antigenic epitopes that cross-react with the self structures of the body. Under certain circumstances, those epitopes can lead to a T or B cell-mediated immunological attack on corresponding self components, such as myelin.

More recently, new mechanisms by which infectious pathogens affect autoreactive lymphocytes have been analyzed in detail. These mechanisms include the modulation of the cytokine phenotype of autoreactive cells during T cell subset development and the enhancement of antigen spreading in CNS lesions during chronic inflammation. In recent years, superantigens, a class of pathogen-derived molecules that are strong stimulators of T lymphocytes, have been intensively studied for their role in autoimmunity. Superantigens might contribute to pathological events by stimulating autoreactive T cells, both locally and systemically.

Structural aspects of superantigens

Endogenous superantigens

The endogenous superantigens include murine Mls antigens. The Mls loci have been found to be genetically inseparable from endogenous mouse mammary tumor virus (MMTV) loci (for review see [14]). The role of endogenous superantigens in autoimmunity is unclear at present.

Exogenous superantigens

Certain gram-positive cocci (*Staphylococcus aureus* and *Streptococcus pyogenes*) and *Mycoplasma arthritidis* secrete biologically active proteins that have been referred to as superantigens because of their powerful stimulatory activity for T lymphocytes. Many staphylococci strains produce one of five serologically distinct enterotoxins, designated SEA through SEE. Some isolates secrete toxic shock syndrome toxin 1 (TSST-1) as well as the exfoliative toxins (ExFT) [13, 14]. Interestingly, several staphylococcal species, i.e, not only *S. aureus*, but also strains of *S. epidermidis* have been reported to produce nanogram quantities of enterotoxins [1]. *S. pyogenes* secretes the streptococcal pyrogenic exotoxins SPE A-C that have superantigenic properties. All these superantigens share significant structure homologies. Based an amino acid sequence comparison, at least three homology groups of staphylococcal toxins have been differentiated: (1) SEA, SED and SEE, (2), SEB and the SEC toxins (3) TSST-1 and ExFT [15].

There are some similarities in the mechanism of T cell activation by superantigens and conventional protein antigens, amongst them autoantigens. Peptide antigens are presented to T cells bound to MHC class II molecules. Similarly, superantigens bind to class II molecules, forming a binary complex that stimulates T lymphocytes through defined V region β chains (Vβ) of the T cell receptor (TCR). Characterizing the ternary complex of SEB, a soluble TCR, and a soluble MHC class II molecule DR1, and the three binary complexes TCR-SEB, SEB-DR1, and the pepide-specific complex DR1-TCR, Seth et al. [18] found that superantigens are able to bind directly with very low affinity to the TCR. The observation that a superantigen can form complexes with TCR in both the absence and presence of MHC class II may provide a mechanism for its ability to induce anergy in some circumstances and activation in others [18]. In contrast to conventional protein antigens, superantigens do not need to undergo degradation prior to binding to MHC class II molecules.

SEB is the first superantigen for which detailed structural information became available [23]. In brief, the SEB molecule contains two domains. The first domain is composed of residues 1–120, the second comprises residues 127–239. Both domains contain β-sheets and several α-helices that are partially exposed to the solvent. Several regions have been identified on the SEB molecule that affect binding to MHC class II and/or T cell actvation [10]. Mutations of residues 9–23 in region 1 influence both MHC and TCR binding. Region 2, comprising residues 40–53, is probably involved in binding of SEB to MHC class II molecules. Mutations in region 3, amino acids 60 and 61, affect cell activation without having an influence on MHC class II binding. Elucidation by X-ray crystallography of the structure of the MHC class II molecule HLA-DR1 complexed with SEB [8] or TSST-1 [12] revealed two overlapping binding sites on DR1 with different binding modes of non-peptide forms of antigens. SEB binds primarily to one edge of the peptide binding site of DR1, interacting with the side of the $\alpha 1$ domain helix and with the loops joining the first and third β-strands. In the co-crystal structure, the peptide binding site of DR1 is occupied with endogenous peptide, indicating that superantigen binding does not prevent peptide binding [8]. In a recent study random point mutations in the gene encoding TSST-1 were generated and functionally impaired mutants were identified based on their lack of murine and human T cell stimulatory activities [7]. The results showed that the great majority of mutations are clustered in two distinct regions of TSST-1. The MHC class II binding site is located in the hydrophobic region of the NH_2-terminal domain, and the TCR binding site is primarily in the major central groove of the COOH-terminal domain. Interestingly, SEA and SEB share a common structural motif for binding MHC class II molecules [24]. Although T cell responses to SEA, and probably other baterial superantigens, could be mediated by variations of a structurally conserved HLA-DR α chain binding motif, differences between various superantigen binding modes exist. For instance, unlike SEB, SEA requires two separate binding sites on HLA-DR1 for optimal T cell stimulatory activity. A high-affinity interaction is mediated by a zinc atom coordinated by residues H187, H225 and H227 in SEA and H81 in the polymorphic DR1 β chain [6]. The same authors described a second low-affinity interaction with the DR1 α chain mediated by residue F47 of the SEA molecule. This binding site is similar to the DR binding site of SEB [6].

The possible role of superantigens in demyelinating disease

Superantigens have been implicated in several pathological conditions, including acute responses such as severe fever, toxic shock, food poisoning and multiple-organ failure. In addition, there is increasing evidence for a key role of superantigens in the pathogenesis of autoimmune diseases. Bacterial superantigens can specifically trigger classes of T cells through interaction with their TCR Vβ region. MS may arise from pathogenic T cells that have somehow evaded mechanisms promoting selftolerance. These pathogenic cells may rearrange a restricted number of TCR Vα and Vβ genes [21]. This conclusion stems mainly from animal studies analyzing T cell clones that recognize MBP causing EAE and has recently been confirmed in studies of some T cell clones from the peripheral blood in man that recognize MBP. In EAE, relapsing paralysis can be induced or exacerbated through injection of low doses of the superantigens SEA and SEB [2, 3, 17, 20]. This effect is in part due to the action of TNF-α [2]. Recent studies show that exposure to various bacterial superantigens can induce MBP-specific T cells with different TCR Vβ chain expression to mediate relapsing paralysis in EAE. Similarly, since autoimmune conditions such as arthritis, diabetes and MS are characterized by heterogeneous T cell infiltrates during active disease, a broad spectrum of superantigens could precipitate exacerbation [20]. In general, the disease exacerbating effect of superantigens depends on the presence of encephalitogenic cells in vivo after immunization with myelin antigens or after transfer of specific T cells [2, 3, 17, 20].

Some earlier reports showed that treatment of mice with SEB prior to the induction of EAE or lupus nephritis prevented disease [9, 11, 16, 19] or reduced the severity of the disease symptoms [5]. These observations have led to interest in the potential of superantigen-based specfic therapies for human autoimmune diseases. However, the presence of autoreactive T cells in humans and the possibility of exacerbating preexisting autoimmune disease limit the feasibility of this therapeutic approach. Indeed, activated autoreactive T cells appear to be resistant to superantigen-mediated anergy induction [17, and Brocke and Steinman, unpublished observations].

Recently, the first direct evidence was published that bacterial superantigens play a role in the activation and expansion of T cells from MS patients. Two common superantigens, SEB and TSST-1, were tested for activation of $\gamma\delta$ TCR expressing T cells. All the $\gamma\delta$ T cell clones tested showed a substantial reactivity in vitro to low amounts of the superantigens. Moreover, the T cell clones lysed target cells pulsed with the superantigens, indicating a cytotoxic response elicited by SEB and TSST-1 [22]. Although the evidence for a role of superantigens in autoimmune disease and MS is largely circumstantial, an increasing number of observations point in that direction. The potent stimulation of the immune response by superantigens can lead to acute disease and could reactivate autoreactive cells both systemically as well as inside organ lesions in autoimmune disease.

References

1. Bergdoll MS (1995) Importance of staphylococci that produce nanogram quantities of enterotoxin. Int J Med Microbiol 282: 1
2. Brocke S, et al (1993) Induction of relapsing paralysis in experimental autoimmune encephalomyelitis by bacterial superantigen. Nature 365: 642
3. Brocke S, et al (1994) Infection and multiple sclerosis: a possible role for superantigens? Trends Microbiol 2: 250
4. Burns J, Rosenzweig A, Zweiman B, Lisak RP (1983) Isolation of myelin basic protein-reactive T cell lines from normal human blood. Cell Immunol 81: 435
5. Gaur A, et al (1993), SEB induced anergy: modulation of immune response to T cell determinants of myoglobin and myelin basic protein. J Immunol 150: 3062
6. Hudson KR, et al (1995) Staphylococcal enterotoxin has two cooperative binding sites on major histocompatibility complex class II. J Exp Med 182: 711
7. Hurley JM, et al (1995) Identification of class II major histocompatibility complex and T cell receptor binding sites in the superantigen toxic shock syndrome toxin 1. J. Exp Med 181: 2229
8. Jardetzky TS, et al (1994) Three-dimensional structure of a human class II histocompatibility molecule complexed with superantigen. Nature 368: 711
9. Kalman B, et al (1993) Effects of staphylococcal enterotoxin B on T cell receptor V beta utilization and clinical manifestations of experimental allergic encephalomyelitis. J Neuroimmunol 45: 83
10. Kappler JW, et al (1992) Mutations defining functional regions of the superantigen staphylococcal enterotoxin B. J Exp Med 175: 387
11. Kim C, Siminovitch KA, Ochi A (1991) Reduction of lupus nephritis in MRL/lpr mice by a bacterial superantigen treatment. J Exp Med 174: 1431
12. Kim J, et al (1994) Toxic shock syndrome toxin-1 complexed with a class II major histocompatibility molecule HLA-DR1. Science 266: 1870
13. Kotzin BL (1994) Superantigens and their role in disease. Hosp Prac (Off Ed) 29: 59
14. Kotzin BL, et al (1993) Superantigens and their potential role in human disease. Adv Immunol 54: 99
15. Marrack P, Kappler J (1990) The staphylococcal enterotoxins and their relatives. Science 248: 705
16. Rott O, Wekerle H, Fleischer B (1992) Protection from experimental allergic encephalomyelitis by application of a bacterial superantigen. Int Immunol 4: 347
17. Schiffenbauer J, et al (1993) Staphylococcal enterotoxins can reactivate experimental allergic encephalomyelitis. Proc Natl Acad Sci 90: 8543
18. Seth A, et al (1994) Binary and ternary complexes between T-cell receptor, class II MHC and superantigen in vitro. Nature 369: 324

19. Soos JM, Schiffenbauer J, Johnson HM (1993) Treatment of PL/J mice with the superantigen staphylococcal enterotoxin B prevents development of experimental allergic encephalomyelitis. J Neuroimmunol 43: 39
20. Soos JM, et al (1995) Accelerated induction of experimental allergic encephalomyelitis in PL/J mice by a non-V beta 8-specific superantigen. Proc Natl Acad Sci USA 92: 6082
21. Steinman L (1994) Specific motifs in T cell receptor V beta D beta J beta gene sequences in multiple sclerosis lesions in brain. Behring Inst Mitt 94: 148
22. Stinissen P, et al (1995) Increased frequency of $\gamma\delta$ T cells in cerebrospinal fluid and peripheral blood of patients with multiple sclerosis. J Immunol 154: 4883
23. Swaminathan S, et al (1992) Crystal structure of staphylococcal enterotoxin B, a superantigen. Nature 359: 801
24. Ulrich RG, Bavari S, Olson MA, (1995) Staphylococcal enterotoxins A and B share a common structural motif for binding class II major histocompatibility complex molecules. Nature Struct Biol 2: 554
25. Vinken PJ, et al (eds) (1989) Handb Clin Neurol vol 56, Elsevier Science Publishers

Pathophysiology of the blood-brain barrier

Krzysztof Selmaj

Department of Neurology, Medical Academy of Lodz, 22, Kopcinskiego Street, PL-90-153 Lodz, Poland

Introduction

The blood-brain barrier (BBB) is a unique feature of the central nervous system (CNS), which isolates the brain from the external environment and prevents access of the majority of circulating substances in the blood to the brain. The concept of complete isolation of the brain from the rest of the body was coined at the beginning of the 20th century [60], and was based mainly on studies on xeno- and allogeneic tissue transplants into the brain. Until recently, the brain was regarded to be an 'immunologically privileged organ'. Current studies showed that this is only partially true. The 'immune privilege' of the brain is in fact determined at the level of the BBB. Several elements should be considered when discussing the role of BBB in the pathophysiology of the CNS: (1) the unique anatomy and histology of the BBB that constitute the basis for BBB function; (2) the transportation systems in the BBB; (3) adhesion molecule events on brain endothelial cells which play a crucial role in directing and controlling the blood-derived cell movement into the CNS; and (4) the BBB cell function as producer of and responder to soluble mediators which can, in concert with adhesion molecules, provide mechanisms for BBB opening during pathophysiogical conditions.

Anatomy and histology of the BBB

The BBB forms a unique anatomical structure very different from other blood-organ barriers in the body. The major difference is the impermeable construction of endothelial cells. The fenestrations in endothelial cells, which are commonly present in other organs, are absent in the CNS except in some locations (see below and [63]). Additionally, endothelial cells adhere strongly to each other forming structures specific for CNS called 'tight junctions'. Ultrastructurally, tight junctions comprise two opposing plasma membranes which form a membrane fusion with cytoplasmic densities on either side [81]. These tight junctions prevent cell migration between endothelial cells and transcapillary movement of other compounds. Brain endothelial cells are also coated with a glycocalyx that maintains a negative charge on the luminal surface of the CNS vasculature. Surrounding the endothelial cells on the CNS side is a basal lamina, the major components of which are type IV collagen and the glycoprotein laminin [29]. The basal lamina sometimes encloses contractile cells called pericytes. These cells represent an intermittent layer. The function of pericytes is not yet fully understood and is still under investigation. They are claimed to be involved in endothelial cell contraction and demonstrate some phagocytic activity. Pericytes may also play an important role in controlling BBB permeability [74]. Another special feature of the BBB is a layer of astrocytic end feet; almost the entire abluminal surface of brain microvessels is covered with astrocytic foot processes. Astrocytic end feet (glia limitans) represent an important part of the BBB and secure brain involvement in mechanisms controlling access of external factors into the CNS. Recent studies suggest that astrocytic end feet contribute to the structural integrity of the BBB [44]. It has been shown that the formation of tight junctions is induced by astrocytic foot processes interacting with endothelial cells, and that this process is mediated by soluble factors produced and released by astrocytes [4]. The encountered space between the basal lamina of endothelial cells and that of the glia limitans is called the Virchow-Robin space. Recent studies on BBB function also implicate perivascular macrophages and microglia as constitutive cells of the BBB. At the microscopic level, a striking feature of the BBB is the high number of mitochondria in endothelial cells [64]. To maintain BBB function, a high level of energy, generated in mitochondria, is required. Under normal conditions, the BBB is almost impermeable to blood-borne cells, proteins and the majority of other compounds which cannot cross the BBB by diffusion. Molecules larger than 20 kDa cannot cross the BBB. It should be noted that not all areas of the CNS contain a fully developed BBB. Selective non-BBB regions are present in the pituitary gland, median eminence, area postrema, pineal gland and most importantly in the choroid plexus [87]. During inflammation or an immune reaction, the organized BBB structure undergoes changes which allow access of inflammatory cells and other compounds into the CNS. Leukocytes migrate across the BBB at sites of inflammation through the specialized endothelium formed by large cuboidal cells similar to those seen in lymphoid organs, which are defined as high endothelial venules (HEV) [70]. Although it is still somewhat controversial, it seems that several intra- and extracellular pathways can contribute to the opening of BBB (see below).

Membrane transport across the BBB

The transport of non-cellular elements across the BBB is determined by diffusion, pinocytosis, carrier-mediated transport and receptor-mediated transcytosis. Diffusion across the BBB is only possible for lipid-soluble compounds. Diffusion is the primary mechanism of transportation through the BBB for respiratory gases and some CNS-active drugs such as diazepam, phenobarbital and phenytoin. Ethanol also crosses the BBB by diffusion. The BBB is highly permeable to water which readily enters the CNS by diffusion. Pinocytosis seems to play a less significant role in transportation through the BBB under normal conditions. However, in a variety of pathological states, increased numbers of pinocytotic vesicles have been observed in the BBB endothelial cells [61]. Pinocytosis is defined as engulfment of extracellular fluid by cell membranes leading to formation of intracellular vesicles. These vesicles move from the luminal surface of endothelial cells to the opposite side where the content of the vesicles is released. A variant of pinocytosis is called absorptive-mediated transcytosis. In this transport system, protein is attached to the membrane based on charge and the fragment of membrane with attached protein undergoes a similar process to receptor-mediated transcytosis (see below). Carrier-mediated transport plays an important role as a mechanism which enables compounds with low lipid solubility to cross the BBB. Glucose, the major energy substrate for the brain, is transported across the BBB by carrier-mediated transport. Five isoforms, GLUT 1–5, of glucose transporters, which are insulin independent, have been identified. Two of them, GLUT 1 and GLUT 3, have been localized in the CNS. GLUT 1 is exposed on endothelial cells and is believed to facilitate glucose transport through the BBB [45]. Essential amino acids, i.e., phenyalanine, leucine, tyrosine, isoleucine, valine, tryptophan, methionine and histidine, enter the CNS by carrier-mediated transport using the L system [11]. Similarly, choline and vitamins are transported into the brain by carrier-mediated transport, albeit at a slower rate. For proteins, the major mechanism for crossing of the BBB is receptor-mediated transcytosis [12]. In this system, after binding of the protein to the receptor on the luminal surface of brain capillaries, the complex is internalized and forms intracellular vesicles similar to pinocytotic vesicles. As a result of receptor-mediated transcytosis, the protein reaches the abluminal surface of the BBB. Receptor-mediated transcytosis is particularly important from the immunological point of view since immunoglobulins enter the CNS by this system [107].

The presence of a limited number of mechanisms of transportation across the BBB demonstrates why many blood-borne compounds with low lipid solubility and lack of specific transport carriers or receptors are prevented from entering the CNS. Additionally, several enzymes, such as transaminases, aminopeptidases, endopeptidases and monoamine oxidase [34], are present on the luminal surface of endothelial cells. These enzymes contribute to the protection of the brain by preventing access by many compounds which serve as their substrates. The brain vasculature also contains enzymes metabolizing neurotransmitters, such as GABA transaminase, and enzymes metabolizing several toxins, drugs and exogenous compounds [59]. The enzyme system of brain capillaries has been termed the 'enzymatic brain barrier'.

Immune cell trafficking through the CNS

Under physiological conditions, immune cells are not present in the CNS. For a long time, it was believed that there was no lymphatic drainage from the CNS. This argument was widely used to support the notion that the brain was an 'immune-privileged organ'. More recent studies [67], however, have shown connections between the CNS and the lymphatic system through cervical lymph nodes. Although these connections do not form conventional lymphatic vessels, several authors have demonstrated specific immune responses in the cervical lymph nodes after immunization into the brain [98]. There is now also substantial evidence that T lymphocytes can regularly, albeit not in great numbers, cross the BBB and migrate through the CNS [36]. After entering the CNS, T lymphocytes reacting to neural antigens remain in the brain and initiate inflammation, whereas T lymphocytes which do not recognize CNS antigens exit from the CNS quite rapidly [4]. A necessary early step in the development of T lymphocyte – endothelial cell interaction is the adhesion of immune cells to endothelial cells.

This process depends on the interaction of a series of specific adhesion molecules on endothelial cells with their ligands on the leukocytes (see below). Interestingly, when lymphocyte binding by cerebral endothelium is compared with lymphocyte binding by extracerebral endothelium, it appears that the base level of lymphocyte binding to brain endothelium is much lower [51]. Adhesion to brain endothelial cells by lymphocytes can be greatly increased by pretreatment of brain endothelium with the cytokines, interferon-γ (IFN-γ) and tumor necrosis factor (TNF) [40]. Similarly, lymphocyte activation significantly increases adhesion of lymphocytes to endothelial cells of the BBB. Interestingly, it has been demonstrated that lymphocytes activated with CNS antigen do not show a more effective adhesion than cells activated non-specifically [38]. Therefore, it is believed that the state of activation is more important in inflammatory cell homing to the brain than antigen specificity. There are, however, differences in the kinetics of antigen-specifically and non-specifically activated cells homing to the CNS. In chronological studies involving radiolabeled myelin basic protein (MBP)-reactive cells, the trafficking pattern of MBP-specific lymphocytes in the CNS showed that antigen-specific T cells migrated to the CNS early in inflammatory lesion formation and remained mainly in the perivascular areas [25]. Conversely, nonspecific T lymphocytes entered the CNS later and did not remain in perivascular areas but migrated into the parenchyma. These observations may suggest that antigen-specifically and non-specifically activated cells use different pathways to enter the CNS.

There are also differences in effectiveness of lymphocyte homing to the CNS and phenotypic expression. The phenotypic analysis of cells trafficking through the CNS has revealed a predominance of T lymphocytes [33]. B cells migrate to the CNS much less frequently [92]. Observations on T lymphocyte subset migration into the CNS showed that, as in other organs, $CD4^+$ T lymphocytes also expressing CD29 but not CD45R, represented the major cell population passing into the brain [85]. This phenotype defines helper/inducer memory T lymphocytes, which respond to recall antigens [1]. These results suggest that the CNS is under constant immune system surveillance and that the process is dependent on T lymphocytes.

Adhesion molecules and the BBB

Leukocyte homing to the CNS is orchestrated by adhesion molecules on endothelial cells of the BBB and their counter receptors on immune cells. The adhesion molecules on brain vasculature are basically the same as those present on other organ endothelial cells. Adhesion molecules can be divided into at least three groups according to structural similarities: the selectin, integrin and immunoglobulin (Ig) superfamilies, plus two additional groups, the addressins and cadherins [105]. The selectins can be divided into P-selectins, E-selectins and L-selectins [88]. The selectins are a group of adhesion molecules which are involved in leukocyte 'rolling' along the blood vessels, a phenomenon involving rapid attachment to and detachment from endothelium [39]. Leukocyte 'rolling' is believed to be initially mediated by P-selectin, and occurs early in the cascade of vascular events leading to leukocyte brain homing. Expression of P-selectin is short-lived and E-selectin, expressed on endothelium, with L-selectin, expressed on leukocytes, mediate the 'rolling' phenomenon for longer periods of time. The selectins bind to carbohydrate ligands but E-selectin and L-selectin can act as counter receptors for themselves [65]. Following initial, but not firm, cell attachment to endothelium mediated by selectins, the second phase of adhesion events is mediated by the integrins expressed on activated leukocytes and the Ig expressed on endothelial cells [76]. The Ig superfamily contains intracellular adhesion molecules (ICAM), vascular CAM (VCAM) and LFA-3 (CD58). ICAM-2 is constitutively expressed in the CNS, but ICAM-1 is induced by cytokine stimulation. At present, two adhesion molecules belonging to the Ig family, ICAM-1 and VCAM-1, are thought to play a primary role in mediating lymphocyte and monocyte binding to the brain endothelial cells [69]. The integrins define a group of adhesion molecules which frequently serve as counter receptors for adhesion molecules of the Ig superfamily. The integrins can be divided into three groups according to common β chains, β_1, β_2 and β_3. The molecule belonging to integrin β_1, VLA-4, binds to VCAM-1. This binding, which occurs late after T cell activation, seems to be of particular importance to T cell homing into the brain. The CD4 Th1 clones that are highly positive for VLA-4 enter brain parenchyma in abundance, while VLA-4-low-positive clones do not [5]. The second integrin subfamily, β_2, contains the adhesion molecules LFA-1 (CD11a/CD18) and CR3 (CD11b/CD18-complement receptor type 3). LFA-1, which is expressed on many activated leukocytes, interacts mainly with another member of the Ig superfamily, ICAM-1, but can also interact with ICAM-2. CR3 binds to the C3b component of the alternative complement cascade. Anti-CR3 has been shown to block immune cell recruitment during CNS inflammation [72]. The third subfamily, β_3 integrins, is involved in platelet binding. There is also a calcium-dependent integrin, LPAM-1, which is believed to contribute to lymphocyte homing to HEV. An important category of adhesion molecules is represented by the addressins [10]. These molecules are strongly associated with activated HEV. They are believed to be the ligand for lymphocyte homing receptors, MEL-14/LAM-1 and LPAM-1. In mice, the monoclonal antibody (mAb) MECA-367 (mice endothelial cell antigen) defines a 58–66 kDa molecule that is selectively expressed by endothelial cells involved in lymphocyte homing into mucosal lymphoid sites and was the first addressin found to interact with mice lymphocyte homing receptor, MEL-14. Another mAb, MECA-79, defines a 90-kDa molecule that is selectively expressed on peripheral lymph node HEV. Both addressins are characterized by interaction with mAb MECA-367 and MECA-97 and can mediate lymphocyte extravasation at inflammatory sites [89]. The mAb MECA-325 defines the HEV differentiation antigen at lymphoid tissue and inflammatory sites. In humans, addressins have been designated as HECA (human endothelial cell antigen). Most importantly in the lymphoid system, lymphocyte binding is mediated by addressins specific for different regions. Therefore, addressins seem to serve as tissue-specific adhesion molecules. Future studies might characterize addressins specific for other organs including CNS. Another category of adhesion molecules with tissue specificity are the cadherins [27]. Cadherins that are functionally associated with CNS are called N-cadherins. However, they are believed to be mostly involved in tissue development and differentiation. Studies on adhesion molecule expression on the BBB are shedding new light on leukocyte brain homing. In normal CNS tissue, adhesion molecules are not expressed or expressed at a very low level. However, there are several reports of increased endothelial CAM expression during CNS inflammation. Increased expression of ICAM-1 was described in multiple sclerosis (MS) [86] and experimental autoimmune encephalomyelitis (EAE), the animal model of MS, [6, 99]. Increased expression of VCAM-1 during the course of EAE strongly correlated ($r > 0.8$) with infiltration of T cells, B cells and monocytes into the CNS [6]. Similarly, increased expression of addressins, MECA-325, correlated with inflammatory cell influx into the CNS in EAE [17]. In this regard, interesting data have been provided by the EAE preventive experiments in which antibodies against adhesion molecules have been applied. Anti-ICAM therapy appeared to be very effective in inhibiting EAE induced by active immunization [3]. However, in the passive transfer model of EAE, which is more relevant to MS, anti-ICAM-1 therapy did not show any significant effect [18, 100]. More striking data have been obtained with antibody against VLA-4 ($\alpha_4\beta_1$ integrin) in EAE. This antibody effectively prevented development of EAE, suggesting that VCAM-1/VLA-4 interaction is of particular importance in immune cell brain homing in inflammatory autoimmune conditions [106]. Investigations in vitro have shown that there may also be species differences in the interactions between the brain vasculature and immune cells. Paradoxically, when the adhesion of splenocytes to brain endothelial cells was compared between SJL/J mice, a strain considered autoimmune prone, and BALB/c mice, considered immunologically normal, there was paradoxically lower adhesion in SJL/J mice [96]. In BALB/c mice, the splenocyte adhesion to endothelial cells was partially dependent on LFA-1 interaction with ICAM-1. In SJL/J mice, adhesion was unaffected by EDTA, suggesting that LFA-1/ICAM-1 interaction does not occur and implies the existence of calcium-independent molecules.

BBB function and cytokines

Cytokines are soluble glycoproteins, non-immunoglobulin in nature, released by living cells of the host, which act nonenzymatically in picomolar to nanomolar concentrations to regulate host cell function. Cytokines play a crucial role in immune processes. In the CNS, cytokines contribute, usually by enhancing immune and inflammatory events, to an array of immunopathological conditions [77]. Since the BBB is the primary site of inflammatory events in the CNS, cytokines can significantly influence its function. In general, cytokines enhance the process of leukocyte homing into the CNS [58]. Cytokines like TNF-α and IFN-γ activate cerebral endothelial cells and enhance lymphocyte adhesion [52]. This enhanced adhesion is not blocked by antibodies to the major histocompatibility complex (MHC) class I and class II, suggesting that intercellular adhesion is enhanced. At the same time, lymphocyte activation increases the binding of lymphocytes to cerebral endothelial cells. This might indicate a role for cytokines in the enhancement of lymphocyte binding that might be of particular importance within the CNS, where the constitutive adhesion of lymphocytes to endothelial cells is lower than in other organs.

Studies with radiolabelled albumin have shown that interleukin (IL)-1β and TNF-α induced an increase in BBB permeability [68]. A single microinjection of TNF-α and IFN-γ into the lumbosacral spinal cord of the rat was shown to produce an inflammatory response [82], whereas injection of IL-2 had no effect [104]. The pathological changes were characterized by mononuclear cell adherence to vessels and inflammatory cell accumulation in the meninges and spinal cord. TNF-α was several times more potent than IFN-γ in evoking inflammation within the CNS when measured by the presence of ^{51}Cr-labeled splenocytes in the spinal cord. In another study, the effect of TNF-α on BBB function was investigated by implantation of genetically engineered cells that continuously secreted TNF into the brains of mice. The most striking change was gliosis and neovascularization in the vicinity of the TNF-producing tumor cells. Focal margination of inflammatory cells was also present

[91]. The effect of cytokines on the BBB and inflammation in vivo was investigated in the rabbit eye model [54]. The injection of IL-1 into the vitreous humor of the eye induced enhanced attachment of inflammatory cells to the luminal surface of endothelial cells of the epiretinal blood vessels, and an increased permeability of endothelial cells for labeled albumin. Electrophysiological recordings of epidural-evoked potentials demonstrated that the intraocular injection of IFN-γ, TNF-α and IL-1 increased the peak latency [13]. These results suggest that cytokines are capable of inducing acute but reversible slowing of neuronal conduction that appears to be primarily associated with vascular events.

The cytokine effect on brain vasculature has been studied in vitro in models of cultured monolayers of endothelial cells. Adhesion of lymphocytes to brain endothelial cells increased three- to fourfold after incubation of endothelial cells with IL-1 and IL-6 [26]. Preincubation of endothelial cells with IFN-γ and TNF-α also significantly enhanced lymphocyte adherence [51]. Cytokine-induced lymphocyte adherence to endothelial cells seems to be due to enhanced expression of adhesion molecules, since adherence of lymphocytes to endothelial cells was prevented by mAb against these molecules. In fact, ICAM-1 expression was up-regulated two- to fourfold after human brain endothelial cell treatment with cytokines and lipopolysaccharides in vitro [102]. In endothelial cell monolayers treated with IFN-γ, endothelial cells became spindle-shape and formed whorls. The morphological changes were associated with increased permeability [42].

Additional data on the importance of cytokine-induced adhesion molecule expression on brain vasculature during inflammation have been provided by experiments in EAE. Neutralization of the proinflammatory cytokines TNF-α, TNF-β and IL-1 led to the prevention of EAE. Two studies demonstrated that the administration of either anti-TNF-α or anti-TNF-α/TNF-β antibody after the adoptive transfer of MBP-reactive T cells to induce EAE, effectively prevented the development of EAE both at the clinical and pathological levels [75, 79]. Similarly, in a study in which soluble IL-1 receptor, which binds IL-1, was applied, significant delay of the onset of EAE and shortening of disease duration was observed [43]. A recent study by Barten and Ruddle [6] has provided a link between these earlier studies on anti-cytokine therapy in EAE and adhesion molecules. It has been demonstrated that, during anti-TNF-α antibody treatment, the up-regulation of VCAM-1 is markedly reduced or prevented. Decreased VCAM-1, but not ICAM-1 or fibronectin expression correlates with decreased lymphocyte and monocyte spinal cord homing. These findings once again underscore the importance of a VCAM-1/VLA-4 interaction for inflammatory changes induced in BBB.

The architecture of the BBB, with astrocytic processes in close contact with endothelial cells, implicates an astrocyte role in maintaining BBB function. In this regard, it is of interest that cytokines express a profound effect on astrocytes. IL-1 has been implicated in astrocyte proliferation in vitro,and in vivo, [32]. IL-1 can increase the expression of ICAM-1 on human astrocytes [30] and is also a strong inducer of cytokine production, TNF-α and IL-6, by astrocytes [9, 19]. It has also been demonstrated that IL-1β can induce astrocytes to serve as a source of complement components within the CNS [48]. Activation of the complement cascade by an alternative pathway has been implicated in the destruction of the BBB in inflammatory conditions of the CNS [23]. TNF-α also has a mitogenic effect on astrocytes [78] parallel to its inhibitory effect on glial fibrillary acidic protein gene expression, suggesting that it might induce dedifferentiation of these cells]80]. Like IL-1, TNF-α enhances

ICAM-1 expression on astrocytes [30] and is a potent inducer for other cytokines, IL-6, granulocyte-colony-stimulating factor (G-CSF), granulocyte/macrophage (GM)-CSF, by these cells [9, 53]. In this way, TNF-α might significantly contribute to the attraction of inflammatory cells to the CNS and promote inflammatory processes.

The BBB cells not only respond to cytokines but can also be induced to produce cytokines by themselves. These observations underscore the active role of the BBB in pathological conditions. Brain microvessels and pericytes produce IL-1α, IL-1β, IL-6 and GM-CSF [7]. Astrocytes produce TNF-α, IL-1α, IL-1β, IL-6, transforming growth factor β and GM-CSF [49]. Significant differences can occur in cytokine gene expression depending on the stimulation factor. A similar pattern of cytokines can be produced by perivascular macrophage/microglial cells.

BBB involvement in antigen presentation in the CNS

MHC molecule expression by antigen-presenting or target cells is essential for antigen recognition by CD4$^+$ and CD8$^+$ lymphocytes. One element which is listed as contributory to 'immunological privilege' in the brain is low expression of MHC molecules within the CNS. However, recent studies have clearly demonstrated that antigen presentation can occur in the CNS. Brain endothelium, pericytes, astrocytes and microglia in perivascular areas appear to be involved in antigen recognition and its presentation to T lymphocytes. Expression of MHC molecules by cells of the BBB may suggest that circulating blood lymphocytes can be exposed to CNS antigens. Consequently, it may suggest that antigen recognition which leads to an autoimmune reaction can occur at the level of the BBB. MHC class I and class II molecules may also have a role in cell adhesion, implicating MHC antigens in vascular adhesion events [28]. Upon treatment with IFN-γ astrocytes, oligodendrocytes and microglial cells can express class I molecules [103]. Microinjection of IFN-γ into the brain also induces MHC class I expression. In addition, IFN-γ induces class II molecule expression on endothelial cells and promotes the presentation of myelin antigens [56]. Similarly, IFN-γ can induce MHC class II molecules expressed by astrocytes [95]. It has been suggested that antigen presentation by astrocytes and microglia in the perivascular areas might serve as a secondary line of antigen recognition, leading to further activation of T lymphocytes. TNF-α enhances IFN-γ-induced class II expression on astrocytes [8]. IFN-γ acts at the transcriptional level to induce class II expression by activating class II molecule promoters. Studies in vivo using CNS samples from inflammatory diseases confirmed the expression of MHC molecules on the cells of the BBB, although the exact site of expression is still controversial. MHC class I expression has been demonstrated on brain endothelial cells and its presence correlates with the course of EAE [55]. Sobel et al. [84] localized class II molecules on endothelial cells in EAE and the expression of Ia molecules increased prior to inflammatory cell infiltration. Other authors have observed the presence of MHC class II molecules mostly on perivascular cells, astrocytes [47], and microglial cells [37]. Although TNF-α increases MHC class I expression on astrocytes, it has no influence on class II expression [46]. A recent report demonstrated that TNF-α induces the expression of ICAM-1 and other adhesion molecules on human astrocytes [41]. It has also been demonstrated that astrocytes and endothelial cells can present MBP to T lymphocytes [90], resulting in the lysis of astrocytes by antigen-specific T cells. However, although astrocytes could process and present antigen in the context of appropriate MHC molecules, they were unable to induce T cell proliferation [97]. These results suggest that astrocytes are only partially competent antigen presenting cells. This might be related to the failure of astrocytes to express the B7 molecule. B7 belongs to the group of co-stimulatory molecules required for complete T cell activation. Similar partial T cell activation characterized by activation of the cytotoxic pathway without induction of proliferation has been previously described in brain endothelial cells [73]. The incomplete function of astrocytes and endothelial cells as antigen-presenting cells has led to the interesting hypothesis that the BBB may play a role in inhibiting immune events in the CNS. The results of studies on antigen presentation by astrocytes and brain endothelial cells are in contrast with the microglia study which demonstrated that human microglial cells can act as fully competent antigen-presenting cells [101].

BBB during inflammatory processes

During inflammatory processes, the highly specialized function of the BBB is partially or totally lost and inflammatory cells and other blood-borne compounds have greater access to the CNS. Factors known to disrupt the BBB during inflammation include arachidonic acid or eicosanoids, bradykinin, histamine and cytokines. The increased permeability of the BBB during inflammation may be dependent on several mechanisms, one of which is the opening of tight junctions. In the context of inflammatory/autoimmune processes, the studies which provided most information on BBB function concerned EAE. In EAE, it has been elegantly demonstrated by ultrastructural studies and studies with special tracers that tight junctions are opened in endothelial venules [21, 71]. Lymphocytes accumulate near to opened tight junctions and migrate through parajunctional channel-like structures. Additional structures involved in increased permeability of the BBB are the gaps across endothelial cells [21]. Both the opening of tight junctions and the formation of gaps across endothelial cells seem to be induced by inflammatory cells since they are always associated with the presence of these cells. Although one of the most important differences between the BBB and the systemic vasculature is the lack of fenestrae in brain endothelial cells, such pores can appear during CNS inflammation. This concept was, again, specifically investigated in EAE. Pores were first seen in CNS blood vessels in chronic EAE [83]. It has also been demonstrated that canalicular profiles and tracer-positive intracellular vesicles can be identified in animals with EAE (50). These results have been interpreted as evidence for re-appearance of immature features of the BBB since similar changes have been observed in the developing brain. Ultrastructural analysis of early lesions in EAE revealed the presence of numerous lymphocytes closely attached to brain vasculature (Fig. 1). At points of attachment, a number of unique membrane punctate interactions were observed. Structural analysis of the points of attachment suggested a receptor-ligand interaction. This was characterized by the presence of lymphocyte pseudopodia embedded in the endothelium as well as associations between lymphocytes and clathrin-coated pits on endothelial cells [71]. Longitudinal analysis suggests that lymphocytes entered the CNS directly through endothelial cells and not along tight junctions (Fig. 2). The migration into the CNS is facilitated by the secretion by lymphocytes of the enzyme heparanase [62]. Heparanase mediates degradation of the heparan sulfate proteoglycans of the extracellular matrix. Another way for inflammatory cells to reach the CNS during severe inflammatory/autoimmune conditions might be through the transient destruction of endothelial cells allowing them rapid passage into the CNS. It has been demonstrated that MBP-reactive encephalitogenic T cell lines are capable of damaging MHC class II-positive endothelial cells [57]. Depending on the state of activation, T cell-mediated lysis of endothelial cells might also be non-MHC restricted. Neutrophils may also contribute to endothelial cell destruction by the production and release of oxygen radicals and proteolitic enzymes. With regard to non-cellular transport through BBB during inflammation, two mechanisms are being considered: increased fluid-phase transcytosis and increased receptor-mediated transcytosis. The first mechanism is directly associated with a decrease in mitochondrial content [20]. This process is also highly correlated with the clinical course of EAE [35]. Using horseradish peroxidase as a tracer, it has been demonstrated that the number of transcytotic vesicles increases fivefold when EAE is at its peak. Receptor mediated-transcytosis provides the mechanism for many proteins to enter the brain (see above). A morphological equivalent of receptor-mediated transcytosis is represented by coated vesicles in capillary endothelium. Although in EAE the number of coated vesicles in brain capillaries does not seem to change, in several other inflammatory models an increased incidence of coated vesicles has been described [92]. In recent years, magnetic resonance imaging (MRI) has become an important technique for studies on BBB permeability in CNS diseases. Using the T_1 and T_2-weighted images for change in water content in connection with the MR contrast reagents, BBB damage has been demonstrated in many inflammatory conditions. Although MRI appeared to be extremely useful in quantitative studies on BBB permeability in situ, it provided relatively little information on the mechanism of BBB dysfunction. Interesting sources of information on BBB dysfunction during inflammatory processes have been provided by studies of MS, neuroviral infections and AIDS. Immunocytochemical analysis demonstrated a well-defined ring of plasma protein leakage around capillaries and small veins in MS plaques [31]. Endothelial cells of vessels within inflammatory plaques contain many vesicles which are immunopositive for fibrinogen and α_2 macroglobulin. These results complement earlier findings on excessive micropinocytosis described in MS biopsies [14]. The density of pinocytotic vesicles in endothelium in biopsied CNS tissue from chronic-progressive MS was increased twofold in gliotic areas. Mitochodria were decreased in the same respective samples. Thus, an inverse correlation between vesicular and mitochodrial content occurs in MS [22]. Recent studies demonstrate an overexpression of adhesion molecules on inflamed veins in active MS lesions (Fig. 3). Immunocytochemical studies have provided evidence that the VCAM-1/VLA-4 adhesion pathway is expressed at higher levels in chronic MS lesions, while ICAM-1/LFA-1 is used more uniformly. The expression of most adhesion molecules and some cytokines is negligible in normal CNS tissue [15]. However, no adhesion or cytokine molecule or pattern of expression specific for MS is apparent. In AIDS, a significant increase in BBB permeability, measured by extravasation of fibrinogen and IgG, has been described [66]. With regard to bacterial infections of the CNS, a hypothesis has been proposed that anti-bacterial antibodies reversibly open the BBB by interacting with an antigen which mimics the native ligand for the CR3 integrin on endothelial cells [94]. In CNS viral infections, serious dysfunction of the BBB is mediated by Thy-1^+, $CD8^+$, class I-restricted T lymphocytes [2]. All these results imply a substantial involvement of immune mechanisms in inducing BBB dysfunction in inflammatory diseases.

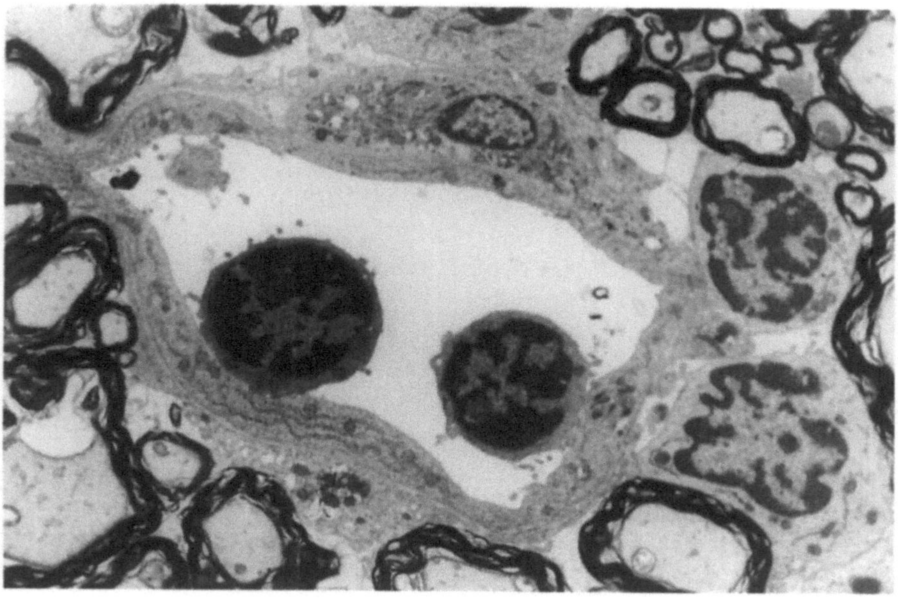

Fig. 1. Electron micrograph of blood-brain barrier in acute experimental allergic encephalomyelitis (EAE) in mice; two lymphocytes attached to endothelial cells; note punctate adhesion interactions between lymphocyte membrane and surface of endothelial cells (Courtesy of Prof. Raine). ×7000

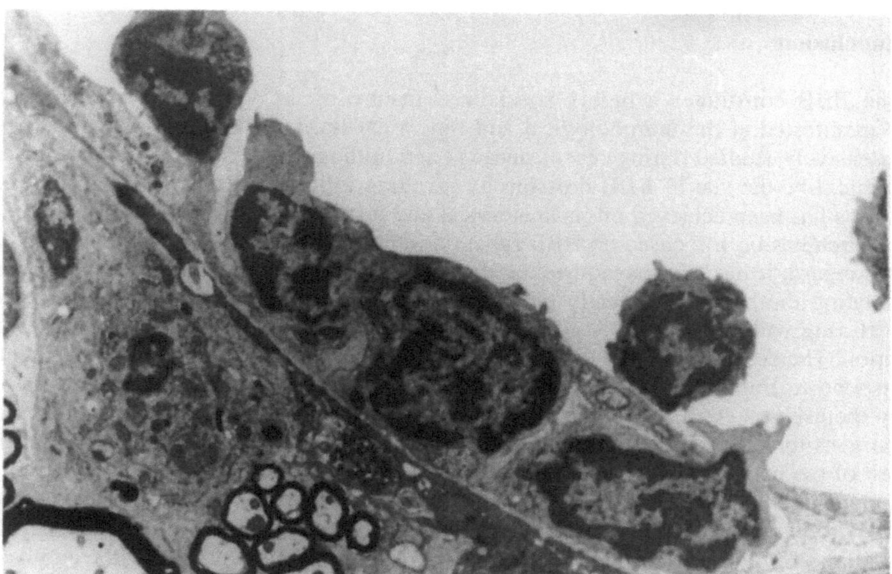

Fig. 2. Electron micrograph of blood-brain barrier in acute EAE in mice; note a lymphocyte that cuts across the cytoplasm of endothelial cell (Courtesy of Prof. C.S. Raine). ×7000

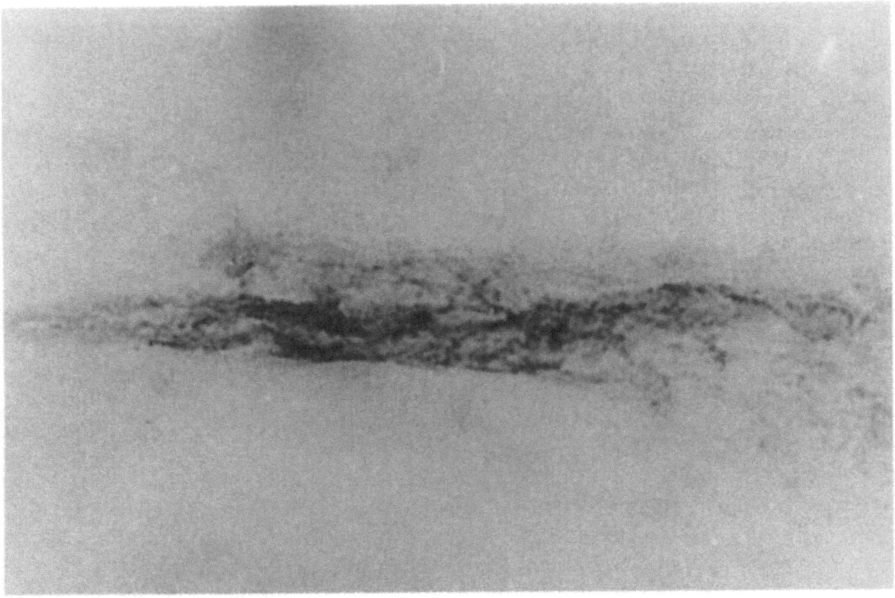

Fig. 3. A frozen section from an active multiple sclerosis lesion; ICAM-1-positive endothelial cells demonstrated by avidin-biotin-complex technique (Courtesy of Prof. C.S. Raine)

Conclusion

The BBB constitutes a highly specialized feature of the CNS. This specialization is manifested at the morphological and functional level. BBB morphology has been extensively studied during recent decades and, although there are still some unclear points, like the site of BBB crossing by invading inflammatory cells, a general consensus has been achieved on its anatomical and histological components. More recent research has been focused on BBB function in normal and pathological conditions. In this regard, it has been demonstrated that the BBB is not a passive protective barrier isolating the brain,but actually plays an active role in pathophysiological processes occurring within the CNS. A good example is the role of the BBB in antigen presentation. The cells of the BBB can express unique molecules of the immune system, can respond to immune mediators and can even be induced to produce these mediators by themselves. As a result, the BBB may participate in the presentation of neuronal antigens to blood-derived immune cells. Since the BBB separates the brain from the rest of the body, it also plays a crucial role in controlling access of blood-borne compounds and cells to the CNS. One of the most important challenges for the future of BBB research, is the characterization of the molecular mechanisms involved in the homing of blood cell to the brain. In the era of molecular immunology, the emphasis of BBB research has moved towards investigations of adhesion receptors on BBB cells and their interaction with counter receptors on blood-borne cells. Several adhesion events have already been discovered, but undoubtedly more will appear in the near future. The exciting possibility exists that an adhesion molecule, an addressin, specific for the BBB will be discovered. This should allow for selective intervention in immune cell homing into the brain. In that context, new therapeutic approaches for inflammatory and autoimmune conditions of the CNS would be possible.

Acknowledgement. I would like to express my appreciation to Prof. Cedric S. Raine, Albert Einstein College of Medicine, New York, for his helpful advice and assistance in the editing of the manuscript.

References

1. Akbar AN, Terry L, Timms A, Beverley PCL, Janossy G (1988) Loss of CD45R and gain of UCHL-1 reactivity is a feature of primed T cells. J Immunol 140:2170
2. Andersen IH, Marker O, Thomsen AR (1991) Breakdown of blood-brain barrier function in the murine lymphocytic choriomeningitis virus infection mediated by virus-specific $CD4^{+}$T cell. J Neuroimmunol 31:155
3. Archelos JJ, Jung S, Maurer M, Schmied M, Lassmann H, Tamatani T, Miyasaka M, Toyka KV, Hartung H-P (1993) Inhibition of experimental autoimmune encephalomyelitis by an antibody to the intercellular adhesion molecule ICAM-1. Ann Neurol 34:145
4. Arthur FE, Shivers RR, Bowman PD (1987) Astrocyte-mediated induction of tight junctions in brain capillary endothelium:an efficient in vitro model. Dev Brain Res 36:155
5. Baron JL, Madri JA, Ruddle NH, Hashim G, Janeway CA (1993) Surface expression of a4 integrin by CD4 T cells is required for their entry into brain parenchyma. J Exp Med 177:57
6. Barten DM, Ruddle NH (1994) Vascular cell adhesion molecule-1 modulation by tumor necrosis factor in experimental allergic encephalomyelitis. J Neuroimmunol 51:123
7. Benveniste E (1992) Inflammatory cytokines within the central nervous system:sources, function, and mechanism of action. Am J Physiol 263:C1
8. Benveniste EN, Sparacio SM, Bethea JR (1989) Tumor necrosis factor-α enhances interferon-γ mediated class II antigen expression on astrocytes. J Neuroimmunol 25:209
9. Benveniste EN, Sparacio SM, Norris JG, Grnett HE, Fuller GM (1990) Induction and regulation of interleukin-6 gene expression in rat astrocytes. J Neuroimmunol 30:201
10. Berg EL, Goldstein LA, Jutila MA, Nakache M, Picker LJ, Streeter PR, Wu NW, Zhou D, Butcher EC (1989) Homing receptors and vascular addressins, cell adhesion molecules that direct lymphocyte traffic. Immunol Rev 108:5
11. Betz AL, Goldstein G (1978) Polarity of the blood-brain barrier:Oeutral amino acid transport into isolated brain capillaries. Science 202:225
12. Broadwell RD (1989) Transcytosis of macromolecules through the blood-brain barrier:a cell biological perspective and critical appraisal. Acta Neuropathol 79:117
13. Brosnan CF, Litwak MS, Schroeder C, Selmaj K, Raine CS, Arezzo JC (1989) Preliminary studies of cytokine-induced functional effects on the visual pathways in the rabbit. J Neuroimmunol 25:227
14. Brown WJ (1978) The capillaries in acute and subacute multiple sclerosis plaques:a morphometric analysis. Neurology [Suppl] 28:84
15. Cannelia B, Raine CS (1995) The adhesion molecule and cytokine profile of multiple sclerosis lesions. Ann Neurol 37 (in press)
16. Cannelia B, Cross AH, Raine CS (1990) Up-regulation and co-expression of adhesion molecules correlate with relapsing autoimmune demyelination in the central nervous system. J Exp Med 172:1521
17. Cannelia B, Cross AH, Raine CS (1991) Relapsing autoimmune demyelination:a role for vascular addressins. J Neuroimmunol 35:295
18. Cannelia B, Cross AH, Raine CS (1993) Anti-adhesion molecule therapy in experimental autoimmune encephalomyelitis. J Neuroimmunol 46:43
19. Chung IY, Benveniste EN (1990) Tumor necrosis factor-alpha production by astrocytes:induction by lipoplysaccharide, interferon-gamma and interleukin-1. J Immunol 144:2999
20. Claudio L, Kress Y, Norton WT, Brosnan CF (1989) Increased vesicular transport and decreased mitochondrial mitochondrial content in blood-brain barrier endothelial cells during experimental autoimmune encephalomyelitis. Am J Pathol 135:446
21. Claudio L, Kress Y, Factor J. Brosnan CF (1990) Mechanisms of edema formation in experimental autoimmune encephalomyelitis. The contribution of inflammatory cells. Am J Pathol 137:1033

22. Claudio L, Raine CS, Brosnan CF (1995) Evidence of generalized bloodbrain barrier abnormalities in chronic progressive multiple sclerosis. Acta Neuropathol (in press)
23. Compston DAS, Morgan BP, Campbell AK, Wilkins P, Cole G, Thomas ND (1989) Immunocytochemical localization of the terminal complement complex in multiple sclerosis. Neuropathol Appl Neurobiol 15:307
24. Cross AH (1992) Immune cell traffic control and the central nervous system. Semin Neurosci 4:213
25. Cross AH, Canneila B, Brosnan CF, Raine CS (1990) Homing to central nervous system vasculature by antigen specific lymphocytes. I. Localization of ^{14}C-labeled cells during acute, chronic and relapsing experimental allergic encephalomyelitis. Lab Invest 63:162
26. De Vries HE, Moor ACE, Blom-Roosemalen MCM, Boer AG de, Breimer DD, Berkel TJC van, Kuiper J (1994) Lymphocyte adhesion to brain capillary endothelial cells in vitro. J Neuroimmunol 52:1
27. Donalies M, Cramer M, Ringwald M, Starzinski-Powitz A, (1991) Expression of M-cadherin, a member of the cadherin multigene family, correlates with differentiation of skeletal muscle cells. Proc Natl Acad Sci USA 88:8024
28. Doyle C, Strominger JL (1987) Interaction between $CD4^{+}$and class II MHC molecules mediates cell adhesion. Nature 330:256
29. Essner E, Lin W-L (1988) Immunocytochemical localization of laminin, type IV collagen and fibronectin in rat retinal vessels. Exp Eye Res 47:317
30. Frohman EM, Frohman TC, Dustin ML, Vayuvegula B, Choi B, Gupta A, Noort S van den, Gupta S (1989) The induction of intracellular adhesion molecule 1 (ICAM-1) expression on human fetal astrocytes by interferon-γ, tumor necrosis factor-α, lymphotoxin, and interleukin-1: relevance to intracerebral antigen presentation. J Neuroimmunol 23:117
31. Gay D, Esiri M (1991) Blood-brain Barrier damage in acute multiple sclerosis plaques. Brain 114:557
32. Giulian D, Lachman LB (1985) lnterieukin-1 stimulation of astroglial proliferation after brain injury. Science 228:497
33. Hailer DA, Weiner HL (1987) In vivo labeling of blood T cells:rapid traffic into cerebrospinal fluid in multiple scierosis. Ann Neurol 22:89
34. Hardebo JE, Owman C (1979) Barrier mechanisms for neurotransmitter monoamines and their precursors at the blood-brain barrier. Ann Neurol 8:1
35. Hawkins CP, Mackenzie F, Tofts P, Du Boulay EPGH, McDonald WI (1991) Patterns of blood-brain barrier breakdown in inflammatory demyelination. Brain 114:801
36. Hickey WF, Kimura H (1987) Graft-vs-host disease elicits expression of class I and class II histocompatibility antigens and the presence of scattered T lymphocytes in rat central nervous system. Proc Natl Acad Sci USA 84:2082
37. Hickey WF, Kimura H (1988) Perivascular microglial cells of the CNS are bone marrow-derived and present antigen in vivo. Science 239:290
38. Hickey WF, Hsu GL, Kimura H (1991) T-Iymphocyte entry into the central nervous system. J Neurosci Res 28:254
39. Hogg N (1992) Roll, roll, roll your leukocyte gently down the vein. Immunol Today 13:113
40. Hughes CCW, Male DK, Lantos PL (1988) Adhesion of lymphocytes to cerebral microvascular cells:effects of interferon-γ, tumor necrosis factor and interleukin-1. Immunology 64:677
41. Hurwitz AA, Lyman WD, Guida MP, Calderon TM, Berman JW (1992) Tumor necrosis factor α induces adhesion molecule expression on human fetal astrocytes. J Exp Med 176:1631
42. Huynh HK, Dorovini-Zis K (1993) Effects of interferon-gamma on primary cultures of human brain microvessels endothelial cells. Am J Pathol 142:1265
43. Jacobs CA, Baker PE, Rous ER, Picha KS, Toivola B, Waugh S, Kennedy MK (1991) Experimental autoimmune encephalomyelitis is exacerbated by IL-1α and suppressed by soluble IL-1 receptor. J Immunol 146:2983
44. Janzer RC, Raff MC (1987) Astrocytes induce blood-brain barrier properties in endothelial cells. Nature 325:253
45. Malaria RN, Gravina SA, Schmidley JW, Perry G, Harik SI (1988) The glucose transporter of the human brain and blood-brain barrier. Ann Neurol 24:75
46. Lavi E, Suzumura A, Murasko DM, Murray EM, Silberberg H, Weiss SR (1988) Tumor necrosis factor induces expression of MHC class I antigens in mouse astrocytes. J Neuroimmunol 18:245

47. Lee SC, Moore GRW, Golenowsky G, Raine CS (1990) Multiple sclerosis:a role for astrogiia in active demyelination suggested by class II MHC expression and ultrastructural study. J Neuropathol Exp Neurol 49:122
48. Levi-Strauss M, Mallat M (1987) Primary cultures of murine astrocytes produce C3 and factor B, two components of the alternative pathway of complement activation. J Immunol 139:2361
49. Lieberman AP, Pitha PM, Shin HS, Shin ML (1989) Production of tumor necrosis factor and other cytokines by astrocytes stimulated with lipopolysaccharide or a neurotropic virus. Proc Natl Acad Sci USA 86:6348
50. Lossinsky AS, Badmajew V, Robson JA, Moretz RC, Wisniewski H (1989) Sites of egress of inflammatory cells and horseradish peroxidase transport across the blood-brain barrier in a murine model of chronic relapsing experimental allergic encephalomyelitis. Acta Neuropathol 78:359
51. Male DK, Pryce G, Hughes C, Lantos P (1990) Lymphocyte migration into brain modelled in vitro:control by lymphocyte activation, cytokines and antigen. Cell Immunol 127:1
52. Male D, Pryce G, Linke A, Rahman J (1992) Lymphocyte migration into the CNS modeled in vitro. J Neuroimmunol 40:167
53. Malipiero UV, Frei K, Fontana A (1990) Production of hemopoietic colony-stimulating factors by astrocytes. J Immunol 144:3816
54. Martiney JA, Litwak MS, Berman JW, Arezzo JC, Brosnan CF (1990) Pathophysiologic effect of IL-1β in the rabbit retina. Am J Pathol 137:1411
55. Matsumoto Y, Fujiwara M (1986) In situ detection of class I and II major histocompatibility complex antigens in the rat cerebral nervous system during experimental allergic encephalomyelitis:an immunohistochemical study. J Neuroimmunol 12:256
56. McCarron RM, Kempsi O, Spatz M, McFariin DE (1985) Presentation of myelin basic protein by murine cerebral vascular endothelial cells. J Immunol 134:3100
57. McCarron R, Racke M, Spatz M, McFarlin DE (1991) Cerebral vascular endothelial cells are effective targets for in vitro lysis by encephalitogenic T lymphocytes. J Immunol 147:503
58. McCarron RM, Wang L, Racke MK, McFarlin DE, Spatz M (1993) Cytokine-regulated adhesion between encephalitogenic T lymphocytes and cerebrovascular endothelial cells. J Neuroimmunol 43:23
59. Minn A, Ghersi-Egea JF, Perrin R, Leininger B, Siest G (1991) Drug metabolizing enzymes in the brain and cerebral microvessels. Brain Res Rev 16:65
60. Murphy JB, Sturm E (1923) Conditions determining the transplantability of tissue in the brain. J Exp Med 38:183
61. Nag S, Robertson DM, Dinsdale HB (1979) Quantitative estimate of pinocytosis in acute experimental hypertension. Acta Neuropathol (Berl) 46:1
62. Naparstek Y, Cohen IR, Fuks Z, Vlodarsky I (1984) Activated T lymphocytes produce a matrix-degrading heparin sulphate endoglycosidase. Nature 310:241
63. Olendorf WH (1977) The blood-brain barrier. Eye Exp Res [Suppl] 25:177
64. Olendorf WH, Cornford ME, Braun WJ (1977) The large apparent work capability of the blood-brain barrier:a study of the mitochondrial content of capillary endothelial cells in brain and other tissues of the rat. Ann Neurol 1:409
65. Osborn L (1990) Leukocyte adhesion to endothelium in inflammation. Cell 62:3
66. Petito CK, Cash KS (1992) Blood-brain barrier abnormalities in the acquired immunodeficiency syndrome:immunohistochemical localization of serum proteins in postmortem brain. Ann Neurol 32:658
67. Prineas JW (1979) Multiple sclerosis:presence of lymphatic capillaries and lymphoid tissue in the brain and spinal cord. Science 203:1123
68. Quagliarello VJ, Wispelwey B, Long WJ, Scheld WM (1991) Recombinant human interleukin-1 induces meningitis and blood-brain barrier injury in the rat. J Clin Invest 87:1360
69. Raine CS, Cannelia B (1992) Adhesion molecules and central nervous system inflammation. Semin Neurosci 4:201
70. Raine CE, Lee S, Scheinberg LC, Duijvestijn A, Cross AH (1990) Adhesion molecules on endothelial cells in the central nervous system:an emerging area in the neuroimmunology of multiple sclerosis. Clin Immunol Immunopathol 57:173
71. Raine CS, Cannelia B, Duijvestijn AM, Cross AH (1990) Homing to central nervous system vasculature by antigen-specific lymphocytes. II. Lymphocyte/endothelial cell adhesion during the initial stages of autoimmune demyelination. Lab Invest 63:476

72. Reid DM, Perry VH, Andersson PB, Gordon S (1993) Mitosis and apoptosis of microglia in vivo by an anti-CR3 antibody which crosses the blood-brain barrier. Neuroscience 56:529
73. Risau W, Engelhardt B, Wekerle H (1990) Immune function of the bloodbrain barrier:incomplete presentation of protein (auto-)antigens by rat microvascular endothelium in vitro. J Cell Biol 110:1757
74. Risau W, Dingier A, Albrecht U, Dehouck MP, Cecchelli R (1992) Blood-brain barrier pericytes are the main source of gamma-glutamyltranspeptidase activity in brain capillaries. J Neurochem 58:667
75. Ruddle NH, Bergman CH, McGrath C, Lingenheld EL, Grunnet ML, Padula SJ, Clark RP (1990) An antibody to lymphotoxin and tumor necrosis factor prevents transfer of experimental allergic encephalomyelitis. J Exp Med 172:1193
76. Ruoslahti E (1991) integrins. J Clin Invest 87:1
77. Selmaj K (1992) The role of cytokines in inflammatory conditions of the central nervous system. Semin Neurosci 4:221
78. Selmaj K, Farooq M, Norton WT, Raine CS, Brosnan CF (1990) Proliferation of astrocytes in vitro in response to cytokines. A primary role for tumor necrosis factor. J Immunol 144:129
79. Selmaj K, Raine CS, Cross AH (1991) Anti-tumor necrosis factor therapy abrogates autoimmune demyelination. Ann Neurol 30:694
80. Selmaj K, Shafit-Zagardo B, Farooq M, Aquino D, Norton WT, Raine CS, Brosnan CF (1991) TNF-induced proliferation of astrocytes from mature brain is associated with down-regulation of GFAP mRNA. J Neurochem 51:823
81. Simionescu M, Simionescu N, Palade GE (1978) Segmental differentiationes of cell junctions in the vascular endothelium. Cell Biol 68:705
82. Simmons RD, Willenborg DO (1990) Direct injection of cytokines into the spinal cord causes autoimmune encephalomyelitis-like inflammation. J Neurol Sci 100:37
83. Snyder D, Hirano A, Raine CS (1975) Fenestrated CNS blood vessels in chronic experimental allergic encephalitis. Brain Res 100:645
84. Sobel RA, Blanchette BE, Bhan AK, Colvin RB (1984) The immunopathology of experimental allergic encephalomyelitis. I. Endothelial cells. Ia increases prior to inflammatory cell infiltration. J Immunol 132:2402
85. Sobel RA, Hailer DA, Castro EE, Morimoto C, Weiner HL (1988) The 2H4 ($CD4^+R$) antigen is selectively decreased in multiple sclerosis lesions. J Immunol 140:2210
86. Sobel RA, Mitchell ME, Fondren G (1990) Intercellular adhesion molecule-1 (ICAM-1) in cellular immune reactions in the human central nervous system. Am J Pathol 136:1309
87. Spector R, Johnson CE (1989) The mammalian choroid plexus. Sci Am 261:68
88. Springer TA (1990) Adhesion receptors of the immune system. Nature 346:425
89. Streeter PR, Rouse BTN, Butcher EC (1988) Immunohistologic and functional characterization of a vascular addressin involved in lymphocyte homing into peripheral lymph nodes. J Cell Biol 107:1853
90. Sun D, Wekerle H (1986) Ia restricted encephalitogenic T lymphocytes mediating EAE lyse autoantigen-presenting astrocytes. Nature 320:70
91. Tracey KJ, Morgello S, Kopkin B, Fahey TJ, Fox J, Aledo A, Manogue KR, Cerami A (1990) metabolic effects of cachectin/tumor necrosis factor are modified by site of production. J Clin Invest 86:2014
92. Triguero D, Buciak JB, Yang J, Partridge WM (1989) Blood-brain barrier transport of canonized immunoglobulin G:enhanced delivery compared to native protein. Proc Natl Acad Sci USA 86:4761
93. Trotter J, Steinman L (1984) Homing of $Lyt\text{-}2^+$ and $Lyt\text{-}2^-$ T cell subsets and B lymphocytes to the central nervous system of mice with acute experimental allergic encephalomielitis. J Immunol 132:2919
94. Tuomanen EI, Prasad SM, George JS, Hoepeiman AI, Ibsen P, Heron I, Starzyk RM (1993) Reversible opening of the blood-brain barrier by antibacterial antibodies. Proc Natl Acad Sci USA 90:7824
95. Vidovoc M, Sparacio SM, Elowitz M, Benveniste EN (1990) Induction and regulation of class II MHC mRNA expression in astrocytes by IFN-γ and TNF-α J Neuroimmunol 30:189
96. Waldschmidt MM, Fabry Z, Keiner J, Love-Homan L, Hart MN (1991) Adhesion of splenocytes to brain microvascular endothelium in the BALB/c and SJL/j mouse systems. J Neuroimmunol 35:191
97. Weber F, Meini E, Aloisi F, Nevinny-Stickel C, Albert E, Wekerle H, Hohlfeld R (1994) Human astrocytes are only partially competent antigen presenting cells. Brain 117:59
98. Widner H, Moller G, Johansson BB (1988) Immune response in deep cervical lymph nodes and spleen in the mouse after antigen deposition in different intracerebral sites. Scand J Immunol 28:563

99. Wilcox CE, Ward AMV, Evans A, Baker B, Rothlein R, Turk JL (1990) Endothelial cell expression of the intercellular adhesion molecule-1 (ICAM-1) in the central nervous system of guinea pigs during acute and chronic relapsing experimental allergic encephalomyelitis. J Neuroimmunol 30:43
100. Willenborg DO, Simmons RD, Tamatani T, Miyasaka M (1993) ICAM-1-dependent pathway is not critically involved in the inflammatory process of autoimmune encephalomyelitis or in cytokine-induced inflammation of the central nervous system. J Neuroimmunol 45:147
101. Williams K, Bar-Or A, Ulvestad E, Olivier A, Antel JP, Young VW (1992) Biology of adult human microglia in culture:comparison with peripheral blood monocytes and astrocytes. J Neuropathol Exp Neurol 51:538
102. Wong D, Dorovini-Zis K (1992) Upregulation of intercellular adhesion molecule 1 (ICAM-1) in primary cultures of human brain microvessel endothelial cells by cytokines and lipopolysaccharide. J Neuroimmunol 39:11
103. Wong GHW, Bariett PF, Clark-Lewis I, Battye F, Schrader JW (1984) Inducible expression of H-2 and Ia antigens on brain cells. Nature 310:688
104. Yamasaki T, Kikuch H, Pain JT, Yamashita J, Miyatake S, Iwasaki K, Koboyashi H, Namba Y, Hanaoka M (1989) Murine intracerebral interleukin-2 injection pathological and immunological effects. J Neurosurg 71:732
105. Yednock TA, Rosen SD (1989) Lymphocyte homing. Adv Immunol 44:313
106. Yednock TA, Cannon C, Fritz LC, Sanchez-Madrid F, Steinman L, Karin N (1992) Prevention of experimental autoimmune encephalomyelitis by antibody against $\alpha4/\beta1$ integrin. Nature 356:63
107. Zlokovic BV, Skundric DS, Segal MB, Lipovic MN, Mackie JB, Davson H (1990) A saturable mechanism for transport of immunoglobulin G across the blood-brain barrier of the guinea pig. Exp Neurol 107:263

Peripheral neuropathy in polyarteritis nodosa

Gérard Said

Service de Neurologie, CHU de Bicêtre, F-94275 Le Kremlin Bicêtre, France

Introduction

Vasculitis which has come to be applied to vascular inflammation syndromes can be observed as a primary feature in connective tissue disorders, or as a secondary feature in a number of inflammatory, infective and immune-mediated disorders of the peripheral nervous system (PNS). Primary vasculitides are assumed to have an autoimmune pathogenesis, but the treatment of primary vasculitis, in spite of attempts to use innovative treatments based on presumed pathophysiological mechanisms empirically rests on corticosteroids. In primary vasculitis of the type observed in polyarteritis nodosa, the nerve blood vessels are often affected and nerve and muscle biopsies represent a useful tool for diagnosis.

Definitions: vasculitis and nerve lesions

Classification of vasculitis

In spite of a number of substantial efforts, the nomenclature of the systemic vasculitides remains enigmatic [19, 23, 43], and vasculitides are often classified according to the size of vessels predominantly affected, but the overlaps are common. The mechanism of the nerve lesions associated with vasculitis is rather uniform, with nerve ischemia as the common consequence of the occlusion of blood vessels in the different patterns of necrotizing arteritis (NA). In polyarteritis nodosa (PAN), the size of the blood vessels affected corresponds to that of most epineurial arteries, thus lesions and occlusion of a large number of such vessels induce nerve ischemia with subsequent axonal degeneration of the vast majority of nerve fibers.

Immunological aspects

It is widely accepted that NA of the type observed in PAN is related to the formation of soluble, circulating immune complexes, which is a logical consequence of the large-scale synthesis of antibodies by plasma cells. In spite of the evidence for an autoimmune mechanism, the diagnosis of PAN remains histological. Histological criteria include: transmural infiltration of small arteries with polymorphonuclear cells, often admixed with lymphocytes and eosinophils; leukocytoclasia; fibrinoid necrosis; destruction of the internal elastic lamina; occlusion of the lumen; and the usual sparing of adjacent venules. The pathophysiological significance of immune complexes contrasts with the contestable usefulness of their demonstration for disease description and therapy [28]. The immune complexes are phagocytosed through the interaction of neutrophil Fc receptors with the Fc portion of the complexed antibody [18]. Proteolytic enzymes, free radicals and various inflammatory peptides may be released by neutrophilic polymorphonuclear cells, which damage the vessel wall [26, 40]. The complement attack complex (C5b-9) may also contribute to the vessel wall injury [21, 40]. Deposits of immunoglobulins and fractions of complement in the epineurial vessels of nerve biopsy samples from patients with vasculitic neuropathy are often found in this setting [17, 20, 30], but their diagnostic value is not established.

Other humoral mechanisms may be involved in the pathogenesis of vasculitis, including anti-endothelial cells antibodies found in several connective tissue and systemic disorders, but their presence is very variable. They are more often found in conditions that are seldom associated with PNS involvement, such as scleroderma, Kawasaki's disease or systemic lupus erythematosus. Thus, the role played by such antibodies and the usefulness of their detection in sera from patients with vasculitic neuropathy remain to establish [12]. In patients with Wegener's granulomatosis anti-neutrophil cytoplasmic autoantibodies (ANCA) are commonly detected, and their serum level seems to correlate with disease activity [16, 32]. ANCA are able to induce neutrophil degranulation and inactivate protease inhibitors [13], as well as damage endothelial cells through activation of polymorphonuclear cells and release of reactive oxygen species [10]. However, the identification of this important marker of disease activity in Wegener's granulomatosis is not a reliable marker for polyarteritis nodosa, which is responsible for the vast majority of vasculitic neuropathy.

Cellular factors should also be considered in vasculitis. In this respect macrophages and cytotoxic T lymphocytes are likely to play a role in vasculitic lesions [17, 21, 30]. Cytotoxic lymphocytes specific for major histocompatibility complex class II molecules, after activation by interferon-γ, also seem able to lyse endothelial cells in vitro [34]. Endothelial cells can serve as antigen-presenting cells and can be induced to secrete a number of cytokines, including IL-6, IL-1 and intercellular adhesion molecules, which can recruit and activate lymphocytes and increase cell damage [31].

Pathophysiology of nerve lesions associated with vasculitis

NA is the most common primary vasculitis responsible for vasculitic neuropathy. This process, which is related to local or systemic vascular inflammation, involves large or small vessels, and produces vessel and tissue necrosis with typical leukocytic infiltration, leukocytoclasia, hemorrhage and fibrinoid change [43]. This pattern of vasculitis is commonly associated with symptomatic or subclinical peripheral nerve involvement due to nerve ischemia [22, 24]. The same pattern of blood vessel lesions occurs in allergic angiitis.

In this group of vasculitis, NA of the PAN type, which represents the most homogeneous entity, occurs *de novo* in PAN or as a secondary feature in diseases such as rheumatoid arthritis and, occasionally, in other connective tissue disorders. The consequences of vascular inflammation and occlusion depend on the size and number of blood vessels affected. Clinical neuropathy occurs in 50–75% of the patients with systemic vasculitis of the PAN group [27]. The ischemic neuropathy induced by NA can be observed in apparent isolation [8, 20, 37, 41] or in the context of a multisystemic disorder.

We have recently reviewed the clinical and pathological data of 200 patients in whom we found NA in nerve and/or in muscle biopsy specimens (Said and Lacroix, in preparation). NA occurred in the context of multisystem disorders in 131 patients (65%), including 116 patients with connective tissue disorders.

NA and multisystemic disorder

Patients in this group commonly manifested loss of weight, fever, asthenia, myalgia, in association with multisystemic involvement like arthritis and cutaneous vasculitis; renal lesions were seen in 10% of these patients.

Classic polyarteritis nodosa

Classic PAN affected 36% of our patients. In these patients multisystem involvement was present, with cutaneous vasculitis as the most common non neurological manifestation. Specific skin involvement including livedo, cutaneous necrosis and nodules, were present in 22 patients. In 17 patients, nonspecific edema, usually affecting one limb extremity, was present. In patients with cutaneous vasculitis and neuropathy, a nerve biopsy should be performed for diagnostic purpose only in those in whom the skin biopsy did not provide specific vasculitic lesions. Renal involvement was observed in 10% of the patients. Asthma was present in 12.6% of patients of another large series [15].

The much-debated Churg and Strauss variant of PAN [4] was diagnosed in 11 patients of our series. In 1951, Churg and Strauss [4] reported a study of 14 cases of a form of disseminated necrotizing vasculitis occurring frequently among asthmatic patients, with fever, eosinophilia, and a fulminant multisystem disease showing a pathology of NA, eosinophilic infiltration and extravascular granulomas. Fauci et al. [11] classify this syndrome with PAN, whereas Chumbley et al. [3] clearly differentiate their cases from PAN. Pulmonary infiltrates can be transient or nodular and chronic; the asthma may be of short or long duration, peripheral eosinophilia is observed routinely. Neurological involvement, usually mononeuritis multiplex, was found in 19 of 30 cases [3]. In our experience, the vascular and nerve lesions observed in nerve biopsy samples from patients with this syndrome are similar to those observed in PAN, and peripheral eosinophilia and eosinophilic infiltration also occur in classic PAN. The response of the so-called Churg-Strauss allergic granulomatosis to corticosteroid therapy does not differ from that observed in classic PAN. Although pANCA seem to be commonly found in this syndrome [5], the number of patients studied is not large enough to make of this syndrome more than a variant of PAN, as far as the PNS and the muscles are concerned.

NA and neuropathy in patients with rheumatoid arthritis

Of our series 43 patients (21.5%) fitted into this group, which represents the second largest group of patients with NA and multisystemic connective tissue disorder. The occurrence of NA in the context of rheumatoid arthritis is classically associated with a poor outcome. Of the 32 patients with rheumatoid arthritis and neuropathy due to histologically proven necrotizing vasculitis in muscle and/or in nerve biopsy specimens [35], 15 had a sensory and motor neuropathy, the others a purely sensory neuropathy. Two thirds of the patients had a multifocal pattern of neurological deficit, the other third manifested distal symmetrical sensory neuropathy. In this group of patients, Sjögren's syndrome was present in 22%, rheumatoid factor in 97%, nodules in 31% and articular erosions in all of them. NA, which was demonstrated in the nerve and/or muscle biopsy specimens of all of these patients, was more often found in the muscle (86%) than in the nerve biopsy specimens (64%). Axonal degeneration affected an average 77% of nerve fibers isolated by teasing the endoneurial content, while segmental demyelination was virtually absent. The survival rate was 57% at 5 years. Low CH50 and C4 complement levels, seen in 60% of the patients, were associated with a poorer outcome. Cryoglobulinemia was present in 13% of the patients of this series. Search for hepatitis B antigens was negative in all cases.

NA associated with other connective tissue disorders

Necrotizing arteritis was associated with other connective tissue disorders including isolated Sjögren's syndrome in 4% of our patients, systemic lupus erythematosus in 1%; systemic sclerosis in 1%. In 16 patients (8%) there was more than one connective tissue disorder associated with NA and neuropathy, including the 12 patients with Sjögren's syndrome and rheumatoid arthritis.

NA and polysystemic non-connective tissue disorder

NA occurs in retroviral infection [1, 37] in patients with chronic hepatitis and in association with hepatitis B infection. Infection with hepatitis B virus was found in 19% of our patients, including two with chronic hepatitis. These figures confirm that infection with the hepatitis B virus is more common in patients with NA than in controls and that it may play a role in the onset of NA [7, 9, 38, 39].

NA and isolated neuropathy

In a large proportion of patients with peripheral neuropathy related to NA, there is no organ other than the PNS clinically involved [41]. In a series of patients seen in our Department, who presented with necrotizing arteritis and neuropathy, peripheral neuropathy was the only manifestation of NA in 70 of 200 patients (35%).

Is there a syndrome such as vasculitis restricted to the PNS? This was suggested by Dyck et al. [8], who consider that these patients have a non-systemic, organ-specific, vasculitis affecting only the PNS. In fact, silent involvement of other organs is common in such patients, as shown by the frequent finding of NA in muscle biopsy specimens. In our series of patients with isolated vasculitic neuropathy, the muscle biopsy was as often diagnostic for vasculitis as the nerve biopsy, demonstrating that silent lesions of medium-sized muscle arteries are common in this context, and that vasculitis is not restricted to the nervous system. A similar question arises for the so-called cutaneous vasculitis, for which 52% of the patients supposed to have a cutaneous variety of periarteritis nodosa, actually manifest associated neuromuscular involvement [6]. It is, thus, more appropriate to assume that patients with apparently isolated vasculitic neuropathy are affected by a milder form of vasculitis, symptomatic in nerves only.

In our series of patients, the mean age was 61 years with 61% females. General signs or symptoms, usually minor, including fever and loss of weight were present in half of the patients. Many complained of spontaneous pains of neurogenic or muscular origin (59%). The erythrocyte sedimentation rate was normal in 33% of the patients. Patients with cryoglobulinemia were not included in this group. From a neurological standpoint, approximately one fourth of the patients presented with a distal symmetrical sensory or sensorimotor neuropathy, and the diagnosis of NA had seldom been considered before the results of the nerve and muscle biopsies were evaluated.

We were able to follow 29 patients over an average 6 years (extremes 2–14 yrs); 37% of them developed systemic manifestations; 37% died after an average 3.3 years after the onset of the neuropathy, including 8 from systemic manifestations and 3 from infection; 24% had one or more relapses of neuropathy (the mean interval between the first and the second episodes of neuropathy was 3.5 years, in contrast to an interval of 6 months to 1 year in patients who had multisystemic signs from the beginning); 31% did not relapse. These findings are in keeping with the better prognosis observed in the 'non-systemic' vasculitis than in the polysystemic forms by Dyck et al. [8]. In all subgroups of NA, the neuropathy has common clinical, electrophysiological and pathological characteristics.

The peripheral neuropathy of necrotizing arteritis

Clinical aspects

Focal (mononeuritis) neuropathy affected 16.5% of the patients and multifocal neuropathy (mononeuritis multiplex) 56.5%. Distal symmetrical sensory or sensorimotor neuropathy was present in 50 patients (25%). The peroneal nerve was affected unilaterally in 62.5% and bilaterally in one third of the patients. Unilateral involvement of the popliteal nerve was present in 27.5% of the patients and on both sides, in 5%. Unilateral involvement of the ulnar nerve was present in 25.5% of the patients and bilateral in 8%. The median nerve was affected on one side in 21.5% and on both sides in 3%. Unilateral involvement of the radial nerve was present in 8% of the patients, and in 2% on both sides. Unilateral involvement of the femoral nerve was observed in 6% of the patients. Proximal sciatic nerve or sciatic root involvement was observed in 2% of the patients. Involvement of the fifth and seventh cranial nerves was observed once. Signs of central nervous system involvement were observed in 3% of the patients.

In typical cases, the onset of the neuropathy is abrupt and the deficit severe, but in many cases only partial deficit is observed. However, the course of the neuropathy may be slowly progressive in some cases, especially in the elderly. The CSF is usually normal, although a mild elevation of the protein content occurs.

Recovery from motor deficit due to ischemic neuropathy takes months because of the axonal lesions. Residual pains are common and may be difficult to differentiate from relapses of the neuropathy.

Electrophysiological data

The electrophysiological studies show an axonal neuropathy in all cases. Some authors have suggested the use of electrophysiological data to orientate the site of the nerve biopsy [42]. An early ischemic conduction block has occasionally, been observed [40], but a conduction block is more suggestive of nerve entrapment or of acute focal demyelination, as observed in Guillain-Barré syndrome or in other forms of acquired demyelinative neuropathy, than of an ischemic nerve lesion.

Morphological aspects

Demonstration of NA in biopsy specimens. A diagnosis of NA needs histological confirmation, which can be usually achieved by examining biopsy samples of a specific skin lesion. If this does not bring confirmation, nerve and/or muscle biopsies may be advisable to search for characteristic lesions of muscular or epineurial arteries [25, 29, 33, 42]. Biopsy of the kidney rarely aids diagnosis, even in patients with nephropathy. It is not possible to estimate how often the diagnosis of NA cannot be achieved histologically, and it should be borne in mind that a negative biopsy does not rule out a diagnosis of NA; however, repetition of biopsies seldom helps in our experience.

In patients with vasculitic neuropathy due to NA, vasculitis can be demonstrated in muscle biopsy specimens, in the nerve biopsy specimen, or in both of them. In our most recent evaluation, in 27% of our patients with vasculitic neuropathy, the specific vascular lesion was in the muscle specimen only, in 35% in the nerve specimen only, and in 27% of the patients, both in the nerve and muscle specimens. It is, thus, wiser to sample both the nerve and the muscle during the same procedure [37], especially when the peroneal nerve is affected. In such cases, we take a biopsy sample from the superficial peroneal nerve and the peroneus brevis muscle during the same procedure, under local anesthesia. Biopsy specimens must be studied on *serial sections*, because arteritis is segmental, and characteristic lesions may be present on segments of the arteries as short as 50 μm in our experience. Total nerve biopsy must be performed instead of fascicular biopsy because, almost invariably, only the epineurial arteries are affected, sometimes in association with inflammatory infiltration of perineurial vessels. Epineurial and perineurial inflammatory infiltrates are common in the vicinity of characteristic arterial lesions, but they are seldom observed in the endoneurial space, except in the patients with NA occurring in the context of HIV infection.

Lesions of nerve fibers – the ischemic neuropathy. Nerve ischemia, as observed in vasculitic neuropathy, induces acute axonal degeneration in most cases. In our series,

Wallerian degeneration affected an average (± SD) of 65 ± 35% of the isolated fibers, with simultaneous degeneration of fibers in many cases (normal value: less than 1%), a pattern which fits well with an ischemic lesion and may suggest vasculitis when other characteristic arterial lesions have not been found. The average incidence of fibers showing segmental abnormalities of the myelin sheath was 1.9% (extremes 0–15%; normal < 1%). Asymmetry of lesions between and within fascicles is also common in vasculitic neuropathy. Sometimes axon loss predominates in the centrofascicular area, which suggests an ischemic origin.

Quantitative studies of Epon-embedded specimens and teased fiber preparations showed that: (1) the intensity of axon loss and the incidence of degenerating fibers varied greatly from case to case, and between fascicles of individual nerves; (2) axon loss predominated on myelinated fibers larger than 7 μm diameter; (3) the unmyelinated fibers were less affected; (4) occasionally, even the Schwann cells seemed to be degenerating – this may preclude axonal regeneration and clinical recovery; and (5) in most cases, the nerve lesions appeared to result from the summation of lesions that have occurred at different time points [14] of nerve blood vessels.

Indication of nerve biopsy for diagnosis of vasculitis

Since vasculitis is a neuropathy that can be easily treated, its diagnosis should not be missed. Its presence is obvious when an acute or subacute multifocal neuropathy occurs in the context of multisystem involvement, in a patient suffering from asthma, or presenting other manifestations including renal insufficiency and multiple ischemic lesions over the skin, especially of the lower extremities, plus eosinophilia. In such patients the diagnosis is easily suspected clinically and confirmed by nerve and muscle biopsy. At the other end of the clinical spectrum, an elderly patient may present with a progressive, painful, distal, roughly symmetrical sensory or sensorimotor polyneuropathy without any clinical or biological sign of inflammation. The patient may be in good general condition, and the electrophysiological investigations may detect signs of a severe axonopathy. In such patients with a moderately disabling isolated sensory polyneuropathy, nerve and muscle biopsies are the only way to diagnose PAN, a diagnosis which may come as a surprise to many. In a recent retrospective study we presented the findings from nerve biopsies of 100 patients over the age of 65 years who presented with a disabling polyneuropathy [2]. The neuropathy was considered disabling because of either pains or a progressive sensorimotor deficit. In this group of patients, we found evidence of NA in 23%. In 10 additional patients the diagnosis of vasculitic neuropathy was considered as very likely on the basis of the presence of inflammatory infiltration in the vicinity of nerve and muscle blood vessels associated with axonal degneration of nerve fibers; however, specific vascular lesions could not be demonstrated in spite of examination of serial nerve and muscle sections. Thus, vasculitic neuropathy accounted for 33% of the disabling neuropathies in this population. It is our opinion that elderly people should be investigated in the same way as younger patients, providing that they are in relatively good general condition and are not demented.

Treatment

The treatment of vasculitis neuropathy rests mainly on corticosteroids. It is widely admitted that one should start with prednisone at 1 mg/Kg/day. High doses of corticosteroids may not be well tolerated in elderly. Simultaneous treatment with cyclophosphamide, 2 mg/Kg/day, may help reducing the doses of corticosteroids. It is difficult to tell how long this treatment should be maintained at full dose. This depends on the reaction to treatment, on the course and form of the disease and on the tolerance of the treatment. We usually give full dose of steroids for approximately 6–8 weeks and then start decreasing the dose of steroids very gently. It is necessary to control the sedimentation rate (ESR) periodically and to increase prednisone if the ESR increases, if this increase in ESR is not due to another factor, especially to infection.

In cases without clinical evidence of polysystemic involvement, in cases with normal sedimentation rate, and in elder patients, the benefit of prolonged treatment with high doses of steroids is less certain.

Indomethacin (100 mg/day) may be associated to steroids when fever persists or when the sedimentation rate remains abnormal in spite of large doses of steroids. It has been recently advocated that in patients with systemic vasculitis and antibodies to antineutrophil cytoplasm had a 51 percent decrease in these antibodies, that persisted during follow-up [44]. Besides the cost of this treatment, it should be evaluated against corticotherapy. In the evaluation of the efficiency of the treatment of the different forms of NA, it must be remembered that there is a wide range of modalities of evolution in NA, with spontaneous remissions of several years of the disease in some cases.

In conclusion, necrotising vasculitis of the polyarteritis nodosa type, is not a rare cause of neuropathy, especially of the multifocal type. Our series show that vasculitic neuropathy of this type is much more common in the elderly. Nerve and muscle biopsy should be performed whenever this diagnosis is suspected, but it is important to remember that not finding vasculitis in a biopsy specimen does not rule out the diagnosis.

References

1. Chia L, Planté V, Adams D, Fernandez A, Lacroix C, Said G. (1996) High incidence of vasculitis in the elderly with disabling neuropathy – a study of 100 consecutive patients. Brain (in press)
2. Calabrese LH, Estes M, Yen-Lieberman B, et al (1989) Systemic vasculitis in association with human immunodeficiency virus infection. Arthritis Rheum 32:569–576
3. Chumbley LC, Harrison EG Jr, Deremee RA (1977) Allergic granulomatosis and angiitis (Churg-Strauss syndrome). Proc Mayo Clin 52:477
4. Churg J, Strauss L (1951) Allergic granulomatosis, allergic angeitis, and periarteritis nodosa. Am J Pathol 27:277–301
5. Cohen-Tervaert JW, Limburg PC, Elema JD, et al (1991) Detection of autoantibodies against myeloid lysosomal enzymes: a useful adjunct to classification of patients with biopsy-proven necrotising arteritis. Am J Med 91:1991
6. Diaz-Perez J, Winkelmann RK (1980) Cutaneous periarteritis nodosa: a study of 33 cases. In: Wolff K, Winkelmann RK (eds) Lloyd-Luke, London, pp 273–284
7. Drueke T, Barbanel C, Jungers P, et al (1980) Hepatitis B antigen-associated periarteritis nodosa in patients undergoing long-term hemodialysis. Am J Med 68:86
8. Dyck PJ, Benstead TJ, Conn DL, et al (1987) Nonsystemic vasculitic neuropathy. Brain 110:843
9. Editorial (1985) Systemic vasculitis. Lancet I:1252

10. Ewert BH, Jennette JC, Falk RJ (1991) The pathogenetic role of antineutrophil cytoplasmic antibodies. Am J Kidney Dis 18:188
11. Fauci AS, Haynes BF, Katz P (1978) The spectrum of vasculitis. Ann Intern Med 89:660
12. Ferraro G, Meroni PL, Tincani A, et al (1990) Anti-endothelial cell antibodies in patients with Wegener's granulomatosis and micropolyarteritis. Clin Exp Immunol 79:47
13. Frampton G, Jayne DRW, Perry GJ, et al (1990) Autoantibodies to endothelial cells and neutrophil cytoplasmic antigens in systemic vasculitis. Clin Exp Immunol 79:47
14. Fujimura H, Lacroix C, Said G (1991) Vulnerability of nerve fibers to ischaemia. Brain 114:1929
15. Godeau P, Guillevin L (1982) Périartérite noueuse systémique. In: Kahn M-F, Peltier AP (eds) Maladies dites systémiques. Flammarion, Paris, pp 414–445
16. Goeken JA (1991) Antineutrophil cytoplasmic antibody – a useful serological marker for vasculitis. J Clin Immunol 11:161
17. Hawke SHB, Davies L, Pamphlett R, et al (1991) Vasculitic neuropathy. A clinical and pathological study. Brain 114:2175
18. Hogg N (1988) The structure and function of Fc receptors. Immunol Today 9:185
19. Jennette JC, Falk RJ, Andrassy K, et al (1994) Nomenclature of systemic vasculitides. Arthritis Rheum 37:187
20. Kissel JT, Mendell JR (1994) Peripheral neuropathy due to vasculitis: immunopathogenesis, clinical features and treament. In: Holhlfeld R (ed) Immunology of neuromuscular disease. Kluwer, Dordrecht, pp 105–121
21. Kissel, JT, Riethman JL, Ormeza J., et al (1989) Peripheral nerve vasculitis: immune characterization of the vascular lesions. Ann Neurol 93:129
22. Kussmaul A, Maier R (1866) Über eine bisher nichtbeschriebene eigenthümliche Arterienerkrankung (Periarteritis nodosa), die mit Morbus Brightii und rapid fortschreitender allgemeiner Muskellähmung einhergeht. Dtsch Arch Klin Med 1:484
23. Lie JT (1994) Nomenclature and classification of vasculitis. Arthritis Rheum 2:181
24. Lovshin LI, Kernohan JW (1948) Peripheral neuritis in periarteritis nodosa. Arch Intern Med 82:321
25. Maxeiner SR, MCDonald JR, Kirlklin JW (1952) Muscle biopsy in the diagnosis of periarteritis nodosa: an evaluation. Surg Clin North Am 32:1225
26. Moore PM (1989) Immune mechanisms in the primary and secondary vasculitides. J Neurol Sci 93:129
27. Moore PM, Cupps TR (1983) Neurologic complications of vasculitis. Ann Neurol 14:155
28. Nydegger UE (1985) A place for soluble immune complexes in clinical immunology? Immunol Today 6:80
29. Pages M, Pages AM (1984) La biopsie nerveuse au cours de la périartérite noueuse. Son intêret diagnostique. Sem Hop Paris 60:3295
30. Panegyres PK, Blumbergs PC, Leong AS-Y, Bourne AJ (1990) Vasculitis of peripheral nerve and skeletal muscle: clinicopathological and immunopathic mechanisms. J Neurol Sci 100:193
31. Panegyres PK, Faull RJ, Russ GR, et al (1992) Endothelial cell activation in vasculitis of peripheral nerve and skeletal muscles. J Neurol Neursurg Psychiatry 55:4
32. Parlevliet KJ, Henzen-Logmans SC, Oe PL, et al (1988) Antibodies to components of neutrophil cytoplasm: a new diagnostic tool in patients with Wegener's granulomatosis and systemic vasculitis. Q J Med 66:55
33. Parry GJ, Brown MJ, Asbury AK (1981) Diagnostic value of nerve biopsy in mononeuritis multiplex. Neurology 31:129
34. Pober JS, Cotran RS (1990) Cytokines and endothelial cell biology. Physiol Rev 70:427
35. Puechal X, Said G, Hilliquin P, et al (1995) Peripheral neuropathy with necrotizing vasculitis in rheumatoid arthritis. A clinicopathologic and prognostic study of thirty two patients. Arthritis Rheum 38:1618
36. Ropert A, Metral S (1990) Conduction block in neuropathies with necrotizing arteritis. Muscle Nerve 13:102
37. Said G, Lacroix C, Fujimura H, et al (1988) The peripheral neuropathy of necrotizing arteritis: a clinicopathologic study. Ann Neurol 23:461
38. Sergent JS, Lockshin MD, Christian CL, Gocke DJ (1976) Vasculitis with hepatitis B antigenemia: long term observations in nine patients. Medicine 55:1
39. Shusterman N, London WT (1984) Hepatitis B and immune complex disease. N Engl J Med 313:43
40. Smiley JD, Moore SE (1989) Immune complex vasculitis: role of complement and IgG-Fc receptor functions. Am J Med Sci 298: 267

41. Torvik A, Berntzen A (1968) Necrotizing vasculitis without visceral involvement. Acta Med Scand 184:69
42. Wees SJ, Sunwood LN, Oh SJ (1981) Sural nerve biopsy in systemic necrotizing vasculitis. Am J Med 71:525
43. Winkelmann RK (1980) Classification of vasculitis. In: Wolff K, Winkelmann RK (eds) Lloyd-Luke, London, pp 1–24
44. Jaynes DRW, Davies MJ, Fox CJV, Black CM, Lockwood CM (1991) Treatment of systemic vasculitis with pooled intravenous immunoglobulin. Lancet 337:1137–1139

Neurological paraneoplastic syndromes

Josep Dalmau[1,2], **Jerome B. Posner**[1,2]

[1] Department of Neurology, Memorial Sloan-Kettering Cancer Center, 1275 York Ave, New York, N.Y., 10021, USA
[2] Department of Neurology and Neurosciences, Cornell University Medical College, New York, N.Y., USA

Introduction

Cancer arising outside of the nervous system (systemic cancer) often causes neurological disability without spreading to the nervous system [1] (Table 1). Such indirect or paraneoplastic effects of systemic cancer on the nervous system include vascular disorders (hemorrhages and/or infarction), infections, metabolic and nutritional disorders, as well as several disorders called by Brain and Norris [2] "remote effects of cancer on the nervous system". For the neurologist, the term paraneoplastic syndrome and remote effects of cancer are synonymous and can be defined as "neurological disorders pathogenetically related to a systemic cancer but not ascribable to nervous system metastases or to destruction of vital systemic organs by the tumor or by treatment of the tumor". Paraneoplastic syndromes affect any portion of the nervous system; they either mainly involve a single cell type, such as the Purkinje cell in paraneoplastic cerebellar degeneration (PCD), or the cholinergic synapse, in the Lambert-Eaton myasthenic syndrome (LEMS), or they may involve multiple areas of the nervous system simultaneously, as in neuromyopathy or encephalomyelitis associated with cancer (Table 2). Neurological disorders that are clinically and pathologically identical to paraneoplastic syndromes also occur in the absence of cancer but at substantially decreased incidence [3]. For example, LEMS is rare in the general population but affects about 3% of patients with small cell lung cancer [4]. About 60% of patients with LEMS have small cell lung cancer [5]. Nevertheless, 30–40% of patients with LEMS will have neither identifiable cancer nor develop that disease within several years following the diagnosis and treatment of LEMS. Table 3 gives our estimates of the likelihood that a given neurological syndrome is paraneoplastic in origin.

Correspondence to: J.B. Posner, (address see [1] above)

Table 1. Indirect (non-metastatic) neurologic complications of cancer

A. Vascular disorder	
B. Infections	
C. Metabolic and nutritional disorders	
D. Side effects of therapy:	(1) chemotherapy
	(2) radiation
	(3) surgery
	(4) immunologic therapy
E. "Remote effects" or paraneoplastic syndromes	

Table 2. Paraneoplastic syndromes affecting the nervous system[a]

I. Brain and cranial nerves
A. Subacute cerebellar degeneration[b]
B. Opsoclonus-myoclonus[b]
C. Limbic encephalitis[b] and other dementias
D. Brain stem encephalitis
E. Optic neuritis
F. Photoreceptor degeneration
II. Spinal cord and dorsal root ganglia
A. Necrotizing myelopathy
B. Subacute motor neuronopathy[b]
C. Motor neuron disease
D. Myelitis
E. Sensory neuronopathy[b]
III. Peripheral nerve
A. Subacute or chronic sensorimotor peripheral neuropathy
B. Acute polyradiculoneuropathy (Guillain-Barré syndrome)
C. Mononeuritis multiplex and microvasculitis of peripheral nerve
D. Brachial neuritis
E. Autonomic neuropathy
F. Peripheral neuropathy with islet cell tumors
G. Peripheral neuropathy associated with paraproteinemia
IV. Neuromuscular junction and muscle
A. Lambert-Eaton myasthenic syndrome[b]
B. Myasthenia gravis
C. Dermatomyositis[b], polymyositis
D. Acute necrotizing myopathy
E. Carcinoid myopathies
F. Myotonia
G. Cachectic myopathy
V. Multiple levels of central and peripheral nervous system or unknown site
A. Encephalomyelitis[b]
B. Neuromyopathy
C. Stiff-man syndrome

[a] Modified from [1]
[b] Neurological disorders strongly associated with cancer (or $\geq$ 10%)

Diagnosis

A diagnosis of paraneoplastic syndrome is made when:
1. The patient has one of the typical neurological syndromes and an appropriate underlying cancer (Tables 2 and 4). Paraneoplastic syndromes are usually associated

Table 3. Estimated likelihood that a given neurological disorder is a paraneoplastic syndrome[a]

Syndrome	% paraneoplastic
Lambert-Eaton myasthenic syndrome	60
Subacute cerebellar degeneration	50
Subacute sensory neuronopathy	20
Opsoclonus/myoclonus (children)	50
Opsoclonus/myoclonus	20
Sensorimotor peripheral neuropathy	10
Encephalomyelitis	10
Dermatomyositis	10
Myasthenia gravis	10

[a]Modified from [1]

with specific cancers, although any cancer can cause any paraneoplastic syndrome (e.g., LEMS with small cell lung cancer, myasthenia gravis with thymoma, PCD with gynecological cancers).
2. The patient has a neurological disorder associated with the presence of a typical anti-neuronal antibody in the serum (Table 4) [1, 6, 7].
3. The patient has a neurological disorder which resolves when a cancer is discovered and is adequately treated [8].

When encountering a patient with a neurological disorder, the physician should first determine if the disorder is caused by cancer (Table 5). In addition, an acute or subacute development of neurological disability should lead one to suspect paraneoplasia. Most, but not all paraneoplastic syndromes develop over a few weeks to a few months and then stabilize, differentiating them from the more chronic and progressive degenerative diseases of middle age and adulthood [9–11]. When the neurological disorder peaks, it is usually severe. Patients with cerebellar degeneration are often unable to walk, feed themselves, or even read because of oscillopsia. Patients with limbic encephalitis may have virtually no memory, although other cognitive functions are usually intact. There are exceptions; in occasional patients, the paraneoplastic syndrome may be mild and non-disabling [12].

Cerebrospinal fluid (CSF) examination may reveal pleocytosis, with an elevated protein concentration and oligoclonal bands. Quantitative measurement of immunoglobulins may indicate intrathecal IgG synthesis, and specific measurement of paraneoplastic autoantibodies may indicate synthesis of paraneoplastic antineuronal antibody within the central nervous system [13]. If the disorder is discovered late in its course, there may be no cells in the CSF, but the IgG usually remains elevated.

If the disorder is one suggesting involvement of intracranial or intraspinal structures, the CSF should be examined for pleocytosis, IgG and oligoclonal bands. If the disorder is one suggesting largely peripheral involvement, electrodiagnostic studies should be performed. These should include conduction velocities to help distinguish demyelination from axonal damage, electromyography to identify motor neuron disorders, and neuromuscular transmission studies to identify LEMS or other disorders of the neuromuscular junction. For all patients, the blood should be examined for anti-neuronal antibodies and, if the suspicion is high, for cancer markers as well (Tables 4 and 5). Cancer markers that may assist in the diagnosis include carcinoembryonic antigen, CA-125 (ovarian cancer), BRCA (breast cancer), PSA (prostate cancer).

Table 4. Some well-characterized autoantibodies associated with paraneoplastic syndromes[a]

Antibody	Antigen	Syndrome(s)	Tumor(s)
Anti-Yo	34 and 62 kDa Purkinje cell cytoplasm	Cerebellar degeneration	Ovary Breast
Anti-Hu	35–40 kDa neuronal nuclei	Encephalomyelitis, sensory neuronopathy, limbic encephalitis, autonomic neuropathy	Small cell lung
Anti-Ri	55 and 80 kDa neuronal nuclei	Opsoclonus/ataxia	Breast
Anti-retinal	23, 65, 145, and 205 kDa photoreceptors and ganglion cells	Cancer-associated retinopathy	Small cell lung
Anti-NMJ	Voltage-gated calcium channel	Lambert-Eaton myasthenic syndrome	Small cell lung
Anti-NMJ	ACh receptor	Myasthenia gravis	Thymoma
Anti-Tr	Purkinje cells	Cerebellar degeneration	Hodgkin's
Anti-Amphiphysin	128 kDa synaptic protein	Stiff-man	Breast

NMJ, neuromuscular junction
[a]Modified from [1] with permission

A careful check of the patient's history to identify risk factors for specific neurological disorders can also be very helpful. Small cell lung cancer is a disease of smokers and the absence of a smoking history strongly suggests that the patient does not have that particular form of cancer. A positive family history of ovarian or breast cancer may lead the physician to consider those disorders as possible underlying causes of the neurological syndrome. Weight loss or malaise may also increase the neurologists suspicion.

If the neurologist suspects cancer, the search should be directed for the tumor most likely to be causal. For example, a woman with paraneoplastic cerebellar degeneration who has anti-Yo antibody in her serum, almost certainly has a breast or gynecological cancer [9]. Nevertheless, there are occasional patients with anti-Yo antibody who have other tumors [9], so the search should not be restricted. If lung cancer is suspected, a computed tomography (CT) of the chest is required. In all women, mammograms and pelvic CT or ultrasound should be carried out. Men and women should be examined carefully for lymph nodes because Hodgkin's disease is a common cause of paraneoplastic syndromes [6, 14, 15]. Careful prostate examination is also sometimes rewarding. There are enough patients with colon cancer presenting with paraneoplastic syndromes that stool guaiac examination is probably useful in every patient. The physician must realize that even if his initial search for an underlying neoplasm is negative, this does not rule out the presence of a cancer. If clinically suspected, and the suspicion of an underlying cancer remains, the patient should be reassessed periodically.

A different approach is required for patients who are known to have cancer but are presented to the neurologist for the diagnosis of a neurological disorder which may not be typical of a paraneoplastic syndrome (Table 6). The first step is to search for a metastasis, either by MRI examining the clinically symptomatic areas, or by

Table 5. Paraneoplastic syndromes: diagnostic approach

- Are neurologic symptoms typical?
  - Subacute onset
  - Rapid evolution
  - Symptoms suggestive of paraneoplasia
- Examine CSF
  - Inflammatory cells
  - IgG and IgG index
  - Oligoclonal bands
  - Anti-neuronal antibodies
- Examine blood
  - Anti-neuronal antibodies
  - Cancer markers
- Electrodiagnostic tests (neuropathy)
  - Sensory potentials
  - Synaptic transmission – (repetitive stimulation)
  - Electromyography (lower motor neuron)
- Cancer risk factors
  - Smoking
- Search for cancer
  - Chest CT
  - Mammograms
  - Pelvic exam, CT, ultrasound
  - Testis exam, ultrasound
  - Lymph node exam
  - Stool guaiac
  - Tumor markers in blood (CEA, PSA, CA-125, BRCA1)

CSF, cerebrospinal fluid; CT, computed tomography

examining the spinal fluid for the presence of malignant cells, or both. Even patients not previously known to have cancer may present with metastases to the brain, epidural spinal space, or the leptomeninges. These examinations will usually establish the diagnosis of metastasis without difficulty. In the absence of evidence of metastasis, a careful review of previous therapy, including radiation portals and dose and specific chemotherapeutic agents used, may indicate that the neurological disorder is a side effect of anti-cancer therapy. Nutritional and metabolic diseases, such as liver or renal failure, or malnutrition may cause neurological disability but no other signs that the patient suffers such an underlying problem. Opportunistic infections are common in patients with lymphoma. The clinical symptoms of these disorders may be restricted to the nervous system, as for example, progressive multifocal leukoencephalopathy and herpes simplex encephalitis, or may be more widespread as in aspergillosis involving lungs and brain, or herpes zoster involving skin and nervous system [1]. Opportunistic infections are generally restricted to patients who suffer from either an immune suppressive tumor, such as lymphoma, or from immune suppression by prior chemotherapy. Steroids are a particular offender. Patients with cancer are hypercoagulable and at risk for vascular thrombosis. They can develop deep vein thrombosis with paradoxical embolism to the brain, dural venous sinus thrombosis, non-bacterial thrombotic endocarditis with cerebral embolization, or disseminated intervascular coagulation with small cerebral vessel occlusion. Furthermore, patients who have received chemotherapy may be hypocoagulable from thrombocytopenia leading to CNS hemorrhage.

Table 6. Approach to paraneoplastic syndromes in a patient with known cancer

• Search for metastases	
	– MRI of symptomatic area(s)
	– CSF for malignant cells
• Review previous therapy	
	– Radiation therapy
	– Chemotherapy
	– Immunotherapy
• Consider metabolic problems	
	– Liver, renal failure, hypoxia
• Check nutrition	
	– Wernicke-Korsakoff
	– Cancer cachexia
• Opportunistic infections (lymphoma, immunosuppression, AIDS)	
	– PML, CMV, herpes
• Vascular disorders	
	– Hypercoagulability
	– Hypocoagulability
• Non-cancer related	
	– Diabetes, alcohol
• Paraneoplastic	

MRI, magnetic resonance imaging; PML, progressive multifocal leukoencephalopathy; CMV, Cytomegalovirus

In a review of a patient with a neurological disorder and cancer, one must always also consider that the disorder may be non-cancer related. Diabetic neuropathy and nutritional neuropathy due to alcohol occur in patients without, as well as with, cancer. This may confound the course of treatment in patients with cancer.

When all of the above metastatic and non-metastatic complications of systemic cancer are ruled out, the neurologist can then consider paraneoplastic syndromes.

Pathogenesis

Several mechanisms have been suggested as a possible pathogenesis for paraneoplastic syndromes. These mechanisms and examples are given in Table 7. Current evidence suggests that most paraneoplastic syndromes are autoimmune. The hypothesis is that antigens normally restricted to the nervous system are aberrantly expressed by a cancer. These antigens may be either normal or mutated proteins. The immune system recognizes the antigen in the cancer as non-self, and an immune response is generated to the antigen. The immune response attacks both the tumor and those nervous system structures that express the antigen. The result may be that the tumor grows more slowly than it would have in the absence of the paraneoplastic disorder, while the nervous system dysfunction results in severe neurological symptoms. The evidence supporting such a hypothesis is best seen in LEMS [16, 17]. In this syndrome, voltage-gated calcium channel proteins are present in both the tumor and the normal presynaptic cholinergic synapse [18]. Antibodies to voltage-gated calcium channels interfering with the release of Acetylcholine are found in the serum of patients with LEMS [19]. These antibodies can be removed by plasma exchange, relieving the patient of the neuromuscular symptoms [20]. Transfer of the patient's plasma or IgG to experimental animals, reproduce the disease [16]. In most other paraneoplastic syndromes, particularly those involving the central nervous system (CNS), the evidence for autoimmunity is less well established. However, suggestive evidence that these disorders are immune mediated include the following: (1) high titers of "paraneoplastic antibodies" correlate with neurological dysfunction. Low-titer antibodies may be found in patients without paraneoplastic symptoms [21, 22] and may be related to indolent growth of the neoplasm; (2) some paraneoplastic antibodies are synthesized intrathecally [13]; (3) intraneuronal deposits of antibodies are found in patients at autopsy [23]; (4) the expression of the paraneoplastic antigen is highly restricted to the tumor and nervous system [24]; and (5) specific inflammatory infiltrates are found in both the tumor and the CNS of patients with paraneoplastic syndromes [25, 26].

Table 7. Possible pathogenesis of paraneoplastic syndromes[a]

Hypothesis	Example
Toxin secreted by tumor	ACTH = Cushing's syndrome PTHRP = hypercalcemia
Competition for essential substrate	Utilization of glucose by sarcomas = hypoglycemia Carcinoid tumors competing with brain for tryptophan = pellagra-like syndrome
Opportunistic infection	Papovavirus = progressive multifocal leukoencephalopathy
Autoimmune process	Lambert-Eaton myasthenic syndrome

PTHRP, parathyroid hormone-related protein; ACTH, adrenocorticotropic hormone
[a]Modified from [1]

Therapy

The therapy of most paraneoplastic syndromes is unsatisfactory. Although anecdotal reports exist of dramatic responses to various therapeutic endeavors [27–32], most disorders do not to respond [10, 33, 34]. With few exceptions, most patients with paraneoplastic syndromes, particularly those involving the CNS, do not respond to either treatment of the tumor or immune suppression. Therapeutic strategies directed at the paraneoplastic syndrome fit into two major categories. The first is to treat the underlying tumor and the second is to treat the paraneoplastic syndrome itself, usually by some form of immune suppression. Table 8 outlines the various approaches to the treatment of paraneoplastic syndromes.

Treating the tumor

Several isolated reports suggest dramatic response of various paraneoplastic syndromes to successful treatment of the underlying tumor [27–31, 35]. Nevertheless, the fact that some paraneoplastic syndromes arise when the tumor is in remission, and that they, for the most part, appear to pursue a course independent from that of the neoplasm, suggests that, except in rare instances, treating the tumor does not assure reversal of the paraneoplastic syndrome. In a relatively large series of patients,

Table 8. Therapy of paraneoplastic syndromes

- Treat neoplasm
- Suppress humoral immunity (B cells, antibody)
  - Plasma exchange
  - Immunoglobulins
  - Protein A immunoadsorption
- Suppress cytotoxic T cells (CD8)
  - Lymphocyte irradiation
  - Corticosteroids
  - Cyclosporin A
- Suppress helper T cells (CD4)
  - Anti-CD4 antibodies
  - β-Interferon
- Interfere with auto-reactive T lymphocyte activation
  - Inhibit MHC recognition with anti-MHC monoclonal antibodies
  - Compete for antigen binding with "design" peptides
  - Inhibit T cell activation
    - Anti-CD4
    - Anti-TCR
- Induce tolerance
  - Feed antigen
- Suppress inflammation
  - Corticosteroids
  - Non-steroidal anti-inflammatory agents

Graus et al. [33] were not able to find paraneoplastic syndromes that responded to neoplastic treatment.

Treating the paraneoplastic syndrome-immunological therapy

Plasma exchange. Plasma exchange is effective in treating LEMS [20]. Anecdotal reports describe that plasma exchange can be effective for paraneoplastic syndromes of the CNS [32]. However, our own patients [6] and those reported by Graus et al. [33] rarely responded to plasma exchange and, when there was an apparent response, it was usually minor and not enduring.

Intravenous immunoglobulin (IVIg). IVIg has been reported to be effective in some patients with paraneoplastic disorders [36–38]. However, Vega et al. [34] reported 18 patients with anti-Hu-positive paraneoplastic dysfunction who received 1–12 cycles of IVIg. None of the severely disabled patients responded. There was some suggestion of mild to modest benefits in patients with only mild to moderate neurological dysfunction. A caveat is necessary in evaluating partial responses of patients with paraneoplastic syndromes to treatment. Since the paraneoplastic disease in most instances is non-progressive, treatment is started when the patient has stabilized. There seems to be an apparent partial response as the patient learns to cope better with the degree of static neurological disability. Thus, unless the response is dramatic, it is very difficult to evaluate. Our experience with IVIg has paralleled that of Vega et al. [34].

Immunoadsorptive protein A columns. A recent report by Cher et al. [39] describes six patients with paraneoplastic syndromes who were treated with the use of a protein

A column that absorbs gammaglobulin from the patient's blood. Three patients with opsoclonus/myoclonus had a complete and enduring response to the treatment. Two patients with paraneoplastic encephalomyelitis had a partial response and one patient with paraneoplastic cerebellar degeneration had no response. Because paraneoplastic opsoclonus/myoclonus is often a remitting and exacerbating disease, one must view these data with caution. Nevertheless, further studies are indicated to determine the usefulness of this technique in paraneoplastic syndromes. Protein A column immunoadsorption probably does not work by the same mechanism as plasma exchange since only small amounts of plasma are removed and treated at each treatment. Some believe that it uncovers antibodies, thus suppressing B cell function.

Immune suppressive drugs. Table 9 lists immunotherapeutic agents in current use. Virtually all of these drugs have been tried in patients with paraneoplastic syndromes. There are isolated reports of corticosteroids being effective in paraneoplastic syndromes. They are reported to be particularly effective in opsoclonus/myoclonus in children [40]. Our own data have failed to show a consistent response of any of the paraneoplastic syndromes to these agents. Therefore, one is left with the problem of immunosuppressing patients who have cancer, raising the theoretical possibility of whether one might promote the growth of the cancer by immunosuppression.

Table 9. Immunotherapeutic agents in current use

Corticosteroids
Azathioprine
Cyclosporin A
Cyclophosphamide
Chlorambucil
Methotrexate
Tacrolimus (FK506)

Current and future therapy

Despite our limited knowledge of the pathogenesis and appropriate treatment of paraneoplastic syndromes, neurologists are presented with patients having severe and ongoing neurological disability that must be dealt with. The best approach is to identify and treat the tumor. Treatment of the tumor may produce, but does not guarantee, improvement of the paraneoplastic syndrome but is likely to prolong the patient's life. Because in many instances the tumors grow more indolently in patients with paraneoplastic syndromes than those free of the neurological disability, it may be possible to treat the tumor using a less vigorous approach than in patients who do not have paraneoplastic disorders. Some evidence suggests that given the same treatment, the tumor may be more successfully treated in patients with paraneoplastic disorders than those without the neurological disease.

A second approach is directed at the paraneoplastic syndrome itself. If an autoantibody can be identified in the serum and spinal fluid, removal of that antibody by plasma exchange, particularly in a patient early in the course of the disease which is still evolving, appears a rational approach. One should attempt to lower the level of the antibody in both serum and CSF. This may require more vigorous exchange than if one is only attempting to remove the antibody from the blood. Although the mechanism of action of IVIg is not known, after plasmapheresis it might be useful to begin IVIg realizing that it has, itself, no direct effect on the antibody levels but certainly has effects on the immune system. Further evaluation of the immunoadsorption with protein A column seems warranted. At the present time, there is no evidence that immunosuppressive agents, such as corticosteroids and cyclophosphamide, have a major role to play in the treatment of paraneoplastic syndromes but some of the agents, such as cyclophosphamide in breast cancer and corticosteroids in lymphoma, may be utilized for treatment of the underlying tumor. Although future therapies are best designed around the development of an animal model, one should consider attempting to develop tolerance with use of the antigen to block autoreactive T cells, and CD4 antibodies [41] as possible endeavors in the treatment of paraneoplastic syndromes.

References

1. Posner JB. (1995) Neurologic complications of cancer. F.A.Davis, Philadelphia
2. Brain WR, Norris FH (1965) The remote effects of cancer on the nervous system. Grune and Stratton, New York
3. Torvik A, Slettebo M (1980) Encephalomyelitis with polyneuropathy. Acta Neurol Scand 61:287
4. Elrington GM, Murray NM, Spiro SG, Newsom-Davis J (1991) Neurological paraneoplastic syndromes in patients with small cell lung cancer. A prospective survey of 150 patients. J Neurol Neurosurg Psychiatry 54:764
5. O'Neill JH, Murray NM, Newsom-Davis J (1988) The Lambert-Eaton myasthenic syndrome. A review of 50 cases. Brain 111:577
6. Posner JB, Furneaux HM (1990) Paraneoplastic syndromes. In: Waksman BH (ed) Immunologic mechanisms in neurologic and psychiatric disease. Raven Press, New York, pp 187–219
7. Dalmau J, Posner JB (1994) Neurologic paraneoplastic antibodies (anti-Yo; anti-Hu; anti-Ri): the case for a nomenclature based on antibody and antigen specificity. Neurology 44:2241
8. Rosenfeld MR, Posner JB (1991) Paraneoplastic motor neuron disease. Adv Neurol 56:445
9. Peterson K, Rosenblum MK, Kotanides H, et al (1992) Paraneoplastic cerebellar degeneration. I. A clinical analysis of 55 anti-Yo antibody positive patients. Neurology 42:1931
10. Dalmau J, Graus F, Rosenblum MK, et al (1992) Anti-Hu-associated paraneoplastic encephalomyelitis/sensory neuronopathy. A clinical study of 71 patients. Medicine 71:59
11. Luque FA, Furneaux HM, Ferziger R. et al (1991) Anti-Ri: an antibody associated with paraneoplastic opsoclonus and breast cancer. Ann Neurol 29:241
12. Graus F, Bonaventura I, Uchuya M, Valls-Sole J, Rene R, Leger JM, Tolosa E, Delattre J (1994) Indolent anti-Hu-associated paraneoplastic sensory neuropathy. Neurology 44:2258
13. Furneaux HM, Reich L, Posner JB (1990) Autoantibody synthesis in the central nervous system of patients with paraneoplastic syndromes. Neurology 40:1085
14. Graus F, Dalmau J (1995) Management of paraneoplastic syndromes. In: Wiley RG (ed) Neurological complications of cancer. Marcel Dekker, New York, pp 167–198
15. Hammack JE, Kotanides H, Rosenblum MK, Posner JB (1992) Paraneoplastic cerebellar degeneration. II. Clinical and immunologic findings in 21 patients with Hodgkin's disease. Neurology 42:1938
16. Lang B, Newsom-Davis J, Wray D, Vincent A, Murray N (1981) Autoimmune aetiology for myasthenic (Eaton-Lambert) syndrome. Lancet II:224
17. Fukunaga H, Engel AG, Osame M, Lambert EH (1982) Paucity and disorganization of presynaptic membrane active zones in the Lambert-Eaton myasthenic syndrome. Muscle Nerve 5:686
18. Chester KA, Lang B, Gill J, et al (1988) Lambert-Eaton syndrome antibodies: reaction with membranes from a small cell lung cancer xenograft. J Immunol 18:97
19. Sher E, Gotti C, Canal E, et al (1989) Specificity of calcium channel autoantibodies in Lambert-Eaton myasthenic syndrome. Lancet II:640
20. Newsom-Davis J, Murray NMF (1984) Plasma exchange and immunosuppressive drug treatment in the Lambert-Eaton myasthenic syndrome. Neurology 34:480

21. Dalmau J, Furneaux HM, Gralla RJ, Kris MG, Posner JB (1990) Detection of the anti-Hu antibody in the serum of patients with small cell lung cancer – a quantitative Western blot analysis. Ann Neurol 27:544
22. Liu S, Mezrich J, Berk J, Federici M, Dalmau J, Posner JB (1995) Expression of Purkinje cell antigens in ovarian tumor, and presence of anti-Purkinje cell antibodies in the serum of patients without paraneoplastic cerebellar degeneration (abstract). Neurology 45[Suppl 4]:A228
23. Dalmau J, Furneaux HM, Rosenblum MK, Graus F, Posner JB (1991) Detection of the anti-Hu antibody in specific regions of the nervous system and tumor from patients with paraneoplastic encephalomyelitis/sensory neuronopathy. Neurology 41:1757
24. Dalmau J, Furneaux HM, Cordon-Cardo C, Posner JB (1992) The expression of the Hu (paraneoplastic encephalomyelitis/sensory neuronopathy) antigen in human normal and tumor tissues. Am J Pathol 141:881
25. Szabo A, Dalmau J, Manley G, Rosenfeld M, Wong E, Henson J, Posner JB, Furneaux HM (1991) HuD a paraneoplastic encephalomyelitis antigen contains RNA binding domains and is homologous to Elav and Sex-Lethal. Cell 67:325
26. Jean WC, Dalmau J, Ho A, Posner JB (1994) Analysis of the IgG subclass distribution and inflammatory infiltrates in patients with anti-Hu associated paraneoplastic encephalomyelitis. Neurology 44:140
27. Paone JF, Jeyasingham K (1980) Remission of cerebellar dysfunction after pneumonectomy for bronchogenic carcinoma. N Engl J Med 302:156
28. Kearsley JH, Johnson P, Halmagyi M (1985) Paraneoplastic cerebellar disease. Remission with excision of the primary tumor. Arch Neurol 42:1208
29. Batson OA, Fantle DM, Stewart JA (1992) Paraneoplastic encephalomyelitis. Dramatic response to chemotherapy alone. Cancer 69:1291
30. Burton GV, Bullard DE, Walther PJ, et al (1988) Paraneoplastic limbic encephalopathy with testicular carcinoma. A reversible neurologic syndrome. Cancer 62:2248
31. Carr I (1982) The Ophelia syndrome: memory loss in Hodgkin's disease. Lancet I:844
32. Cocconi G, Ceci G, Juvarra G (1985) Successful treatment of subacute cerebellar degeneration in ovarian carcinoma with plasmapheresis. A case report. Cancer 56:2318
33. Graus F, Vega F, Delattre J-Y, et al (1992) Plasmapheresis and antineoplastic treatment in CNS paraneoplastic syndromes with antineuronal autoantibodies. Neurology 42:536
34. Vega F, Graus F, Chen QM, Schuller E, Poisson M, Delattre JY (1994) Intravenous (IV) immunoglobulin therapy in paraneoplastic neurologic syndromes (PNS) with antineuronal autoantibodies. Neurology 44:A157
35. Younger DS, Dalmau J, Inghirami G, Sherman WH, Hays AP (1994) Anti-Hu-associated peripheral nerve and muscle microvasculitis. Neurology 44:181
36. Moll JW, Henzen-Logmans SC, Van der Meche FG, et al (1993) Early diagnosis and intravenous immune globulin therapy in paraneoplastic cerebellar degeneration. Letter to the editor. J Neurol Neurosurg Psychiatry 56:112
37. Counsell CE, McLeod M, Grant R (1994) Reversal of subacute paraneoplastic cerebellar syndrome with intravenous immunoglobulin. Neurology 44:1184
38. Bird SJ (1992) Clinical and electrophysiologic improvement in Lambert-Eaton syndrome with intravenous immunoglobulin therapy. Neurology 42:1422
39. Cher LM, Hochberg FH, Teruya J, Nitschke M, Valenzuela RF, Schmahmann JD, Herbert M, Rosas HD, Stowell C (1995) Therapy for paraneoplastic neurologic syndromes in six patients with protein A column immunoadsorption. Cancer 75:1678
40. Dyken P, Kolar O (1968) Dancing eyes, dancing feet: infantile polymyoclonia. Brain 91:305
41. Ahlberg R, Yi Q, Pirskanen R, et al (1994) Treatment of myasthenia gravis with anti-CD4 antibody: Improvement correlates with decreased T-cell autoreactivity. Neurology 44:1732

Autoimmune responses in peripheral nerve

Hans-Peter Hartung[1], Hugh Willison[2], Stefan Jung[1], Martin Pette[1], Klaus V. Toyka[1], Gerhard Giegerich[1]

[1] Department of Neurology, Julius-Maximilians-Universität Würzburg, Josef-Schneider-Strasse 11, D-97080 Würzburg, Germany
[2] University Department of Neurology, Southern General Hospital, Glasgow, G51 4TF, Scotland

Introduction

Students of the pathogenesis of peripheral neuropathies have benefited greatly from recent advances in immunology, molecular biology and neurobiology, both in terms of general concepts and methodology. These have been applied to the investigation of those diseases of peripheral nerve that are thought to result from aberrant immunity. These immune-mediated neuropathies include the Guillain-Barré syndrome (GBS), multifocal motor neuropathy with conduction block, chronic inflammatory demyelinating polyradiculoneuropathy (CIDP), the neuropathies associated with monoclonal gammopathies, vasculitic neuropathies and possibly paraneoplastic neuropathies. Here we attempt to present an overview of recent advances in understanding immune responses in the peripheral nervous system (PNS) that underlie these disorders.

The peripheral nervous system: an immunocompetent organ

Very much like the brain, the PNS is to some extent secluded from the systemic immune compartment by the blood-nerve-barrier (BNB). The anatomical basis of this functional barrier is provided by tight junctions of endoneurial endothelial cells. The endoneurium, the intrafascicular connective tissue excluding perineurial partitions, comprises axons, Schwann cells, pericytes, mast cells, fibroblasts, macrophages, connective tissue matrix and the endoneurial fluid. Immediately external to the Schwann cell is a thin sheath that is inflected at the nodes of Ranvier where fast saltatory impulse propagation takes place. External to this is a layer of collagen fibrils which is not inflected at the nodes, and which constitute the outer endoneurial sheath. Myelinated nerve fibers are invested by a basal lamina that turns in at the nodes and is continuous across them, surrounding the Schwann cell nodal processes. The endoneurial space is enclosed by the perineurium, a multilamellar arrangement of flattened cells separated

Correspondence to: H.-P. Hartung

by layers of collagenous connective tissue. The contiguous perineurial cells overlap or interdigitate with one another and, at their margins, are linked by tight junctions which probably determine the diffusion barrier properties of this structure. Antibodies of the IgG class are not significantly excluded from the endoneurial space, even under normal conditions. The BNB is relatively deficient at the most proximal and distal parts of nerve, i.e. at spinal roots and at the motor nerve terminals [137, 191].

Antigen presentation in peripheral nerve

Activated T cells can penetrate the intact BNB irrespective of antigen specificity and under certain circumstances (e.g., infections), release cytokines that up-regulate the expression of major histocompatibility complex (MHC) class II gene products and adhesion molecules in the PNS [69, 145]. Professional and non-professional antigen-presenting cells are numerous in peripheral nerve tissue. These include resident endoneurial and perivascular macrophages and Schwann cells.

Normal nerves contain macrophages near the endoneurial vessels. These cells usually lie outside the basal lamina of the blood vessels and have a dendritic appearance [48, 126]. Their perivascular distribution at the blood-nerve interface makes them uniquely suited to act as antigen-presenting cells in the PNS. Many of these resident macrophages constitutively express MHC class II molecules, adhesion molecules, complement receptor type 3 (CR3), and CD4, but the level of expression of these molecules can be greatly enhanced in inflammatory conditions [13, 46, 126].

Human and rat Schwann cells in culture express little, if any, MHC class II antigen but can be induced to do so by co-culture with T lymphoblasts, T lymphoblasts and mycobacterial antigen, or by interferon-γ (IFN-γ) and by tumor necrosis factor α (TNF) [6, 40, 98, 162, 163, 206]. Pharmacological inhibitors of IFN-γ synthesis greatly diminish MHC class II antigen expression on cultured rat Schwann cells [193]. IFN-γ also induces expression of intercellular adhesion molecule (ICAM)-1 on rat Schwann cells and TNF-α synergized with IFN-γ in this respect [40]. Schwann cells, exposed to IFN-γ or IFN-γ plus TNF-α, are able to present exogenous P2 protein to P2-specific neuritogenic T cell lines [40]. Furthermore, HLA-DR immunoreactivity was shown to be associated with Schwann cells in human sural nerve biopsy samples in some reports [120, 124, 143, 144]. Whether Schwann cells can act as antigen-presenting cells in vivo is still a matter of debate [28, 47, 167].

Schwann cells also constitutively express low levels of MHC class I molecules which are markedly increased following incubation with IFN-γ [163, 193]. This would mark Schwann cells as targets for T cell-mediated cytotoxicity. In fact, Schwann cells primed with *Mycobacterium leprae* can be lysed in the presence of IFN-γ through antigen-specific interaction with *M. leprae*-reactive $CD8^+$ T lymphocytes [176].

Immune regulation

Rat Schwann cells are endowed with a cytokine-inducible nitric oxide (NO) synthase (iNOS). IFN-γ and TNF-α up-regulate iNOS-specific mRA in Schwann cells and precipitate release of nitrite in a dose-dependent manner [41]. NO production may in part account for a strong suppressive effect on T cell activation by thymic antigen presenting cells in a co-culture model where Schwann cells were present. These observations raise the possibility that Schwann cells exert potent immunoregulatory functions beyond their role as antigen presenters. Whether some constitutive expression of mRNA for the down-regulatory cytokine transforming growth factor (TGF)-β1 observable in Schwann cells is sufficient to contribute to these control mechanisms is unclear at present [97]. Rat Schwann cells also release another possibly immunosuppressive molecule, prostaglandin E, when challenged with lipopolysaccharide (LPS) or cytokines [26]. LPS, TNF-α and interleukin (IL)-1α are able to stimulate rat Schwann cells to produce the cytokine IL-6. Interestingly, Schwann cells respond to exogenous IL-6 by induction of IL-6 mRNA and expression of the IL-6 receptor [12], suggesting the existence of an autocrine pathway. Mouse Schwann cells can produce neurosteroids locally, but their role in immune regulation has not been established [100].

Schwann cells express the complement receptor type 1 (CR1, CD35), which recognizes C3b and C4b and the immunoglobulin Fc receptor (Fcγ R) in vivo. The latter is present only if Schwann cells are in contact with axons. CR1 and FcγR act as immune adherence receptors mediating the uptake of immune complexes. CR1 may, in addition, function as a cofactor for factor I-mediated C3b cleavage and dissociation of the alternative pathway convertase C3b,Bb. Thus, it may play a role in down-regulating complement activation occurring in the vicinity of Schwann cells [199].

Peripheral nerve antigens

The myelin sheath is composed of three major proteins (P0, P1, and P2), with molecular masses of 14–30 kDa, that make up more than 70% of the total protein content [114]. In human PNS myelin, P0 accounts for more than half of the total membrane protein; it does not occur in central nervous system (CNS) myelin. P0 is a 29 kDa, 219 amino acid transmembrane glycoprotein with a larger extracellular domain, a highly hydrophobic membrane-spanning portion and a basic cytoplasmic domain. It belongs to the immunoglobulin gene superfamily and acts as a homophilic cell adhesion molecule [35, 36]. P0 is thought to be involved in starting and stabilizing the compaction of the extracellular apposition of the myelin membrane in the PNS. P1 is identical to myelin basic protein (MBP) of the CNS, and the 14 kDa P2 protein containing 131 amino acids is common to both CNS and PNS, but predominates in the latter. It is concentrated in the paranodal loops and in Schmidt-Lanterman incisures rather than in compact internodal myelin. The conformation of P0 is highly dependent on its association with lipids. Circulating antibodies to P0 and P2 have been detected in Lewis rat experimental autoimmune neuritis (EAN) [4] and in a minority of patients with GBS; however, they have also been detected in patients with other disease and healthy controls [96, 150]. Similarly, T cells reactive with P0 and/or P2 and peptides thereof can sometimes be found in patients with GBS and occasionally in healthy donors [141, 142, 157].

Lipid and glycoconjugate antigens

Lipids make up about 75% of the total nerve dry weight. The major myelin lipids are cholesterol, galactosylceramide (galactocerebroside), galactosylceramide-3-O-sulfate (sulfatide) and, in smaller amounts, gangliosides and complex neutral glycolipids [108, 114, 183, 184, 217].

Gangliosides

Glyco(sphingo)lipids are composed of the long-chain aliphatic amine ceramide, attached to one or more hexose sugars [113]. Some relevant glycolipid structures are shown in Table 1. The ceramide is immersed in the membrane lipid bilayer with the carbohydrate structure exposed extracellularly. They are highly enriched in neuronal and glial membranes where they modulate a diverse array of cellular functions. A single hexose linked to ceramide (i.e., monohexosyl ceramide) is termed cerebroside, the most abundant of which in man is galactocerebroside (GalC). Sulfatide (sulfated galactocerebroside) is a major component of peripheral nerve myelin. Paragloboside is a neutral glycolipid, which when sialylated on the terminal galactose (i.e., sialosylparagloboside, SGP) is a major peripheral nerve glycolipid, also known as LM1. Substitution of the terminal sialic acid on SGP for sulfated glucuronic acid forms sulfated glucuronyl paragloboside (SGPG), which has a closely related homologue termed SGPLG. GalC, sulfatide, LM1 and SGPG have all been identified as antigens for sulfated glucuronyl lactosaminyl paragloboside autoantibodies in subtypes of autoimmune neuropathy.

Gangliosides are complex glycosphingolipids which must contain at least one sialic acid residue (*N*-acetylneuraminic acid, abbreviated to NeuNAc) attached to the internal and/or the terminal galactose of an oligosaccharide core composed of up to four sugars. Again, many gangliosides have been identified as important neuropathy-associated autoantigens. There are four major gangliosides in brain (GM1, GD1a, GD1b, GT1b), and many minor gangliosides in brain and peripheral nerve which are developmentally regulated, spatially segregated and functionally distinct in a wide range of patterns across different species.

The regional distribution and function of glycolipids in different areas of the human nervous system may play a role in defining the pathophysiological features of their autoantibody-associated diseases. For example, GQ1b is enriched in human ocular muscle nerves compared with other sites, which may account for the vulnerability of these nerves to immune-mediated attack in the anti-GQ1b antibody-associated neuropathy, Miller Fisher syndrome [23, 24, 213, 223]. A similar explanation may hold for other regional or fiber-selective syndromes, e.g., in the myelin sheaths GM1 is enriched in motor nerves compared to sensory nerves [129, 130]. However, other factors, such as antibody access and glycolipid membrane topography, are also clearly important. The fine specificity of antibodies will also have an influence on their ability to bind a particular antigen in a physiological membrane. For example, a panel of human anti-GM1 monoclonal antibodies cloned from motor neuropathy patients has been recently described [215]: all react well with GM1 in enzyme-linked immunosorbent assays (ELISA) but the patterns of immunoreactivity with GM1-containing membranes, both in tissue sections and in cultured cells, are entirely different between the various antibodies, some of which showed no immunoreactivity. Thus, anti-glycolipid

Table 1. Glycolipid and ganglioside structures

Chemical name	Carbohydrate sequence
GM1	Gal (β1-3) GalNAc (β1-4) Gal (β1-4) Glc (β1-1) ceramide
	(a2-3)
	NeuNAc
GM2	GalNAc (β1-4) Gal (β1-4) Glc (β1-1) ceramide
	(a2-3)
	NeuNAc
GM3	Gal (β1-4) Glc (β1-1) ceramide
	(a2-3)
	NeuNAc
GD1a	Gal (β1-3) GalNAc (β1-4) Gal (β1-4) Glc (β1-1) ceramide
	(a2-3) (a2-3)
	NeuNAc NeuNAc
GD1b	Gal (β1-3) GalNAc (β1-4) Gal (β1-4) Glc (β1-1) ceramide
	(a2-3)
	NeuNAc
	(a2-8)
	NeuNAc
GT1b	Gal (β1-3) GalNAc (β1-4) Gal (β1-4) Glc (β1-1) ceramide
	(a2-3) (a2-3)
	NeuNAc NeuNAc
	(a2-8)
	NeuNAc
GT1a	Gal (β1-3) GalNAc (β1-4) Gal (β1-4) Glc (β1-1) ceramide
	(a2-3) (a2-3)
	NeuNAc NeuNAc
	(a2-8)
	NeuNAc
GQ1b	Gal (β1-3) GalNAc (β1-4) Gal (β1-4) Glc (β1-1) ceramide
	(a2-3) (a2-3)
	NeuNAc NeuNAc
	(a2-8) (a2-8)
	NeuNAc NeuNAc
GD3	Gal(β1-4Glc(β1-1) ceramide
	(a2-3)
	NeuNAc
	(a2-8)
	NeuNAc
Asialo-GM1	Gal (β1-3) GalNAc (β1-4) Gal (β1-4) Glc (β1-1) ceramide
LM1 (=SPG)	Gal (β1-4) GlcNAc (β1-3) Gal (β1-4) Glc (β1-1) ceramide
	(a2-3)
	NeuNAc
SGPG	Gal (β1-4) GlcNAc (β1-3) Gal (β1-4) Glc (β1-1) ceramide
	(b1-3)
	$SO_4$3GlcUA

SPG, sialosylparagloboside; SGPG, sulfonated glucuronyl paragloboside

antibody standard ELISA methodologies may be poor identifiers of pathogenic antibodies that are capable of binding glycolipids in biological membranes. These caveats need to be taken into account when defining clinical syndromes.

Myelin-associated glycoprotein

Myelin-associated glycoprotein (MAG) is a glycosylated transmembrane glycoprotein and member of the immunoglobulin gene superfamily. It shares homologies with a number of cell adhesion molecules, P0 and a glycoprotein originally described on natural killer cells that is recognized by the monoclonal antibody HNK-1, and PMP-22. MAG, a 100-kDa glycoprotein, occurs in two isoforms of 72 kDa (L-MAG) and 67 kDa (S-MAG) [114], which are generated by alternative splicing. In peripheral nerve more than 95% of the total MAG is S-MAG. Apart from being glycosylated, MAG undergoes a number of post-translational modifications. A very minor component of peripheral myelin (less than 1% of all myelin protein), it is concentrated in membranes adjacent to Schwann cells, i.e., the adaxonal Schwann cell membrane, the paranodal loops and the Schmidt-Lanterman incisures. It is also found in the internal and external mesaxons. MAG is apparently an essential adhesion molecule for Schwann cell-axolemmal interactions and, therefore, important in the process of myelination.

EAN: a model of T cell-mediated neuropathy

The animal disorder EAN mirrors many of the immunological, electrophysiological, and morphological aspects of GBS and, hence, has been widely used as a model to investigate disease mechanisms of immune-mediated nerve damage. EAN is an acute demyelinating inflammatory polyradiculoneuropathy that can be induced in mice, rats, rabbits and monkeys by active immunization with whole peripheral nerve homogenate, myelin, the myelin protein P2, cyanogen bromide fragments of P2, or synthetic peptides from P2 – in particular those comprising the amino acid sequences P2 53–78 and P2 61–70 (minimal neuritogenic epitopes) and P0 (reviewed in [58, 114]). EAN can also be induced by adoptive transfer (AT-EAN) of P2, P2-peptide-specific, P0 and P0 peptide-specific T cell lines in the Lewis rat [116, 118, 133–135, 147, 148, 156, 158, 160] and by adoptive transfer of MBP-specific $CD4^+$ T helper cell clones in BALB/c mice [1]. MBP is present in peripheral myelin where it is designated P1. There it makes up for approximately 2–16% of all myelin proteins. Two T cell clones recognizing epitopes contained in the 21- and 18.5-kDa isoforms induced changes predominantly in the PNS [1].

Demyelinative changes and admixed axonal degeneration occur depending on the amount of myelin or neuritogenic P2 peptides used for immunization [52, 58]. Similarly, depending on the number of autoreactive cells transferred, T cell line-mediated EAN can take the form of a predominantly demyelinating condition (low cell dose) or of a fulminant neuritis morphologically characterized by prominent endoneurial edema and axonal degeneration when a higher cell dose is injected to produce disease. This is reflected in the profound functional deficit in the latter [74, 115, 147, 148]. These observations in both EAN models are important since they suggest that profound axonal damage in hyperacute forms of the GBS [33, 155] would nonetheless be in keeping with an autoimmune reaction to a single myelin antigen, and does not necessarily imply the requirement for different antigens or etiologies in these clinical and morphological variants.

An important role for T cells in the initiation of EAN was suggested by transfer of the disease to naive Lewis rats with lymph node cells obtained from animals that had been actively immunized with myelin or P2 [79]. Formal proof was presented by two groups who showed that injection of P2- or P0-reactive $CD4^+$ T line cells into Lewis rats produced EAN [115, 118, 158]. Conversely, rats rendered T cell deficient by adult thymectomy and lethal irradiation or treated with anti-T cell antibodies were much less susceptible to disease induction by active immunization [17, 77]. Indirect evidence was also advanced by showing that high-dose cyclosporine or administration of a monoclonal antibody directed to the IL-2 receptor on activated T cells or to the α, β chains of the T cell receptor (TCR) prevented the development of transfer EAN [56, 90]. Epitope specificity of neuritogenic T cell lines was found to reside primarily in the amino acid sequences 61–70 in P2, and 180–199 in the P0 protein. These are amphipathic peptide fragments with an alpha-helical configuration [133–135, 156, 160].

Usage of TCR variable (V) gene families has been analyzed in EAN. Clark and colleagues [25, 231] reported that T cell lines and hybridomas generated with a neuritogenic peptide SP26 exclusively utilized the same TCR V gene families, $V\alpha 2$ and $V\beta 8$, for their TCR as T cells that cause experimental allergic encephalomyelitis (EAE). Subsequent studies from our group revealed a broader expression of $V\alpha$ and $V\beta$ genes [91].

Early invasion of the PNS by T lymphocytes has been documented immunocytochemically. Before onset of clinical signs most T cells express the $CD4^+$ helper/inducer phenotype, whereas $CD8^+$ cells prevail after the peak of the disease and during recovery. However, macrophages greatly outnumber lymphocytes in affected peripheral nervous tissue, particularly beyond day 13 post immunization when only few T cells continue to be present in the active EAN model [69, 81, 182].

To access peripheral nerve across the BNB, circulating autoreactive T lymphocytes need to be activated [145]. The process of homing and transmigration is governed by an ensemble of adhesion molecules which are reciprocally expressed on endothelial cells and leukocytes. This is a multistep process occurring in an ordered sequential fashion [11, 175].

In, EAN, shortly before disease manifestation, both invading T cells, macrophages and endoneurial endothelial cells up-regulate their expression of ICAM-1 [180]. Expression of the counterreceptor for ICAM-1, the leukocyte function associated antigen (LFA)-1, is likewise greatly enhanced on infiltrating mononuclear cells [5]. Breakdown of the BNB is one of the earliest morphologically demonstrable events in lesion development in EAN [51, 146, 147]. Local reactivation of immigrated neuritogenic T cells initiates a cascade of inflammatory responses involving cytokines, additional effector cells, and a host of inflammatory effector molecules both inside and outside the BNB.

Among the cytokines that orchestrate cellular interactions during immune responses, IFN-γ is of major importance. IFN-γ is released by $CD4^+$ T cells of the Th1 inflammatory phenotype. It can (1) induce MHC class II antigens on macrophages and their production of toxic oxygen free radicals and other proinflammatory mediators [61, 167], (2) up-regulate expression of adhesion molecules on endothelial cells, macrophages, T cells and Schwann cells [40], and (3) enhance vascular permeability. IFN-γ could, therefore, be instrumental in allowing influx of antibodies, soluble mediators, and inflammatory cells into the endoneurium.

IFN-γ immunoreactivity is expressed by T cells, some macrophages, and neutrophils, which along with the invasion of T lymphocytes enter the nerve early in EAN [168]. IFN-γ^+ cells are present only transiently from day 11 to day 13 after active immunization, i.e., before onset of overt disease. The presence of IFN-γ^+ cells in nerve roots correlates with the number of Ia$^+$ macrophages. In the early phase of the disease many lean (prephagocytic) macrophages are present and carry MHC class II gene products, while most of the postphagocytic macrophages with large ED1$^+$ vacuoles no longer express Ia at later stages of EAN [136, 167]. Lymph nodes taken from animals with EAN contain increased numbers of T cells that, upon exposure to P2, secrete IFN-γ as early as 7 days post immunization with peripheral myelin [232]. Similarly there was augmented mRNA expression of IFN-γ by mononuclear cells stimulated with P2 [234]. Significantly, in vivo administration of recombinant IFN-γ markedly augments disease severity in actively induced EAN and in EAN induced by P2 T cell lines, whereas application in vivo of a monoclonal antibody to IFN-γ suppressed the disease [61]. TNF-α appears in nerve lesions around the time that the first clinical symptoms appear. With clinical recovery, TNF-α immunoreactivity is no longer detectable. Neutralization of endogenously generated TNF-α ameliorates EAN [181]. Injection of TNF-α into sciatic nerve of mice produced predominantly axonal damage in one study, and a mixture of vascular damage, demyelination and degeneration in another [151, 161].

Macrophages are the predominant cell population in affected nerves of animals with EAN [59, 178]. Most ED1$^+$ macrophages also express MHC class II antigen (OX6) and, hence, most likely function as antigen presenters [136, 167]. Macrophages adhere to normal-appearing nerve fibers in EAN, a feature never observed in Wallerian degeneration where macrophages selectively attack degenerating fibers [47]. They strip off myelin lamellae from the axons, causing demyelination. Local complement activation may play a role in this process by providing the ligand for the adhesion molecule and complement type 3 receptor CR3/Mac1. Through this process macrophages may be attracted to the myelin sheath.

Immunopharmacological investigations shed light on the role of macrophages in the amplification and effector phase of the disease. The essential requirement of macrophages for disease expression was highlighted by depletion experiments in which either silica or the macrophage-specific toxin dichlormethylene disphosphonate encapsulated in liposomes were used. Macrophage depletion prevented all clinical, electrophysiological, and histological signs of EAN and AT-EAN [59, 75, 92, 186]. Macrophages are activated early on in disease development. Shortly before onset of clinical symptoms blood monocytes were shown to exhibit a significant increase in spontaneous chemiluminescence. In addition, superoxide anion and hydrogenperoxide release, like chemiluminescence an index of heightened oxidative burst activity in macrophages, were enhanced in peritoneal macrophages collected from animals with EAN [60, 177]. Another family of proinflammatory molecules is liberated from macrophages during the course of EAN, the eicosanoids [59]. The pathogenetic role of these macrophage products was shown using treatment with inhibitors of arachidonc acid conversion, such as dexamethasone, indomethacin, BW755c, or with the oxygen radical scavengers superoxide dismutase and catalase, which prevented myelin-induced EAN when initiated early or greatly attenuated disease expression when administered after onset of clinical signs [59, 60]. Arachidonic acid metabolites possibly synergizing with activated complement could function as chemoattractants and secretagogues for inflammatory cells or enhance permeability of the BNB. Reactive oxygen species can inflict peroxidative injury upon myelin but may also act by generating chemotactic signals and by damaging endothelial cells [22, 54, 101]. Rodent macrophages are endowed with a cytokine-inducible form of NO synthase (iNOS) and, hence, are able, upon appropriate stimulation, to generate toxic NO metabolites. A role for these compounds is suggested by our recent demonstration that competitive NO synthase inhibitors can mitigate Lewis rat EAN [236].

Macrophage-derived neutral proteases and phospholipases have been implicated in myelin damage in vitro [20, 21]. Microinjection of proteinases into the sciatic nerve of rats in vivo produces inflammatory demyelination [204, 208] and treatment of EAN rats with proteinase inhibitors delayed development of the disease [166].

Endoneurial mast cells decrease in number during the course of EAN [16]. Electron microscopically, they can be shown to degranulate, possibly as a consequence of T cell-mediated delayed-type hypersensitivity [88]. These cells may release vasoactive amines (5-hydroxytryptamine, histamine) and arachidonic acid-derived metabolites that could augment vascular permeability. BNB integrity and nerve conduction in rat tibial nerve were greatly disturbed when mast cell degranulation was induced by intraneural injection of compound 48/80 [68]. A contributory pathogenic role of mast cell degranulation is supported by pharmacological experiments in which mast cell-stabilizing drugs like reserpine and nedocromil prevented or attenuated the disease [15].

Termination of the inflammatory response in EAN

Several mechanisms may operate to terminate the acute immunoinflammatory attack in this monophasic disease model. Production of the down-regulatory cytokine TGF-β1 by macrophages and to a lesser extent by T lymphocytes is increased during the course of EAN in Lewis rat [97, 234]. Importantly, systemic administration of recombinant TGF-β1 markedly inhibited the clinical signs and histological changes of actively induced EAN in the Lewis rat [43, 93]. Zhu et al. [234] demonstrated augmented mRNA expression of the down-regulatory cytokine IL-4 in lymph node and spleen cells from EAN rats cultured with peripheral myelin, P2 or P0, which peaked late in the course of the disease.

Apoptosis, i.e., programmed cell death, of activated T cells accumulating in peripheral nerve may be another mechanism for terminating the ongoing immune response and limiting issue damage. Apoptotic T cells can be identified by their condensed nuclear chromatin and the labeling of fragmented DNA in their nuclei in EAN, and this process was greatly enhanced after administration of high-dose glucocorticosteroids [229, 230].

Antibodies in EAN

Antibodies to the P2 protein have been detected in the blood of animals with EAN [79, 157]. Their titers and kinetics, however, did not correlate with disease activity which casts doubt on an essential causative function. Using the ELISA spot assay technique, IgG and IgM antibodies to P2 were demonstrated in lymph nodes and spleens from rats with EAN. Again there was no correlation to stage or activity of the disease [233]. Similar observations were made regarding P0 [233]. Using ELISA and Western-blot analysis we found high-titer circulating antibodies to the extracellular domain of P0 in blood and cerebrospinal fluid of Lewis rats on day 13 after immunization with purified bovine myelin. Titers were maintained for approximately 2 weeks and then declined, which was in sharp contrast to the antibody response to P2. This appeared later and persisted for several weeks despite complete clinical recovery. When EAN was induced by immunization with P2, no significant antibody reactivity against P0 or its extracellular domain was detectable [4]. Strong evidence arguing against an exclusive role for antibodies in the pathogenesis of EAN stems from studies of AT-EAN. Recipient rats develop disease within 4–5 days of T cell transfer, an interval too short for a significant antibody response to be mounted. In fact, circulating antibodies to P2 could not be measured in the sera of these recipient animals [4, 115]. Screening the B cell repertoire in myelin-induced EAN, Zhu et al. [233] also detected increased numbers of B cells in lymph nodes and spleen cells that synthesized antibodies to ganglioside GM1 and MAG peptides.

Complement

Evidence for the involvement of complement in lesion pathogenesis has been adduced by several lines. Immunocytochemically the terminal complement complex (TCC; C5b-9) was found to be deposited on Schwann cells and along the myelin sheath at sites of impending demyelination. It is there that macrophages concentrate [179]. Strikingly similar observations have recently been made in acute inflammatory demyelinating polyneuropathy [119]. Decomplementation of animals with cobra venom factor partly suppressed EAN [34, 202]. Finally, injection of rats with soluble CR1, a regulatory glycoprotein of the complement cascade, mitigated disease [94].

TCC deposition on the surface of Schwann cells and myelin sheaths may promote an influx of calcium, which is considered an important initial step in myelin breakdown. TCC formation in myelin membranes may cause activation of myelin-associated neutral proteases with subsequent hydrolysis of myelin proteins. It is conceivable that TCC-targeted myelin is selectively attacked by macrophages. Macrophages in culture respond to TCC by liberating chemotactically active and proinflammatory eicosanoids, which have been documented to contribute to the pathogenesis of EAN [59]. Taken together, available evidence suggests that complement may be important in recruiting macrophages into the endoneurium, in opsonizing myelin for phagocytosis and in amplifying ongoing inflammatory reactions.

Synergy of T and B cell responses in EAN

The apparent dichotomy of the cell-mediated and humorally mediated immune responses being considered as pathogenic, as reviewed, may be resolved by envisaging a cooperative action of T cells and antibodies in effecting nerve damage. Experimental evidence to corroborate this notion has recently become available [53, 145, 173]. Injection of Lewis rats with an antibody to myelin or to galactocerebroside enhanced disease in P2-peptide T cell transfer EAN [53, 173]. A local breach of the BNB by neuritogenic T cells would facilitate influx of demyelinating antibodies from the circulation [174]. Most importantly, this permissive action of activated T lymphocytes appears not to be related to their antigenic specificity. When ovalbumin-specific T line cells accumulated in the nerve of Lewis rats at sites of ovalbumin injection, intraperitoneal administration of a galactocerebroside antiserum produced local conduction block and demyelination [69, 145].

Guillain-Barré Syndrome

This is the commonest and to date best studied of the immune-mediated neuropathies in man. Currently, GBS is subdivided into: the classic acute demyelinating variant, which represents the great majority of cases in the Western world; the acute motor axonal neuropathy (AMAN) initially described as Chinese paralytic syndrome and the most prevalent form of GBS in China [122]; the acute motor and sensory axonal neuropathy (AMSAN), which is distinguished by significant sensory involvement and apparently associated with a more severe course and poorer prognosis; and the Miller Fisher syndrome [49, 78]. The earliest changes noted in well-preserved autopsy cases of AMAN were found at the nodes of Ranvier of motor fibers where nodal gaps were lengthened and the nodes decorated with activation products of the complement cascade [49, 50, 119]. Macrophages overlied the nodes of Ranvier and penetrated the outer basal lamina of the Schwann cell to eventually enter the periaxonal space of the internode. Activated complement components (C3d, C9 neoantigen) could be localized to the adaxonal Schwann cell plasmalemma. Despite invasion of the periaxonal space by macrophages, the axons were intact. Wallerian degeneration was observed in the most severe cases. Based on these observations, the authors raise the possibility that antibodies to GM1 or related glycoconjugates bind to the axolemma (50). Ensuing complement activation would provide a chemotactic signal for macrophages to invade the periaxonal space. Binding of such antibodies to the nodes of Ranvier could produce partial and reversible conduction block. Two groups have shown that GM1 antibody containing sera from patients with multifocal motor neuropathies and those with GBS can block conduction and reduce or eliminate endplate potentials in the mouse phrenic nerve hemidiaphragm preparation ([18, 154] and A. Vincent, personal communication).

Inn an autopsy of a patient with "classic demyelinating" GBS who succumbed 3 days after onset, pathological changes in the spinal cords were very mild and only few lymphocytes were identified. Activated complement was found deposited on the outermost surface of the Schwann cell (the abaxonal Schwann cell plasmalema) and electron microscopy provided evidence for vesicular dissolution of the myelin sheaths [119]. Activation products of complement (C3a, C5a, C3c, C5b-9) have been detected in serum, plasma and cerebrospinal fluid of Western GBS patients [57, 64, 67, 105, 164]. Earlier immunohistochemical studies in this population showed that C3d was localized to myelin sheaths in sural nerve biopsy samples and C5b-9 deposits were observed in one biopsied case of classic GBS [72, 105].

Cellular mechanisms in GBS

Activated T cells circulate in the blood of GBS patients. This is evidenced by their enhanced expression of HLA-DR, the transferrin receptor [188], the IL-2 receptor, both in its membrane-bound and shed forms [9, 62, 188], by increased frequencies of mutant T cells [197], and by the presence in increased amounts of soluble products of activated T cells, i.e., cytokines such as IL-2 [9, 63], IFN-γ [55] and TFN-α in serum [32, 170]. The $CD4^+$ $CD29^+$ helper/inducer subset was found to be increased, while the $CD4^+$ $CD45RA^+$ suppressor/cytotoxic subpopulation of T cells was decreased in number [29]. Various attempts have been made to identify nerve antigen(s) to which an aberrant T cell response is mounted. Many researchers employing the macrophage migration inhibition test, skin testing for hypersensitivity, or lymphocyte proliferation assays ex vivo failed to demonstrate a specific T cell response to P2, the major neuritogen of EAN, while others succeeded [7, 39, 55, 80, 87, 102, 114, 131, 237]. In one study it was possible to raise a P2-specific T cell line from one GBS patient, but with equal efficiency from three healthy controls [19]. Very recently it was shown that T cells given irradiated autologous mononuclear cells as antigen presenters underwent marked transformation in response to P2 and its peptides P2 14–25 and P2 58–81 in the first few days of GBS, and to P0 (P0 70–85; P0 180–199) in a minor proportion of patients [95, 189]. The T cell response to P0 or its peptides was detectable before that against P2 and P2 peptides. The response was clearly transient subsiding within 2–3 weeks, while the response to P2 persisted for months. There were no apparent clinical correlations other than onset of disease to any of these responses. A note of caution is in order. All these studies employed bovine proteins. We now have evidence that T cells reacting to bovine P2 or P0 may not recognize the human proteins [141, 205].

Results obtained with the P0 protein and its peptides are of particular importance since the extracellular domain of P0 (residues 1–124) is readily available for T cell recognition, in contrast to the cytoplasmically located protein P2. The relevance of these observations is strengthened by findings in EAN. Immunization with P0 or adoptive transfer of T cell lines raised against extracellular portions of P0 indeed produces an inflammatory neuropathy characterized by primary demyelination with preservation of axons [117, 123].

Macrophages feature prominently in the nerve lesion of GBS [169]. They may be derived from circulating monocytes that invade the peripheral nerve or from endoneurial macrophages. They act as professional antigen presenters and are of pivotal importance in the amplification and effector phase of immune-mediated demyelination [7]. Mechanisms that are operative include phagocytosis and the release of proinflammatory cytokines (such as IL-1, and TNF-α) and other highly active mediators that act at short range (proteases, complement components, oxygen radicals, NO, eicosanoids) [65].

Monocytes/macrophages are activated in GBS. Monocytes isolated from peripheral blood of GBS patients in the early phase of their disease produce increased amounts of toxic oxygen species when stimulated with phorbol diester, and elevated levels of neopterin, a biosynthetic product generated in response to IFN-γ, have been measured in GBS sera. Macrophages are activated by IFN-γ, and neopterin production mirrors the presence of this cytokine [10, 55].

Myelin antibodies in GBS

A large body of evidence implicates myelin antibodies in the pathogenesis of classic demyelinating GBS and this is reviewed here. Complement-fixing IgM antibodies to human myelin were originally identified in 90% of classic GBS cases [103, 104, 112]. On serial determinations no immunoglobulin isotype switching to IgG was noted, which would be compatible with the (auto)antigen being a thymus-independent antigen [104, 106]. These IgM antibodies were claimed to bind to carbohydrate residues similar to those of the Forssman antigen, a ceramide with five carbohydrate residues distributed widely in many mammalian species and infectious agents [106]. However, Ilyas et al. [84] were unable, using a different assay, to detect increased serum titers of antibodies to neutral glycolipids, particularly the Forssman antigen, in a larger series of 54 GBS cases.

Glycolipid antibodies in GBS

Recent research efforts have focused on humoral immune responses to the glycoconjugate family of antigens. The reported frequency of anti-glycolipid antibodies ranged from as few as 5% in one study to 60% in another, which may reflect the variable sensitivity of assays to detect these antibodies [183] and problems with classifying low titer antibodies [235]. Antibodies were frequently directed to one or more of the gangliosides GM1, GD1a, GD1b, GT1b, asialo-GM1, and LM1 (reviewed in [66]).

Different epitopes on glycoconjugates are recognized by these antibodies but the common denominator is the *Gal(β1-3) GalNac* carbohydrate moiety. Titers of these glycolipid antibodies tend to decline after the acute phase of the disease. Several investigators found an association of GM1 antibodies with a more severe and predominantly axonal type of GBS, and some reported that these antibodies occurred more frequently in patients that had a preceding *Campylobacter jejuni* infection [44, 127, 152, 195, 198, 200, 203, 218], but this remains disputed [31, 172, 201]. This lack of correlation with clinical features has now been confirmed in a large prospective case-control study [152]. Yuki et al. [222] claimed that high-titer GD1a antibodies of IgG or IgM class were associated with more severe disease. They never observed isotype class switching in the IgM anti-GD1a^{+} group in serial determinations. Some GBS patients have antibodies against rare ganglioside species; for example the sera from 6 out of 50 GBS cases were recently reported to contain antibodies to *N*-aceytl-galactosaminyl GD1a [108]. Most of the Japanese patients with severe GBS, ganglioside antibodies, and evidence of a preceding *C. jejuni* infection reported by Yuki et al. [218, 219, 221] were found to have the HLA-B35 haplotype, while Kuroki et al. [107] noted an HLA-B35 haplotype association in only 25% of his stool culture-positive Japanese patients, making an immunogenetic basis of the variations reported unlikely. Rees et al. [153] looking at MHC class II alleles in 97 patients with GBS noted a significant association between HLA-DQB1*03 and preceding *C. jejuni* infection.

Fredman et al. [37] found antibody activity mostly of the IgG isotype predominantly to LM1 in 43% of GBS patients, but also in 20% of blood donors. Anti-sulfatide antibodies were detected in 65% of GBS cases. IgM and IgA class antibodies have also been observed [37, 82–86].

Anti-SGPG antibodies of IgM or IgG class were detected in 13 of 53 (25%) of GBS sera [83] in one series, but in only 1 of 21 GBS sera in another [196]. Antibodies to galactocerebroside were not found more frequently in GBS [159] compared to controls, but were detected in 5 patients who developed GBS after infection with *Mycoplasma pneumoniae* [109].

Ho et al. [76] termined anti-glycolipid antibodies in the serum of Chinese patients with GBS. They found anti-GM1 IgG antibodies with a frequency of 42% versus 6% in controls; 17% had antibodies to GM2, 26% to GD1a and 18% to GD1b. Antibodies to LM1 notably occurred less frequently in GBS patients versus controls, and were markedly less common in those cases of GBS which had evidence of an antecedent *C. jejuni* infection. There was no significant correlation with the pattern of the disease (AMAN versus classic GBS) or with the presence of *C. jejuni* antibodies.

In the Miller Fisher variant of GBS, patients almost invariably harbor serum IgG antibodies reactive with the ganglioside GQ1b [23, 213, 223]. This antibody was also detected in some GBS patients with pronounced ophthalmoplegia [24]. The anti-GQ1b antibodies usually cross-react with GT1a and in about half the cases with GD1b and/or GD3 [214]. Two groups reported that the serum GQ1b antibodies were of IgG1 and IgG3 isotype, invoking a primary T cell-dependent antigen [211, 228].

Other clinical syndromes associated with glycolipid antibodies

GBS and the other main clinical syndromes with a well-defined anti-glycolipid antibody specificity are listed in Table 2. Historically, the first syndrome in which anti-glycolipid antibodies were identified was the IgM paraproteinemic neuropathy associated with anti-MAG antibodies [110, 111, 149], which also react with the acidic glycolipid, SGPG and sulfated glucuronyl lactosaminyl paragloboside (SGLPG). The clinical syndrome is of a relatively benign, late-onset, chronic sensori-motor demyelinating neuropathy, often associated with tremor. Nerve biopsy samples may show widely spaced myelin. The anti-MAG specificity accounts for 50% of patients with IgM paraproteinemic neuropathy, most of the remainder react with other glycolipids or gangliosides. Some anti-MAG paraproteins cross-react with sulfatide and, in addition, sulfatide can also be a specific and sole antigen in patients with a IgM paraproteinemic, predominantly sensory, neuropathy.

Another IgM paraproteinemic neuropathy syndrome is a chronic large-fiber sensory neuropathy with prominent ataxia, and in some cases, ophthalmoplegia, but few other motor features; this phenotype is reminiscent of Miller Fisher syndrome [2, 14, 30, 128, 212, 220]. The IgM paraprotein reacts with gangliosides bearing NeuNAc(a2-8) NeuNAc-linked disialosyl groups, including GD3, GD1b, GT1b and GQ1b. Some of these patients also have cold agglutinin disease by virtue of the presence of sialylated glycoprotein epitopes on human red blood cells.

Anti-GM1 antibodies and their related clinical syndromes have also been thoroughly investigated in chronic motor neuropathies. In multifocal motor neuropathy (MMN) with demyelinating conduction block, IgM antibodies to GM1 ganglioside

Table 2. Clinical syndromes associated with specific anti-glycolipid autoantibodies

Clinical syndrome	Autoantibody specificity and isotype
Chronic sensori-motor demyelinating neuropathy	Monoclonal IgM Ab to SGPG, SGLPG, MAG and P0
Chronic axonal sensory neuropathy	IgM Ab to sulfatide
Chronic large fiber sensory neuropathy with ataxia.	Monoclonal IgM Ab to gangliosides containing NeuNAc(a2-8)NeuNAc epitopes
Multifocal motor neuropathy	IgM Ab to GM1 and related Gal(b1-3)GalNAc bearing glycolipids
Miller Fisher syndrome	IgG Ab to GQ1b and GT1a
Guillain Barré syndrome	IgG, IgM and IgA Ab to GM1, GD1a, LM1, sulfatide and other glycolipids

SGPG, sulfated glucuronyl paragloboside; SGLPG, sulfated glucuronyl lactosaminyl paragloboside; MAG, myelin associated glycoprotein

are found in about 50% of cases [215]. Many patients with MMN and anti-GM1 antibodies have a stereotyped clinical picture comprising a chronic asymmetric motor syndrome, usually with distal onset in an upper limb. Atypical cases may occur without focal conduction block, in which the phenotype resembles chronic spinal muscular atrophy. In a large series of cases, high titers of IgM anti-GM1 ganglioside antibodies were found in 51 of 2072 sera, all of whom had pure motor neuropathy, with or without generalized demyelination or focal conduction block and without sensory or upper motor neuron features [99]. Despite evidence of definite sensory involvement in some reported cases, anti-GM1 antibodies are clearly a markiar for predominantly motor nerve syndromes.

The origin of anti-glycolipid antibodies

Anti-glycolipid IgM antibodies of differing specificities are frequently found at low levels in normal sera. In addition, umbilical cord B cells, which are naive to foreign antigen, and adult peripheral blood and tonsillar B cells can be induced to secrete anti-GM1 IgM antibodies from donors with low or absent serum antibody titers [209]. These data suggest that some anti-carbohydrate antibodies form part of the naturally occurring autoantibody repertoire that comprises low-affinity, polyreactive IgM antibodies encoded by immunoglobulin V region genes in the germ-line configuration. Such IgM antibodies may have an immunoregulatory role in shaping immune maturation and may act as a first line defence against invading microorganisms. Microbial polysaccharide antigens, such as LPS containing repeating sugar sequences can activate B cells, particularly those of the Ly1/CD5 subset, by cross-linking their surface immunoglobulin receptors in a T cell-independent (TI) but antigen-specific manner, or alternatively can polyclonally activate B cells. This results in the predominant production of long-lived IgM responses: it is possible that the monoclonal and polyclonal low-affinity IgM anti-glycolipid antibodies seen in chronic neuropathies arise through such TI mechanisms. Affinity maturation, class switching from IgM to IgG and recruitment of B cells into the memory compartment are typical features of cognate, T cell-dependent interactions. There is no evidence that complex carbohydrate molecules such as gangliosides can be processed by antigen-presenting cells in this way. However, in GBS, high-affinity IgG subclass 1 and 3 anti-ganglioside antibody responses are seen, suggesting that T cell help has been recruited in the immune response; the mechanism by which this takes place is unknown [211]. Whether the B cell subset(s) that give rise to the chronically elevated IgM antibodies is CD5/Ly1$^+$ and distinct from that which gives rise to the IgG antibodies is not known [42, 73].

Evidence suggests that the source of the carbohydrate complex that drives the antiglycolipid response in both acute and chronic neuropathies is bacterial LPS. Molecular mimicry, whereby infectious agents share antigenic epitopes common to neural tissues has been demonstrated for certain strains of *C. jejuni* LPS and anti-ganglioside antibodies including anti-GM1, GD1a, GT1a, GD3 and GQ1b. In GBS and Miller Fisher syndrome following *C. jejuni* enteritis, the LPS fractions of the infecting bacteria have been shown to react with anti-GM1 and anti-GQ1b antibodies and to contain the appropriate oligosaccharide structures [8, 89, 107, 138, 224–227]. Anti-glycolipid (GM1 and GQ1b) antibodies associated with chronic neuropathies also react with LPS from specific strains of *C. jejuni*. It is possible that other microorganisms may bear carbohydrate epitopes that mimic glycolipid structures [45, 66, 132].

In contrast to the prediction that anti-glycolipid IgM antibodies, along with IgM antibodies in general, would be encoded by immunoglobulin V region genes in the germ-line configuration, several sequencing studies have shown that anti-glycolipid IgM antibodies contain extensive somatic mutation. This indicates that they have arisen through antigen-driven mechanisms, possibly stimulated by cross-reactive bacterial LPS [121, 140, 207]. No specific or unifying patterns of V gene usage have so far been described and it, thus, seems that anti-glycolipid antibodies can be encoded by diverse V genes from different heavy and light chain families. One exception to this may be the anti-disialosyl IgM antibodies that cross-react with the red blood cell Pr_2 antigens [2, 14, 30, 128, 212, 220]; both anti-Pr_2 antibodies that have been sequenced to date make use of the VH1 gene segment V1-18 [38, 216]. The repertoire of VH segments used in disease-associated antibodies against another red cell carbohydrate antigen, the i/I antigen, is highly restricted to the VH4-21 gene [139, 171, 192]. Whether the use of V1-18 by anti-Pr2 antibodies is similarly restricted remains to be seen.

The pathogenic relevance of anti-glycolipid antibodies

There is evidence from some studies, contradicted in others, in a variety of experimental animal models to support a direct role for anti-ganglioside antibodies in causing neuropathy. Two major contentious issues should be outlined. Firstly, once formed and present in the circulation, can anti-glycolipid antibodies induce neuropathy, in isolation or are other factors such as perturbations in cytokine profiles or also T cells required for full expression of the pathophysiology? Assuming that anti-glycolipid antibodies are involved in causation, are their effects predominantly pro-inflammatory or are they capable of acting pharmacologically to block ganglioside-mediated physiological processes? It is likely that there is some truth in all of these views but this is clearly a complex issue. Interpretation of data derived from animal studies is hampered by species differences in many areas including glycolipid function and distribution in nerve, immunological background and physiological performance.

Experimental studies on the effects of anti-SGPG/anti-MAG IgM paraproteins were initially conducted in the cat [71, 210] and then the chicken [187] and rabbit [125] with positive results. Intraneural injection studies showed evidence of inflammatory demyelination and, in passive transfer to the newborn chicken, widely spaced myelin was observed; this pathological feature is also seen in the human disease. There is, thus, general agreement that anti-SGPG antibodies are capable of causing peripheral nerve demyelination.

A recent passive transfer study of anti-disialosyl IgM antibodies from a chronic neuropathy patient [216] to the mouse has shown electrophysiological evidence of impaired nerve excitability and a reduction in quantal release of neurotransmitter in the absence of clinical or morphological abnormalities. This highlights the need for refined analysis of experimental models to obtain evidence of nerve dysfunction. Collectively these data indicate that the motor nerve terminal and neuromuscular junction may be an important site of action of anti-glycolipid antibodies, particularly since their position outside the BNB allows for easy antibody access.

Anti-GM1 antibody-immunoreactive material is present at nodes of Ranvier [27] and motor nerve terminals, two sites crucial to efficient nerve function. Many studies have now been performed in an attempt to establish whether anti-GM1-containing sera can cause neuropathic effects. Two studies [165, 194] demonstrated conduction block upon intraneural injection of anti-GM1 antibody into rat sciatic and tibial nerve, with or without an additional source of complement. In a rat sciatic nerve preparation in vitro, both human and rabbit anti-GM1 antibody-containing sera induced partial conduction block [3]. Immunization of rabbits with GM1 has been reported to induce serum anti-GM1 antibodies, immunoglobulin deposits at nodes of Ranvier and conduction block in one study [190]. Using the mouse hemidiaphragm preparation as a model system, Roberts et al. [154] have shown electrophysiologically both direct effects in vitro and passive transfer effects of anti-GM1 antibody-containing sera. In some of these studies the effects appeared to be complement independent, in others the converse was seen. The mechanism(s) by which these effects are induced is unknown but has been speculated to involve Na^+ channels; in support of this, high-titer rabbit anti-GM1 antisera have been shown to block Na^+ channels in an isolated single myelinated fiber preparation in the presence of complement [185]. In contrast to studies reporting positive effects of anti-GM1 antibodies in animal models, Harvey et al. [70] failed to show any morphological or electrophysiological effects of IgG or IgM anti-GM1 antisera when injected with complement into rat tibial nerves, despite presence of immunoglobulin at nodes of Ranvier in the experimental animals.

These issues cannot be easily reconciled. Differences in experimental design and conduct must be the major factors in producing these discrepancies: the choice of anti-GM1 antisera, species and electrophysiological site to be studied seem to be critical. Further experimentation is clearly needed.

Conclusions

The PNS is an immunocompetent organ. The participating cellular and humoral elements in the local immune circuitry have been identified. A number of acute and chronic neuropathies appear to result from disturbed immune homeostasis. The model disorder EAN has been useful in examining the induction, amplification and effector phase of autoimmune responses to peripheral nerve antigens. Potential autoantigens contained in the myelin sheath and on the axolemma have been characterized. Thus, the recent years have seen a rapid growth of information concerning the pathogenesis of the important group of immune-mediated neuropathies. It can be anticipated that the better understanding of the principal mechanisms of autoimmunity in the PNS will aid in the development of more specific and efficacious treatments for these crippling diseases.

Acknowledgement. Work from the authors laboratory cited in this review was supported by grants from BMBF, Deutsche Forschungsgemeinschaft and Gemeinnützige Hertie-Stiftung.

References

1. Abromson-Leeman S, Bronson R, Dorf ME (1995) Experimental autoimmune peripheral neuritis induced in BALB/c mice by myelin basic protein-specific T cell clones. J Exp Med 182:587
2. Arai M, Yoshino H, Kusano Y, et al (1992) Ataxic polyneuropathy and anti-Pr2 IgM kappa M proteinemia. J Neurol 239:147
3. Arasaki K, Kusunoki S, Kudo N, Kanazawa I (1993) Acute conduction block in vitro following exposure to antiganglioside sera. Muscle Nerve 16:587
4. Archelos JJ, Roggenbruck K, Schneider-Schaulies J. et al (1993) Detection and quantification of antibodies to the extracellular domain of P0 during experimental allergic neuritis. J Neurol Sci 117:197
5. Archelos JJ, Mäurer M, Jung S, et al (1994) Inhibition of experimental autoimmune neuritis by an antibody to the lymphocyte function-associated antigen-1. Lab Invest 70:667
6. Armati PJ, Pollard JD, Gatenby P (1990) Rat and human Schwann cells in vitro can synthesize and express MHC molecules. Muscle Nerve 13:106
7. Arnason BGW, Soliven B (1993) Acute inflammatory demyelinating polyradiculoneuropathy. In: Dyck PJ, Thomas PK, Griffin JW, et al (eds) Peripheral neuropathy. 3rd edn. Saunders, Philadelphia, pp 1437–1497
8. Aspinall GO, McDonald AG, Pang H (1994) Lipopolysaccharides of *Campylobacter jejuni* serotype O:19: structures of O antigen chains from the serostrain and two bacterial isolates from patients with the Guillain-Barré syndrome. Biochemistry 33:250
9. Bansil S, Mithen FA, Cook SD, et al (1991) Clinical correlation with serum-soluble interleukin-2 receptor levels in Guillain-Barré syndrome. Neurology 41:1302
10. Bansil S, Mithen FA, Singhal BS, et al (1992) Elevated neopterin levels in Guillain-Barré syndrome. Further evidence of immune activation. Arch Neurol 49:1277
11. Bevilacqua MP (1993) Endothelial-leukocyte adhesion molecules. Annu Rev Immunol 91:379
12. Bolin LM, Verity N, Silver JE et al (1995) Interleukin-6 production by Schwann cells and induction in sciatic nerve injury. J Neurochem 64:850
13. Bonetti B, Monaco S, Ciannini C, et al (1993) Human peripheral nerve macrophages in normal and pathological conditions. J Neurol Sci 118:158
14. Brindel 1, Preud'Homme J, Vallat J et al (1995) Monoclonal IgM reactive with several gangliosides in a chronic relapsing neuropathy. Neurosci Lett 181:103
15. Brosnan CF, Tansey FA (1984) Delayed onset of experimental allergic neuritis in rats treated with reserpine. J Neuropathol Exp Neurol 43:84
16. Brosnan CF, Lyman WD, Tansey FA, Carter TH (1985) Quantitation of mast cells in experimental allergic neuritis. J Neuropathol Exp Neurol 44:196
17. Brosnan JV, Craggs RI, King RH, Thomas PK (1987) Reduced susceptibility of T cell-deficient rats to induction of experimental allergic neuritis. J Neuroimmunol 14:267
18. Buchwald BM, Weishaupt A, Toyka KV, Dudel J (1995) lmmunoglobulin G from a patient with Miller-Fisher Syndrome rapidly and reversibly depresses evoked quantal release at the neuromuscular junction of mice. Neurosci Lett 201:163
19. Burns J, Krasner LJ, Rostami A, Pleasure D (1986) Isolation of P2 protein-reactive T-cell lines from human blood. Ann Neurol 19:391
20. Cammer W, Brosnan CF, Bloom BR, et al (1981) Degradation of the P0, P2, and Pr proteins in peripheral nervous system myelin by plasmin: implications regarding the role of macrophages in demyelinating diseases. J Neurochem 36:1506
21. Cammer W, Brosnan CF, Basile C, et al (1986) Complement potentiates the degradation of myelin proteins by plasmin: implications for a mechanism of inflammatory demyelination. Brain Res 364:91
22. Chia LS, Thompson JE, Moscarello MA (1983) Disorder in human myelin induced by superoxide radical: an in vitro investigation. Biochem Biophys Res Commun 117:141
23. Chiba A, Kusunoki S, Shimizu T, et al (1992) Serum IgG antibody to ganglioside GQ1b is a possible marker of Miller Fisher syndrome. Ann Neurol 31:677
24. Chiba A, Kusunoki S, Obata H, et al (1993) Serum anti-GQ-1b IgG antibody is associated with ophthalmoplegia in Miller-Fisher syndrome and Guillain-Barré syndrome: clinical and immunohistochemical studies. Neurology 43:1911
25. Clark L, Heber-Katz E, Rostami A (1992) Shared T-cell receptor gene usage in experimental allergic neuritis and encephalomyelitis. Ann Neurol 31:587

26. Constable AL, Armati PJ, Toyka KV, Hartung HP (1994) Production of prostanoids by Lewis rat Schwann cells in vitro. Brain Res 635:75
27. Corbo M, Quattrini A, Latov N, Hays AP (1993) Localization of GM1 and Gal(b1-3) GalNAc antigenic determinants in peripheral nerve. Neurology 43:809
28. Cowley SA, Butter C, Gschmeissner SE, et al (1989) An immunoelectronmicroscopical study of the expression of major histocompatibility complex (MHC) class II antigens in guinea pig sciatic nerves following induction of intraneural mycobacterial granulomas. J Neuroimmunol 23:223
29. Dahle C, Vrethem M, Ernerudh J (1994) T lymphocyte subset abnormalities in peripheral blood from patients with the Guillain-Barré syndrome. J Neuroimmunol 53:219
30. Duane GC, Farrer RG, Dalakas MC, Quarles RH (1992) Sensory neuropathy with monoclonal immunoglobulin M to GD1b ganglioside. Ann Neurol 31:683
31. Enders U, Karch H, Toyka KV, et al (1993) The spectrum of immune responses to *Campylobacter jejuni* and glycoconjugates in Guillain-Barré syndrome and in other neuroimmunological disorders. Ann Neurol 34:136
32. Exley AR, Smith N, Winer JB (1994) Tumour necrosis factor-alpha and other cytokines in Guillain-Barré syndrome. J Neurol Neurosurg Psychiatry 57:1118
33. Feasby TE, Gilbert JJ, Brown WF, et al (1986) An acute axonal form of Guillain-Barré polyneuropathy. Brain 109:1115
34. Feasby TE, Gilbert JJ, Hahn AF, Neilson M (1987) Complement depletion suppresses Lewis rat experimental allergic neuritis. Brain Res 419:97
35. Filbin MT, Tennekoon GI (1991) The role of complex carbohydrates in adhesion of the myelin protein, P0. Neuron 7:845
36. Filbin MT, Walsh FS, Trapp BD, et al (1990) Role of myelin P0 protein as a homophilic adhesion molecule. Nature 344:871
37. Fredman P, Vedeler CA, Nyland H, et al (1991) Antibodies in sera from patients with inflammatory demyelinating polyradiculoneuropathy react with ganglioside LM1 and sulphatide of peripheral nerve myelin. J Neurol 238:75
38. Friedman DF, Cho EA, Goldman J, et al (1991) The role of clonal selection in the pathogenesis of an autoreactive human B cell lymphoma. J Exp Med 174:525
39. Geczy C, Raper R, Roberts IM, et al (1985) Macrophage procoagulant activity as a measure of cell-mediated immunity to P2 protein of peripheral nerves in the Guillain-Barré syndrome. J Neuroimmunol 9:179
40. Gold R, Toyka KV, Hartung H-P (1995) Synergistic effect of IFN-gamma and TNF-alpha on expression of immune molecules and antigen presentation by Schwann cells. Cell Immunol 165:65
41. Gold R, Zielasek J, Kiefer R, et al (1996) Secretion of nitrite by Schwann cells and its effect on T-cell activation in vitro. Cell Immunol 68:69
42. Graves MC, Ravindranath RM (1992) Do $CD5^+$ cells secrete anti-asialo GM1 antibodies in motor neuron disease? Ann NY Acad Sci 651:570
43. Gregorian SK, Lee WP, Beck LS, et al (1994) Regulation of experimental autoimmune neuritis by transforming growth factor-beta 1. Cell Immunol 156:102
44. Gregson NA, Koblar S, Hughes RAC (1993) Antibodies to gangliosides in Guillain-Barré syndrome: specificity and relationship to clinical features. Q J Med 86:111
45. Griffin JW (1994) Antiglycolipid antibodies and peripheral neuropathies: links to pathogenesis. Prog Brain Res 101:313
46. Griffin JW, Ho TW (1993) The Guillain-Barré syndrome at 75: the Campylobacter connection (editorial; comment). Ann Neurol 34:125
47. Griffin JW, Stoll G, Li CY, et al (1990) Macrophage responses in inflammatory demyelinating neuropathies. Ann Neurol 27 [Suppl]:S64
48. Griffin JW, George R, Ho T (1993) Macrophage systems in peripheral nerves. A review. J Neuropathol Exp Neurol 52:553
49. Griffin JW, Li CY, Ho TW, et al (1995) Guillain-Barré syndrome in Northern China. The spectrum of neuropathological changes in clinically defined cases. Brain 118:577
50. Griffin JW, Li CY, Ho TW, et al (1996) Pathology of the motor-sensory axonal Guillain-Barré syndrome. Ann Neurol 39:17
51. Hahn AF, Feasby TE, Gilbert JJ (1985) Blood-nerve barrier studies in experimental allergic neuritis. Acta Neuropathol (Berl) 68:101

52. Hahn AF, Feasby TE, Wilkie L, Lovgren DT (1991) P2-peptide induced experimental allegic neuritis: a model to study axonal degeneration. Acta Neuropathol (Berl) 82:60
53. Hahn AF, Feasby TE, Wilkie L, Lovgren DT (1993) Anti-galactocerebroside antibody increases demyelination in adoptive transfer experimental allergic neuritis. Muscle Nerve 16:1174
54. Halliwell B (1992) Reactive oxygen species and the central nervous system. J Neurochem 59:1609
55. Hartung HP, Toyka KV (1990) T-cell and macrophage activation in experimental autoimmune neuritis and Guillain-Barré syndrome. Ann Neurol 27 [Suppl]:S57
56. Hartung HP, Schäfer B, Fierz W, et al (1987) Cyclosporin A prevents P2 T cell line-mediated experimental autoimmune neuritis (AT-EAN) in rat. Neurosci Lett 83:195
57. Hartung HP, Schwenke C, Bitter-Suermann D, et al (1987) Guillain-Barré syndrome: activated complement components C3a and C5a in CSF. Neurology 37:1006
58. Hartung HP, Heininger K, Schäfer B, et al (1988) Immune mechanisms in inflammatory polyneuropathy. Ann NY Acad Sci 540:122
59. Hartung HP, Schäfer B, Heininger K, et al (1988) The role of macrophages and eicosanoids in the pathogenesis of experimental allergic neuritis. Serial clinical, electrophysiological, biochemical and morphological observations. Brain 111:1039
60. Hartung HP, Schäfer B, Heininger K, et al (1988) Suppression of experimental autoimmune neuritis by the oxygen radical scavengers superoxide dismutase and catalase. Ann Neurol 23:453
61. Hartung HP, Schäfer B, van der Meide PH, et al (1990) The role of interferon-gamma in the pathogenesis of experimental autoimmune disease of the peripheral nervous system. Ann Neurol 27:247
62. Hartung HP, Hughes RAC, Taylor WA, et al (1990) T cell activation in Guillain-Barré syndrome and in MS: elevated serum levels of soluble IL-2 receptors. Neurology 40:215
63. Hartung HP, Reiners K, Schmidt B, et al (1991) Serum interleukin-2 concentrations in Guillain-Barré syndrome and chronic idiopathic demyelinating polyradiculoneuropathy: comparison with other neurological diseases of presumed immunopathogenesis. Ann Neurol 30:48
64. Hartung HP, Jung S, Stoll G, et al (1992) Inflammatory mediators in demyelinating disorders of the CNS and PNS. J Neuroimmunol 40:197 67.
65. Hartung HP, Archelos JJ, Zielasek J et al (1995) Circulating adhesion molecules and inflammatory mediators in demyelination: a review. Neurology 45 (Suppl):S22
66. Hartung HP, Pollard JD, Harvey GK, et al (1995) Immunopathogenesis and treatment of the Guillain-Barré syndrome Parts I & II. Muscle Nerve 18:137
67. Harvey GK, Pollard JD (1992) Peripheral nervous system demyelination from systemic transfer of experimental allergic neuritis serum. J Neuroimmunol 41:159
68. Harvey GK, Toyka KV, Hartung HP (1994) Effects of mast cell degranulation on blood-nerve barrier permeability and nerve conduction in vivo. J Neurol Sci 125:102
69. Harvey GK, Gold R, Hartung HP, et al (1995) Nonneural-specific T lymphocytes can orchestrate inflammatory peripheral neuropathy. Brain 118:1263
70. Harvey GK, Toyka KV, Zielasek J, et al (1995) Failure of anti-GM1 IgG or IgM to induce conduction block following intraneural transfer. Muscle Nerve 18:388
71. Hays AP, Latov N, Takatsu M, Sherman WH (1987) Experimental demyelination of nerve induced by serum of patients with neuropathy and an anti-MAG IgM M-protein. Neurology 37:242
72. Hays AP, Lee SS, Latov N (1988) Immune reactive C3d on the surface of myelin sheaths in neuropathy. J Neuroimmunol 18:231
73. Heidenreich F, Leifeld L, Jovin T (1994) T cell-dependent activity of ganglioside GM1-specific B cells in Guillain-Barré syndrome and multifocal motor neuropathy in vitro. J Neuroimmunol 49:97
74. Heininger K, Stoll G, Linington C, et al (1986) Conduction failure and nerve conduction slowing in experimental allergic neuritis induced by P2-specific T-cell lines. Ann Neurol 19:44
75. Heininger K, Schäfer B, Hartung HP, et al (1988) The role of macrophages in experimental autoimmune neuritis induced by a P2-specific T-cell line. Ann Neurol 23:326
76. Ho TW, Mishu B, Li CY, et al (1995) Guillain-Barré syndrome in Northern China. Relationship to *Campylobacter jejuni* infection and anti-glycolipid antibodies. Brain 118:597
77. Holmdahl R, Oisson T, Moran T, Klareskog L (1985) In vivo treatment of rats with monoclonal anti-T-cell antibodies. Immunohistochemical and functional analysis in normal rats and in experimental allergic neuritis. Scand J Immunol 22:157
78. Hughes RAC (1995) The concept and classification of Guillain-Barré syndrome and related disorders. Rev Neurol (Paris) 151:291

79. Hughes RA, Kadlubowski M, Gray IA, et al (1981) Immune responses in experimental allergic neuritis. J Neurol Neurosurg Psychiatry 44:565
80. Hughes RAC, Gray IA, Gregson NA, et al (1984) Immune responses to myelin antigens in Guillain-Barré syndrome. J Neuroimmunol 6:303
81. Hughes RAC, Atkinson PF, Gray IA, et al (1987) Major histocompatibility antigens and lymphocyte subsets during experimental allergic neuritis in the Lewis rat. J Neurol 234:390
82. Ilyas AA, Willison HJ, Quarles RH, et al (1988) Serum antibodies to gangliosides in Guillain-Barré syndrome. Ann Neurol 23:440
83. Ilyas AA, Mithen FA, Dalakas MC, et al (1991) Antibodies to sulfated glycolipids in Guillain-Barré syndrome. J Neurol Sci 105:108
84. Ilyas AA, Mithen FA, Chen ZW, et al (1991) Search for antibodies to neutral glycolipids in sera of patients with Guillain-Barré syndrome. J Neurol Sci 102:67
85. Ilyas AA, Mithen FA, Chen ZW, Cook SD (1992) Anti-GM1 IgA antibodies in Guillain-Barré syndrome. J Neuroimmunol 36:69
86. Ilyas AA, Mithen FA, Dalakas MC, et al (1992) Antibodies to acidic glycolipids in Guillain-Barré syndrome and chronic inflammatory demyelinating polyneuropathy. J Neurol Sci 107:111
87. Iqbal A, Oger JJF, Arnason BW (1981) Cell-mediated immunity in idiopathic polyneuritis. Ann Neurol 9 [Suppl]:65
88. Izumo S, Linington C, Wekerle H, Meyermann R (1985) Morphologic study on experimental allergic neuritis mediated by T cell line specific for bovine P2 protein in Lewis rats. Lab Invest 53:209
89. Jacobs BC, Endtz H, Van der Meché FGA, et al (1995) Serum anti-GQ1b IgG antibodies recognize surface epitopes on *Campylobacter jejuni* from patients with Miller Fisher syndrome. Ann Neurol 37:260
90. Jung S, Kramer S, Schluesener HJ, et al (1992) Prevention and therapy of experimental autoimmune neuritis by an antibody against T cell receptors-alpha/beta. J Immunol 148:3768
91. Jung S, Hartung HP, Toyka KV (1993) Shared T-cell receptor gene usage in experimental allergic neuritis and encephalomyelitis (letter). Ann Neurol 34:113
92. Jung S, Huitinga I, Schmidt B, et al (1993) Selective elimination of macrophages by dichlormethylene diphosphonate-containing liposomes suppresses experimental autoimmune neuritis. J Neurol Sci 119:195
93. Jung S, Schluesener HJ, Schmidt B, et al (1994) Therapeutic effect of transforming growth factor-beta 2 on actively induced EAN but not adoptive transfer EAN. Immunology 83:545
94. Jung S, Toyka KV, Hartung HP (1995) Soluble complement receptor type 1 inhibits experimental autoimmune neuritis in Lewis rats. Neurosci Lett 200:167
95. Khalili-Shirazi A, Hughes RAC, Brostoff SW, et al (1992) T cell responses to myelin proteins in Guillain-Barré syndrome. J Neurol Sci 111:200
96. Khalili-Shirazi A, Atkinson P, Gregson N, Hughes RA (1993) Antibody responses to P0 and P2 myelin proteins in Guillain-Barré syndrome and chronic idiopathic demyelinating polyradiculoneuropathy. J Neuroimmunol 46:245
97. Kiefer R, Funa K, Schweitzer T, et al (1996) Transforming growth factor beta1 in experimental autoimmune neuritis: cellular localization and time course. Am J Pathol 148:211
98. Kingston AE, Bergsteinsdottir K, Jessen KR, et al (1989) Schwann cells co-cultured with stimulated T cells and antigen express major histocompatibility complex (MHC) class II determinants without interferon-γ pretreatment: synergistic effects of interferon-gamma and tumor necrosis factor on MHC class II induction. Eur J Immunol 19:177
99. Kinsella LJ, Lange DJ, Trojaborg W, et al (1994) Clinical and electrophysiologic correlates of elevated anti-GM1 antibody titers. Neurology 44:1278
100. Koenig HL, Schumacher M, Ferzaz B, et al (1995) Progesterone synthesis and myelin formation by Schwann cells. Science 268:1500
101. Konat GW, Wiggins RC (1985) Effect of reactive oxygen species on myelin membrane proteins. J Neurochem 45:1113
102. Korn-Lubetzki I, Abramsky O (1986) Acute and chronic demyelinating inflammatory polyradiculoneuropathy. Association with autoimmune diseases and lymphocyte response to human neuritogenic protein. Arch Neurol 43:604
103. Koski CL, Humphrey R, Shin ML (1985) Anti-peripheral myelin antibody in patients with demyelinating neuropathy: quantitative and kinetic determination of serum antibody by complement component 1 fixation. Proc Natl Acad Sci USA 82:905

104. Koski CL, Gratz E, Sutherland J, Mayer RF (1986) Clinical correlation with anti-peripheral nerve myelin antibodies in Guillain-Barré syndrome. Ann Neurol 19:573
105. Koski CL, Sanders ME, Swoveland PT, et al (1987) Activation of terminal components of complement in patients with Guillain-Barré syndrome and other demyelinating neuropathies. J Clin Invest 80:1492
106. Koski CL, Chou DK, Jungalwala FB (1989) Anti-peripheral nerve myelin antibodies in Guillain-Barré syndrome bind a neutral glycolipid of peripheral myelin and cross-react with Forssman antigen. J Clin Invest 84:280
107. Kuroki S, Saida T, Nukina M, et al (1993) *Campylobacter jejuni* strains from patients with Guillain-Barré syndrome belong mostly to Penner serogroup 19 and contain beta-N-acetylglucosamine residues. Ann Neurol 33:243
108. Kusunoki S, Chiba A, Kon K, et al (1994) *N*-Acetylgalactosaminyl CD1a is a target molecule for serum antibody in Guillain-Barré syndrome. Ann Neurol 35:570
109. Kusunoki S, Chiba A, Hitoshi S, et al (1995) Anti-Gal-C antibody in autoimmune neuropathies subsequent to mycoplasma infection. Muscle Nerve 18:409
110. Latov N (1990) Antibodies to glycoconjugates in neurological disease. Clin Asp Autoimmun 4:18
111. Latov N (1995) Pathogenesis and therapy of neuropathies associated with monoclonal gammopathies. Ann Neurol 37 (S1):S32
112. Latov N, Gross BB, Kastelman J, et al (1981) Complement fixing anti-peripheral nerve myelin antibodies in patients with inflammatory polyneuritis and with polyneuropathy and paraproteinemia. Neurology 31:1530
113. Ledeen RW (1989) Biosynthesis, metabolism and biological effects of gangliosides. In: Margolis RU, Margolis RK (eds) Neurobiology of glycoconjugates. Plenum Press, New York, pp 43–83
114. Linington C, Brostoff SW (1993) Peripheral nerve antigens. In: Dyck PJ, Thomas PK, Griffin JW, et al (eds) Peripheral Neuropathy, 3rd edn. Saunders, Philadelphia pp 404–417
115. Linington C, Izumo S, Suzuki M, et 1984) A permanent rat T cell line that mediates experimental allergic neuritis in the Lewis rat in vivo. J immunol 133:1946
116. Linington C, Mann A, Izumo S, et al (1986) Induction of experimental allergic neuritis in the BN rat: P2 protein-specific T cells overcome resistance to actively induced disease. J Immunol 137:3826
117. Reference deleted
118. Linington C, Lassmann H, Ozawa K, et al (1992) Cell adhesion molecules of the immunoglobulin supergene family as tissue-specific autoantige induction of experimental allergic neuritis (EAN) by PO protein-specific T cell lines. Eur J immunol 22:1813
119. Macko CE, Sheikh KA, Li CY et al (1996) Immune attack on the Schwann cell surface in acute inflammatory demyelinating polyneuropathy. Ann Neurol (in press)
120. Mancardi GL, Cadoni A, Zicca A, et al (1988) HLA-DR Schwann cell reactivity in peripheral neuroptthies o different origins. Neurology 38:848
121. Mariette X, Tsapis A, Brouet JC (1993) Nucleotidic sequence analysis of the variable domains of four human monoclonal IgM with an antibody activity to myelin-associated glycoprotein. Eur J immunol 23:846
122. McKhann GM, Cornblath DR, Griffin JW, et al (1993) Acute motor axonal neuropathy: a frequent cause of acute flaccid paralysis in China. Ann Neurol 33:333
123. Milner P, Lovelidge CA, Taylor WA, Hughes RAC (1987) P0 myelin protein produces experimental allergic neuritis in Lewis rats. J Neurol Sci 79:275
124. Mitchell GW, Williams GS, Bosch EP, Hart MN (1991) Class II antigen expression in peripheral neuropathies. J Neurol Sci 102:170
125. Monaco MD, Ferrari S, Bonefti B, et al (1995) Experimental induction of myelin changes by anti-MAG antibodies and terminal complement complex. J Neuropathol Exp Neurol 54:96
126. Monaco S, Gehrmann J, Raivich G, Kreutzberg GW (1992) MHC-positive, ramified macrophages in the normal and injured rat peripheral nervous system. J Neurocytol 21:623
127. Nobile-Orazio E, Carpo M, Meucei N, et al (1992) Guillain-Barré syndrome associated with high titers of anti-GM1 antibodies. J Neurol Sci 109:200
128. Obi T, Kusunoki S, Takatsu M, et al (1992) IgM M-protein in a patient with sensory-dominant neuropathy binds preferentially to polysialogangliosides. Acta Neurol Scand 86:215
128. Ogawa-Goto K, Funamoto N, Abe T, Nagashima K (1990) Different ceramide compositions of gangliosides between human motor and sensory nerves. J Neurochem 55:1486

130. Ogawa-Goto K, Funamoto N, Ohta Y, et al (1992) Myelin gangliosides of human peripheral nervous system: an enrichment of GM1 in the motor nerve myelin isolated from *Cauda equina*. J Neurochem 59:1844
131. Ohno R, Hamaguchi K, Nomura K, et al (1986) Cellular hypersensitivity to nervous antigens in Guillain-Barré syndrome. Neurochem Pathol 4:119
132. O'Leary CO, Willison HJ (1995) Immunological investigation. Curr Opin Neurol 8:349
133. Olee T, Powers JM, Brostoff SW (1988) A T cell epitope for experimental allergic neuritis. J Neuroimmunol 19:167
134. Olee T, Weise M, Powers J. Mostoff S (1989) A T cell epitope for experimental allergic neuritis is an amphipathic alpha-helical structure. J Neuroimmunol 21:235
135. Olee T, Powell HC, Brostoff SW (1990) New minimum length requirement for a T cell epitope for experimental allergic neuritis. J Neuroimmunol 27:187
136. Olsson T, Holmdahl R, Klareskog L, Forsum U (1983) Ia-expressing cells and T lymphocytes of different subsets in peripheral nerve tissue during experimental allergic neuritis in Lewis rats. Scand J Immunol 18:339
137. Olsson Y (1990) Microenvironment of the peripheral nervous system under normal and pathological conditions. Crit Rev Neurobiol 5:265
138. Oomes PG, Jacobs BC, Hazenberg MPH, et al (1995) Anti-GM1 IgG antibodies and campylobacter bacteria in Guillain-Barré syndrome: evidence of molecular mimicry. Ann Neurol 38:170
139. Pascual V, Victor K, Lelsz D, et al (1991) Nucleotide sequence analysis of the V regions of two IgM cold agglutinins. Evidence that the VH4-21 gene segment is responsible for the major cross-reactive idiotype. J Immunol 146:4385
140. Paterson G, Wilson G, Kennedy PGE, Willison Hi (1995) Analysis of anti-GM1 ganglioside antibodies cloned from motor neuropathy patients demonstrates diverse V region gene usage with extensive somatic mutation. J Immunol 155:3049
141. Pette M, Gengaroli C, Hartung HP, et al (1994) Human T lymphocytes distinguish bovine from human P2 peripheral myelin protein: implications for immunological studies on inflammatory demyelinating neuropathies. J Neuroimmunol 52:47
142. Pette M, Linington C, Gengaroli C, et al (1994) T lymphocyte recognition sites on peripheral nerve myelin PO protein. J Neuroimmunol 54:29
143. Pollard JD, McCombe PA, Baverstock J, et al (1986) Class II antigen expression and T lymphocyte subsets in chronic inflammatory demyelinating polyneuropathy. J Neuroimmunol 13:123
144. Pollard JD, Baverstock J, McLeod JG (1987) Class II antigen expression and inflammatory cells in the Guillain-Barré syndrome. Ann Neurol 21:337
145. Pollard JD, Westland KW, Harvey GK, et al (1995) Activated T cells of non;eurai specificity open the blood-nerve barrier to circulating antibody. Ann Neurol 37:467
146. Powell HC, Braheny SL, Myers RR, et al (1983) Early changes in experimental allergic neuritis. Lab Invest 48:332
147. Powell HC, Myers RR, Mizisin et al (1991) Response of the axon and barrier endothelium to experimental neuritis induced by autoreactive T cell lines. Acta Neuropathol 82:364
148. Powell HC, Olee T, Brostoff SW, Mizisin AP (1991) Comparative histology of experimental allergic neuritis induced with minimum length neuritogenic peptides by adoptive transfer with sensitized cells or direct sensitization. J Neuropathol Exp Neurol 50:658
149. Quarles RH, Ilyas AA, Willison HJ (1986) Antibodies to glycolipids in demyelinating diseases of the human peripheral nervous system. Chem Phys Lipids 42:235
150. Quarles RH, Ilyas AA, Willison HJ (1990) Antibodies to gangliosides and myelin proteins in Guillain-Barré syndrome. Ann Neurol 27 Suppl:S48
151. Redford EJ, Hall SM, Smith KJ (1995) Vascular changes and demyelination induced by the intraneural injection of tumour necrosis factor. Brain 118:869
152. Rees JH, Gregson NA, Hughes RAC (1995) Anti-ganglioside GM1 antibodies in Guillain-Barré syndrome and their relationship to *Campylobacter jejuni* infection. Ann Neurol 38:809
153. Rees JH, Vaughan RW, Kondeatis E, Hughes RAC (1995) HLA-class II alleles in Guillain-Barré syndrome and Miller Fisher syndrome and their association with preceding *Campylobacter jejuni* infection. J Neuroimmunol 62:53
154. Roberts M, Willison H, Vincent A, Newsom-Davis J (1995) Multifocal motor neuropathy human sera block distal motor nerve conduction in mice. Ann Neurol 38:111
155. Ropper AH (1986) Severe acute Guillain-Barré syndrome. Neurology 36:429

156. Rostami A, Gregorian SK (1991) Peptide 53–78 of myelin P2 protein is a T cell epitope for the induction of experimental autoimmune neuritis. Cell Immunol 132:433
157. Rostami A, Brown MJ, Lisak RP, et al (1984) The role of myelin P2 protein in the production of experimental allergic neuritis. Ann Neurol 16:680
158. Rostami A, Burns JB, Brown MJ, et al (1985) Transfer of experimental allergic neuritis with P2-reactive T-cell lines. Cell Immunol 91:354
159. Rostami A, Burns JB, Eccleston PA, et al (1987) Search for antibodies to galactocerebroside in the serum and cerebrospinal fluid in human demyelinating disorders. Ann Neurol 22:381
160. Rostami A, Gregorian SK, Brown MJ, et al (1990) Induction of severe experimental autoimmune neuritis with a synthetic peptide corresponding to the 5378 amino acid sequence of the myelin P2 protein. J Neuroimmunol 30:145
161. Said G, Hontebeyrie-Joskowiecz M (1992) Nerve lesions induced by macrophage activation. Res immunol 143:589
162. Samuel NM, Jessen KR, Grange JM, Mirsky R (1987) Gamma interferon, but not *Mycobacterium leprae* induces major histocompatibility class II antigens on cultured rat Schwann cells. J Neurocytol 16:281
163. Samuel NM, Mirsky R, Grange JM, Jessen KR (1987) Expression of major histocompatibility complex class I and class II antigens in human Schwann cell cultures and effects of infection with *Mycobacterium leprae.* Clin Exp Immunol 68:500
164. Sanders ME, Koski CL, Robbins D, et al (1986) Activated terminal complement in cerebrospinal fluid in Guillain-Barré syndrome and multiple sclerosis. J Immunol 136:4456
165. Santoro M, Uncini A, Corbo M, et al (1992) Experimental conduction block induced by serum from a patient with anti-GM 1 antibodies. Ann Neurol 31:385
166. Schabet M, Whitaker JN, Schott K, et al (1991) The use of protease inhibitors in experimental allergic neuritis. J Neuroimmunol 31:265
167. Schmidt B, Stoll G, Hartung HP, et al (1990) Macrophages but not Schwann cells express Ia antigen in experimental autoimmune neuritis. Ann Neurol 28:70
168. Schmidt B, Stoll G, Van der Meide P, et al (1992) Transient cellular expression of gamma-interferon in myelin-induced and T-cell line-mediated experimental autoimmune neuritis. Brain 115:1633
169. Schmidt B, Toyka KV, Kiefer R, et al (1996) Inflammatory infiltrates in sural nerve biopsies in Guillain-Barré and chronic inflammatory demyelinating neuropathy. Muscle Nerve (in press)
170. Sharief MK, McLean B, Thompson EJ (1993) Elevated serum levels of tumor necrosis factor-alpha in Guillain-Barré syndrome. Ann Neurol 33:591
171. Silberstein LE, Jefferies LC, Goldman J. et al (1991) Variable region gene analysis of pathologic human autoantibodies to the related i and I red blood cell antigens. Blood 78:2372
172. Simone IL, Annunziata P, Maimone D, et al (1993) Serum and CSF anti-GM1 antibodies in patients with Guillain-Barré syndrome and chronic inflammatory demyelinating polyneuropathy. J Neurol Sci 114:49
173. Spies JM, Pollard JD, Bonner JG et al (1995) Synergy between antibody and P2-reactive T cells in experimental allergic neuritis. J Neuroimmunol 57:77
174. Spies JM, Westiand KW, Bonner JG, Pollard JD (1995) Intraneural activated T cells cause focal breakdown of the blood-nerve barrier. Brain 118:857
175. Springer TA (1994) Traffic signals for lymphocyte recirculation and leukocyte emigratio: the multiple step paradigm. Cell 76:301
176. Steinhoff U, Kaufmann SHE (1988) Specific lysis by $CD8^{+}T$ cells of Schwann cells expressing *Mycobacterium leprae* antigens. Eur J Immunol 18:969
177. Stevens A, Lang R, Schabet M et al (1990) Spontaneous chemiluminescence activity of peripheral blood cells during experimental allergic neuritis (EAN). J Neuroimmunol 27:33
178. Stoll G, Hartung HP (1992) The role of macrophages in degeneration and immune-mediated demyelinating of the peripheral nervous system. Adv Neuroimmunol 2:163
179. Stoll G, Schmidt B, Jander S, et al (1991) Presence of the terminal complement complex (C5b-9) precedes myelin degradation in immune-mediated demyelination of the rat peripheral nervous system. Ann Neurol 30:147
180. Stoll G, Jander S, Jung S, et al (1993) Macrophages and endothelial cells express intercellular adhesion molecule-1 in immune-mediated demyelination but not in Wallerian degeneration of the rat peripheral nervous system. Lab Invest 68:637

181. Stoll G, Jung S, Jander S, et al (1993) Tumor necrosis factor alpha in immune-mediated demyelination and Wallerian degeneration of the rat peripheral nervous system. J Neuroimmunol 45:175
182. Strigard K, Brismar T, Olsson T (1987) T-lymphocyte subsets, functional deficits, and morphology in sciatic nerves during experimental allergic neuritis. Muscle Nerve 10:329
183. Svennerholm L, Fredman 990) Antibody detection in Guillain-Barré syndrome. Ann Neurol 27 [Suppl]: S26
184. Svennerholm L, Bostrøm K, Fredman P, et al (1994) Gangliosides and allied glycosphingolipids in human peripheral nerve and spinal cord. Biochim Biophys Acta 1214:115
185. Takigawa T, Yasuda H, Kikkawa R, et al (1995) Antibodies against GM, ganglioside affect K^+ and Na^+ currents in isolated rate myelinated nerve fibers. Ann Neurol 37:436
186. Tansey FA, Brosnan CF (1982) Protection against experimental allergic neuritis with silica quartz dust. J Neuroimmunol 3:169
187. Tatum AH (1993) Experimental paraproteinemic neuropathy, demyelination by passive transfer of human IgM anti-myelin-associated glycoprotein. Ann Neurol 33:502
188. Taylor WA, Hughes RAC (1989) T lymphocyte activation antigens in Guillain-Barré syndrome and chronic idiopathic demyelinating polyradiculoneuropathy. J Neuroimmunol 24:33
189. Taylor WA, Brostoff SW, Hughes RAC (1991) P2 specific lymphocyte transformation in Guillain-Barré syndrome and chronic idiopathic demyelinating polyradiculoneuropathy. J Neurol Sci 104:52
190. Thomas FP, Trojabor W, Nagy C, et al (1991) Experimental autoimmune neuropathy with anti-GM1 antibodies and immunoglobulin deposits at the nodes of Ranvier. Acta Neuropathol 82:378
191. Thomas PK, Berthold C, Ochoa J (1993) Microscopic anatomy of the peripheral nervous system. In: Dyck PJ, Thomas PK, Griffin JW, et al (eds) (Peripheral Neuropathy, 3rd edn, Saunders, Philadelphia, pp 28–91
192. Thompson KM, Sutherland J. Barden G, et al (1991) Human monoclonal antibodies against blood group antigens preferentially express a VH4-21 variable region gene-associated epitope. Scand J Immunol 34:509
193. Tsai CP, Pollard JD, Armati PJ (1991) lnterferon-gamma inhibition suppresses experimental allergic neuritis: modulation of major histocompatibility complex expression of Schwann cells in vitro. J Neuroimmunol 31:133
194. Uncini A, Santoro M, Corbo M, et al (1993) Conduction abnormalities induced by sera of patients with multifocal motor neuropathy and antiGM1 antibodies. Muscle Nerve 16:610
195. van den Berg LH, Marrink J deJager AE, et al (1992) Anti-GM1 antibodies in patients with Guillain-Barré syndrome. J Neurol Neurosurg Psychiatry 55:8
196. van den Berg LH, Lankamp CLAM, deJager AEJ, et al (1993) Anti-sulphatide antibodies in peripheral neuropathy. J Neurol Neurosurg Psychiatry 56:1164
197. Van den Berg LH, Mollee I, Wokke JH, Logtenberg (1995) Increased frequencies of HPRT mutant T lymphocytes in patients with Guillain-Barré syndrome and chronic inflammatory demyelinating polyneuropathy; further evidence for a role of T cells in the etiopathogenesis of peripheral demyelinating diseases. J Neuroimmunol 58:37
198. van der Meché FGA, Schmitz PIM, Dutch Guillain-Barré Syndrome Study Group (1992) High-dose intravenous immunoglobulin versus plasma exchange in Guillain-Barré syndrome. N Engl J Med 326:1123
199. Vedeler CA, Matre R (1990) Peripheral nerve CR1 express in situ cofactor activity for degradation in C3b. J Neuroimmunol 26:51
200. Visser LH, Van der Meche FGA, Van Doorn PA et al (1995) Guillain- Barré syndrome without sensory loss (acute motor neuropathy). A subgroup with specific clinical, electrodiagnostic and laboratory features. Brain 118:841
201. Vriesendorp FJ, Mishu B, Blaser MJ, et al (1993) Serum antibodies to GM1, GD1b, peripheral nerve myelin, and *Campylobacter jejuni* in patients with Guillain-Barré syndrome and controls: correlation and prognosis (see comments). Ann Neuroi 34:130
202. Vriesendorp FJ, Flynn RE, Pappolia MA, Koski CL (1995) Complement depletion affects demyelination and inflammation in experimental allergic neuritis. J Neuroimmunol 58:157
203. Walsh FS, Cronin M, Koblar S, et al (1991) Association between glycoconjugate antibodies and Campylobacter infection in patients with Guillain-Barré syndrome. J Neuroimmunol 34:43
204. Watson SL, Westland K, Pollard JD (1994) An electrophysiological and histological study of trypsin induced demyelination. J Neurol Sci 126:116

205. Weishaupt A, Giegerich G, Jung S (1995) T cell antigenic and neuritogenic activity of recombinant human peripheral myelin P2 protein. J Neuroimmunol 63:149
206. Wekerle H, Schwab M, Linington C, Meyermann RT (1986) Antigen presentation in the peripheral nervous system: Schwann cells present endogenous myelin autoantigens to lymphocytes. Eur J Immunol 16:1551
207. Weng NP, Yu Lee LY, Sanz I, et al (1992) Structure and specificities of anti-ganglioside autoantibodies associated with motor neuropathies. J Immunol 149:2518
208. Westland K, Pollard JD (1987) Proteinase induced demyelination. An electrophysiological and histological study. J Neurol Sci 82:41
209. Willison HJ, Kennedy RAC (1993) Gangliosides and bacterial toxin in Guillain-Barré syndrome. J Neuroimmunol 46:105
210. Willison HJ, Trapp BD, Bacher JD, et al (1988) Demyelination induced by intraneural injection of human anti-myelin-associated glycoprotein antibodies. Muscle Nerve 11:1169
211. Willison HJ, Veitch J (1994) Immunoglobulin subclass distribution and binding characteristics of anti-GQ1b antibodies in Miller Fisher syndrome. J Neuroimmunol 50:159
212. Willison HJ, Paterson G, Veitch J, et al (1993) Peripheral neuropathy associated with monoclonal IgM anti-Pr2 cold agglutinins. J Neurol Neurosurg Psychiatry 56:1178
213. Willison HJ, Veitch J. Paterson G, et al (1993) Miller Fisher syndrome is associated with serum antibodies to GQ1b ganglioside. J Neurol Neurosurg Psychiatry 56:204
214. Willison HJ, Almemar A, Veitch J, Thrush D (1994) Acute ataxic neuropathy with cross-reactive antibodies to GD1b and GD3 gangliosides. Neurology 44:2395
215. Willison HJ, Paterson G, Kennedy PGE, Veitch J (1994) Cloning of human anti-GM1 antibodies from motor neuropathy patients. Ann Neurol 35:471
216. Willison HJ, O'Hanlon GM, Paterson G, et al (1996) A somatically mutated human antiganglioside IgM antibody that induces experimental neuropathy in mice is encoded by the variable region heavy chain gene, V1–18. J Clin Invest 97: (in press)
217. Yu RK, Saito M (1989) Structure and localization of gangliosides. In: Margolis RU, Margolis RK (eds) Neurobiology of glycoconjugates. Plenum Press, New York, pp 1–4
218. Yuki N, Yoshino H, Sato S, et al (1990) Acute axonal polyneuropathy associated with anti-GM1 antibodies following Campylobacter enteritis. Neurology 40:1900
219. Yuki N, Sato S, Itoh T, et al (1991) HLA-B35 and acute axonal polyneuropathy following Campylobacter infection. Neurology 41:1561
220. Yuki N, Miyatani N, Sato S, et al (1992) Acute relapsing sensory neuropathy associated with IgM antibody against B-series gangliosides containing a GalNAc beta 1-4(Gal3-2 alpha NeuAc8-2 alpha NeuAc)beta 1 configuration. Neurology 42:686
221. Yuki N, Sato S, Fujimoto S, et al (1992) Serotype of *Campylobacter jejuni*, HLA, and the Guillain-Barré syndrome (letter). Muscle Nerve 15:968
222. Yuki N, Yoshino H, Sato S, et al (1992) Severe acute axonal form of Guillain-Barré syndrome associated with IgG anti-GD1a antibodies. Muscle Nerve 15:899
223. Yuki N, Sato S, Tsuji S, et al (1993) Frequent presence of anti-GQ1b antibody in Fisher's syndrome. Neurology 43:414
224. Yuki N, Taki T, Inagaki F, et al (1993) A bacterium lipopolysaccharide that elicits Guillain-Barré syndrome has a GM1 ganglioside-like structure. J Exp Med 178:1771
225. Yuki N, Taki T, Takahashi M, et al (1994) Penner's serotype 4 of *Campylobacter jejuni* has a lipopolysaccharide that bears a GM1 ganglioside epitope as well as one that bears a GD1a epitope. Infect Immun 62:2101
226. Yuki N, Taki T, Takahashi M, et al (1994) Molecular mimicry between GQ1b ganglioside and lipopolysaccharides of *Campylobacter jejuni* isloated from patients with Fisher's syndrome. Ann Neurol 36:791
227. Yuki N, Handa S, Tai T, et al (1995) Ganglioside-like epitopes of lipopolysaccharides from *Campylobacter jejuni* (PEN 19) in three isolates from patients with Guillain-Barré syndrome. J Neurosci 130:112
228. Yuki N, Ichihashi Y, Taki T (1995) Subclass of IgG antibody to GM1 epitope bearing lipopolysaccharide of *Campylobacter jejuni* in patients with Guillain-Barré syndrome. J Neuroimmunol 60:161
229. Zettl UK, Gold R, Hartung HP, Toyka KV (1994) Apoptotic cell death of T lymphocytes in experimental autoimmune neuritis. Neurosci Lett 176:75

230. Zettl UK, Gold R, Toyka KV, Hartung H-P (1995) Intravenous glucocorticosteroid treatment augments apoptosis of inflammatory T cells in experimental autoimmune neuritis (EAN) of the Lewis rat. J Neuropathol Exp Neurol 54:540
231. Zhang XM, Esch TR, Clark L, et al (1994) Neuritogenic Lewis rat T cells use Tcrb chains that include a new Tcrb-V8 family member. Immunogenetics 40:266
232. Zhu J. Link H, Mix E, et al (1994) Th1-like cell responses to peripheral nerve myelin components over the course of experimental allergic neuritis in Lewis rats. Acta Neurol Scand 90:19
233. Zhu J, Link H, Weerth S, et al (1994) The B cell repertoire in experimental allergic neuritis involves multiple myelin proteins and GM1. J Neurol Sci 125:132
234. Zhu J, Mix E, Olsson T, Link H (1994) Cellular mRNA expression of interferon-gamma, IL-4 and transforming growth factor-beta (TGF-beta) by rat mononuclear cells stimulated with peripheral nerve myelin antigens in experimental allergic neuritis. Clin Exp Immunol 98:306
235. Zielasek J, Ritter G, Magi S, et al (1994) A comparative trial of anti-glycoconjugate antibody assays: IgM antibodies to GM1. J Neurol 241:475
236. Zielasek J, Jung S, Gold R, et al (1995) Administration of nitric oxide synthase inhibitors in experimental autoimmune neuritis and experimental autoimmune encephalomyelitis. J Neuroimmunol 58:81
237. Zweiman B, Rostami A, Lisak RP, et al (1983) Immune reactions to P2 protein in human inflammatory demyelinative neuropathies. Neurology 33:234

Mechanisms of action of interferon-β in multiple sclerosis

Barry G.W. Arnason, Amit Dayal, Zhi Xiang Qu, Mark A. Jensen, Kursad Genç, Anthony T. Reder

Department of Neurology and the Brain Research Institute, The University of Chicago, 5841 S. Maryland Ave. - MC 2030, Chicago, IL 60637, USA

Introduction

Interferon-β-1b (IFN-β-1b, Betaseron) at a dose of 8 million units (8 MU) given subcutaneously (s.c.) every other day to patients with the relapsing-remitting form of multiple sclerosis (MS) lessens the overall frequency of MS attacks by 35%, the frequency of major attacks by 50%, and the number of MS-related hospitalizations by 40% [1, 2]. Benefit persists for at least 5 years, the last assessable time point in the pivotal clinical trial that led to approval of the agent for the treatment of MS [3] by the United States Food and Drug Administration. The agent drastically curtails accumulating disease burden, as assessed by annual magnetic resonance imaging (MRI) scans, again over at least 5 years [4]. Disease activity as measured by serial MRI scans performed at 6-week intervals, is also markedly reduced by IFN-β-1b treatment [4]. IFN-β-1b is currently approved for the treatment of ambulatory MS patients in the United States and Australia; approval in Europe is expected imminently. Trials to determine whether it is effective in progressive MS are underway currently.

The interferons

Interferons comprise a family of proteins (reviewed in [5]). They are divided into two major classes. IFN-β is a type I IFN. It exists as a single protein. The other members of the class are the α interferons (IFN-α), a superfamily of related proteins all evolved from the same common ancestor as IFN-β. At least 18 non-allelic IFN-α genes are known, including 4 pseudo-genes. All the IFN-α genes reside on the short arm of chromosome 9. All contain from 165 to 166 amino acids. None are glycosylated, and there is extensive homology among them. A recombinant IFN-α isoform has shown a beneficial effect on MS attack frequency in a preliminary trial [6].

Correspondence to: B.G.W. Arnason

The IFN-β gene sits next to the IFN-α gene cluster on chromosome 9. IFN-β has 166 amino acids and is glycosylated. IFN-β-1b is a slightly modified form of the native IFN-β molecule. The methionine in position 1 is lacking and serine has been substituted for cysteine at position 17. The protein is generated in *Escherichia coli* and, in contrast to the native molecule, is not glycosylated. A second recombinant form of IFN-β, known as IFN-β-1A, is generated in Chinese hamster ovary cells. IFN-β-1A is glycosylated. IFN-β-1A shares with IFN-β-1b the capacity to lessen both MS attack frequency and MRI-determined disease activity ([7], and C. Pozzilli, personal communication).

IFN-α is induced preferentially in leukocytes by foreign cells, by tumor cells, by bacteria and their products, and by viruses. IFN-β is produced preferentially by fibroblasts, epithelial cells, monocytes, and macrophages. IFN-β can be induced by interleukin (IL)-2, a product of activated T cells, and by IL-1 and tumor necrosis factor (TNF), products of activated macrophages, in addition to viruses. IFN-α and IFN-β bind to the same receptor and share many actions. The IFN-α/β receptor gene is sited on chromosome 21. IFN-α/β receptors are found on all cell types.

IFN-γ is the sole type II IFN. Its gene is found on chromosome 12. It is a glycoprotein, contains 143 amino acids, and forms a homodimer. It has its own receptor which is found on chromosome 9. IFN-γ is made by T cells and natural killer (NK) cells. Its major action is to stimulate the immune system. IFN-γ can induce IFN-α and IFN-β production. Much of its "interferon" activity may depend on its ability to induce IFN-α and IFN-β production. Synthesis of IFN-γ is triggered by antigens, including autoantigens.

That type I IFN preparations are effective in MS, at least to a degree, is clear. What is less evident is how and why they work. Type I IFNs are cytokines. They inhibit viral replication (hence the name), exert antiproliferative effects, and modulate immune responses in a highly selective manner. An understanding of those mechanisms of action of IFN-β-1b that are relevant to its attenuating effect on MS disease activity could pave the way for more effective treatments of the disease.

Clinical features of MS

The relapsing-remitting form of MS is characterized by abrupt worsenings of neurological symptoms known as attacks. Attacks evolve over days to weeks and then fade away, leading into periods of quiescence or seeming quiescence (see below) known as remissions. Attacks may occur two to three times a year, particularly in the early phase of the disease, or may be separated by years or even decades. Overall mean attack frequency in large series is slightly less than one per year but with marked individual variations.

Recovery from attacks may be complete or partial, but as one attack follows another increasing permanent disability becomes the rule. In many patients the disease evolves after a series of attacks, and usually after the elapse of some years, into a slowly but inexorably progressive illness designated as relapsing-progressive or secondarily progressive MS. Other patients, perhaps 10% of the total, are relentlessly progressive from the outset. Such cases are known as primary progressive MS. The disease shortens life expectancy only marginally, but many victims ultimately become wheelchair-bound and 15% end up requiring custodial care. MS affects about 1 in 500 of the adult population of the United States and Northern Europe. Onset is most often at the prime of young adulthood. Incalculable damage occurs to those afflicted, and to their families, in terms of lost productivity and an inferior quality of life.

Frequent MRI scanning at weekly to 6-weekly intervals has provided a far more dynamic sense of the MS disease process than had been appreciated in the past. Areas with increased water content appreciable on T2-weighted MRI images, known colloquially as UBOs (for unidentified bright objects), appear in the cortical white matter with surprising frequency. They enlarge, shrink, and oftentimes disappear over time intervals of from 2 to 8 weeks. The majority of these bouts of activity, in some series up to 80% of them, are silent in the sense that no new deficits are detected on the neurological examination [9]. Evidence for ongoing activity at times of seeming disease quiescence has also been obtained by examination of spinal fluid. Activated T cells are commonly found in the spinal fluid during periods of seeming remission [10]. Despite this evidence for a much brisker disease activity than clinical evaluation would have led one to suspect, the fact remains that MS activity is episodic and that the immune dysregulation that permits activity to occur is similarly episodic.

Biopsy of acute lesions of the type described above reveals a brisk inflammatory response comprised of invading T cells and activated macrophages [11–15]. B cells are rare [14, 15]. Interstitial edema is, as expected, prominent, but sometimes only subtle changes in myelin are observed. In other instances frank demyelination is intermingled with remyelination (i.e., repair). Oligodendrocytes are present in normal or near normal numbers in such lesions.

These features are to be contrasted with those seen at autopsy of evolved MS cases [14, 15]. In established MS lesions demyelination is usually total, oligodendrocytes are absent, inflammation may be sparse or lacking and, when inflammation is present, B cells tend to be more numerous than in acute reversible lesions. Astrogliosis is prominent. Axons are said to be relatively spared in MS, but with emphasis on the word relative. Axonal loss is oftentimes considerable and can even cause shrinkage of the spinal cord. There is evidence that axonal loss and oligodendrocyte loss correlate both with each other and with the severity of the gait disturbances that neurologists take to be a valid measure of disability [16].

Clearly from the above, many MS lesions are reversible. Yet, as one attack follows another, as areas of permanent damage in central nervous system (CNS) white matter (known as plaques) enlarge and coalesce, as repair mechanisms fail, as oligodendrocytes are progressively lost, as myelin disappears, and as axons are depleted, irreversible difficulties with gait, coordination, vision, bladder and bowel control accumulate. It is important to note that reversible lesions may be superimposed on permanent ones in the same patient.

Irreversible plaques can be seen to accumulate on serial annual MRI scans. The increment in permanent "disease burden" averages about 10% a year [17, 18]. Burden of disease as assessed by MRI scanning correlates poorly with locomotion in individual patients, presumably because a single "strategic hit" in an important pathway can do great damage. In a large series, burden of disease was shown to correlate with ambulation and the neurological rating scales used to measure it [3, 4, 19].

The distinction between disease activity (i.e., usually reversible lesions) and disease burden (i.e., accumulating irreversible lesions) is not trivial. Any therapeutic agent that lessens disease activity, whether episodes of activity are perceived as attacks or not, may have a favorable impact on those aspects of MS that correlate with disease activity (see later). If disease burden accumulation can be lessened, further progression of disability may be prevented or slowed, even though little can be expected in terms of previously accumulated fixed deficits.

The clinical manifestations of MS are protean. They can be divided into two broad categories: those that depend on the accumulation of plaques, i.e., disease burden, and those that do not, or do so only in part. Areas of demyelination and of axonal loss interrupt impulse conduction along white matter tracts. If the optic nerve is involved, vision may be compromised; if cerebellar pathways are involved, there may be loss of coordination; if the pyramidal tracts are involved, there may be spasticity, weakness, and ultimately loss of ability to walk; if sensory pathways are involved, there may be dyesthesias or frank loss of sensation; if the pathways that control bladder and bowel function are involved, there may be urinary and fecal incontinence. All of the above are readily quantifiable by the disability rating scales used by neurologists to evaluate disease progression in clinical trials. Such rating scales are heavily weighted towards ambulation on the assumption that gait problems are the major determinant of disability.

In the IFN-β-1b clinical trial of relapsing-remitting MS patients there was less progression of disability, as evaluated by standard disability scales, in the treated arm than in the placebo arm, but the difference between groups was not statistically significant even though the difference between treated and placebo groups in accumulating disease burden seen on MRI scanning was highly significant [1, 3, 4]. This disappointing result may reflect no more than the insensitivity of the rating scales employed. Only a modest progression in disability was observed, on average, in the placebo-treated group despite a large increase in disease burden. This limited change in disability score in the placebo group may have precluded demonstration of a meaningful difference, i.e., the study was probably underpowered in terms of the number of patients required to demonstrate a meaningful difference in disability. There is also the remote possibility that disability does not reflect accumulating disease burden, but such a proposal flies in the face of a considerable body of literature that argues to the contrary [20, 21]. Trials of IFN-β-1b in secondarily progressive MS, currently underway in North America and Europe with larger patient groups than studied heretofore, may resolve the issue.

The disability rating scales in current use capture only one dimension of the MS problem. Cognition, particularly memory function, may be adversely impacted by MS. Depression is seen at some point or another in half the patients. Pain and a general sense of malaise are more often complained of than is generally appreciated. Fatigue is almost universal and inordinate fatigue is more commonly given as the reason for stopping work than is difficulty walking [22]. Problems with thinking, mood, pain, and fatigue may be prominent early even when gait disorders are absent. These difficulties link to disease activity far more than to accumulated fixed lesions [23].

MS patients often note an increase in fatigue during attacks. Sometimes they sleep more and have transient worsening of cognition, such as difficulties with short-term memory recall. Such complaints may be voiced whether the site of disease activity is in the thoracic spine, say, or in the optic nerve. These activity-associated difficulties probably have little to do with demyelination. More likely they depend on actions of cytokines, of free radicals, of proteases, or of other substances released by invading inflammatory cells, that diffuse throughout the CNS and, acting at a distance from their site of generation, adversely affect neuronal function. Cytokines are readily detected in the spinal fluid during MS attacks [24, 25], and neuronal function can be shown to be influenced by them [26]. It follows that these symptoms may abate when activity, whether documented by a reduced number of attacks or by a lesser number of active MRI scans, is lessened.

Both attack frequency and the proportion of active scans were reduced with IFN-β-1b treatment [1, 3, 4]. Patients entered into the IFN-β-1b clinical trial at the University of Chicago showed an improvement in selected aspects of cognitive performance. There was, for example, a statistically significant improvement in delayed visual memory in patients receiving between 2 and 4 years of IFN-β-1b treatment that was not seen in the placebo-treated patients [27]. Mood was also better overall in the treated group. Improved memory has been spontaneously remarked upon by some of our patients. Fatigue is difficult to measure, but patients on IFN-β-1b often offer that this is also, in their estimation and that of their families, improved and that their sleeping habits have changed for the better. Heat intolerance is a common complaint in MS. This, again, has been favorably influenced by IFN-β-1b treatment according to substantial numbers of our MS patients.

We suggest that the products released during inflammation (i.e., disease activity) decrease neuronal excitability and that in the absence of inflammation neuronal excitability may increase. Increased neuronal excitability may explain the improvements mentioned above. Some patients receiving IFN-β-1b note an increase in their spasticity even when other functions remain unchanged or, as in the case of fatigue, improve. Perhaps an increase in the excitability of already overactive motor neurons accounts for this finding. The issue is not trivial since, on first consideration, a worsening of spasticity would suggest an increase in disease activity. This may not always be the case. The rating scales used by neurologists to evaluate disease progression are heavily weighted towards spasticity. If the formulation given above is even partially correct, then the currently employed rating scales may be seriously flawed. One cannot but wonder whether earlier trials with agents abandoned as ineffective may not have floundered because of the preconceived notion that worsening of spasticity equated to ongoing disease activity.

Immunology of MS

During MS attacks, inflammatory cells invade the CNS. The inflammatory infiltrate is comprised of T cells, B cells, and macrophages, these last activated by cytokines released by invading T cells. Macrophages are the final vectors of myelin destruction. The antigen or antigens against which the presumed autoimmune process that underlies MS is directed is unknown. There is good reason to believe that the core mechanism that leads to myelin damage, and hence to a failure of nerve conduction, is one of T cell-mediated delayed-type hypersensitivity (DTH).

DTH responses can be viewed as being comprised of nine overlapping steps: (1) activation of T cells in the lymphoid organs; (2) recruitment of additional cells to the lymphoid organs; (3) expansion of activated clones; (4) migration of activated and expanded T cell clones into the circulation and thence across the CNS endothelial cells into the parenchyma of the CNS; (5) attraction of followers; (6) proliferation in situ in the CNS; (7) tissue destruction; (8) suppression or attenuation of the response; and (9) repair. IFN-β-1b could exert its beneficial effect by impeding any of the first seven steps or by augmenting either of the last two.

What causes attacks?

MS attacks, whether recognized as such or clinically silent, must have a trigger. The trigger acts in the peripheral lymphoid organs to activate T cells that then make their way to the CNS and attack myelin and oligodendrocytes. Attacks run a finite course so that the ending of an attack must also, in some manner, be programmed or controlled.

An established trigger for MS attacks is viral infection [28, 29], though most viral infections in MS patients are not followed by an overt attack. Nonetheless, the frequency of MS attacks as an aftermath of viral infection, usually within 1–4 weeks of the onset of the viral illness, exceeds background attack frequency by some three to fivefold. Whether any viral infection is capable of triggering a MS attack or whether certain ones are more prone to do so than others has not been studied systematically. There is a hint that adenovirus infections may be particularly likely to precipitate a MS attack [30], a point of potential interest inasmuch as adenoviruses inhibit the IFN response normally observed in virally infected cells [31].

MS patients have fewer overt infections than age-matched controls. This suggests that the potential to generate an anti-viral response may be preset at a high level in MS. Those MS patients with the fewest viral infections have the most malignant disease, perhaps pointing, in them, to a persistently overactive immune system [28].

Since viral infections trigger MS attacks, and type I IFNs are anti-viral agents, a facile explanation for the protective effect of IFN-β-1b in MS might be that it lessens the frequency and/or severity of viral illnesses and that this translates into fewer attacks than would otherwise occur. No evidence that viral illnesses were fewer in MS patients receiving IFN-β-1b than in MS patients receiving placebo emerged from the pivotal trial that led to approval of the agent in the United States [32]. Severity of viral infections was not addressed; operationally, this would be a difficult issue to resolve.

An immune response to a newly encountered pathogen such as a virus kindles memory cells specific for previously encountered antigens. Viral infections trigger MS attacks presumably because cytokines released by cells responding to viruses kindle memory cells specific for the elusive MS antigen or antigens. Kindling is likely to occur in the lymphoid organs.

Different infections activate different arms of the immune response. We suspect that generation of DTH responses by viral infections is more germane to the triggering of MS attacks than the cytotoxic CD8 cell-mediated responses that are also triggered by viral infections. Some evidence supports the postulate advanced above. DTH responses are characterized by increased production of IFN-γ by Th1-type CD4 cells. Administration of IFN-γ to MS patients provokes attacks of disease [33]. Conversely, IFN-β-1b augments cytotoxic CD8 cell-mediated responses yet protects in MS.

Th2-type CD4 cells promote antibody production by B cells and the IL-4 and IL-10 that they release cause isotype switching of immunoglobulin produced by B cells. Th2 cell and B cell responses are critical to defense against extracellular bacteria. Bacterial infections are said not to provoke MS attacks [28], suggesting that activation of Th2 cells and antibody production by B cells does not predispose to MS attacks. No effect of IFN-β-1b on IL-4 production by peripheral blood mononuclear cells (PBMC) in vitro has been observed (A.B.C. Noronha, personal communication). The effect of IFN-β-1b on IL-10 production is discussed later. IFN-β-1b has no known role in isotype switching. Whether infections with bacteria that reside intracellularly might provoke MS attacks has not been explored. Intracellular bacteria tend to provoke DTH and cytotoxic T cell responses rather than the antibody response that constitutes the major defense against extracellular bacteria.

In MS, CNS-invading cells come from the blood. The cells that enter the CNS are pre-activated and as a consequence express cell surface adhesion molecules. CNS endothelial cells must similarly express adhesion molecules to permit cell attachment and passage. Interaction between the adhesion molecules expressed by endothelial cells and activated mononuclear cells is essential for passage of cells into the CNS. A detailed discussion of adhesion molecules lies beyond the scope of this review. In simple terms, adhesion molecule expression is induced by circulating cytokines. Circulating IFN-γ, IL-4, TNF, and IL-1 all induce adhesion molecules on CNS endothelial cells [35–39]. IFN-β-1b treatment lowers IFN-γ secretion, though only after a delay (see later), and this action could lessen IFN-γ-induced adhesion molecule expression on CNS endothelial cells. IFN-β-1b also profoundly lowers TNF secretion in vitro [34], although whether it does so in vivo is not known. Similar considerations apply to IL-1. Were IFN-β-1b to lessen TNF and IL-1 secretion in vivo, adhesion molecule expression might again be lessened. IFN-β acts directly on endothelial cells, at least in vitro, to lower IFN-γ binding to them and to reverse IFN-γ-induced up-regulation of adhesion molecules. IFN-α exerts a beneficial effect in MS. It, too, inhibits IFN-γ-induced binding of T cells to human endothelial cells, again in vitro [40]. All of the above suggests, after allowing for the dangers inherent in extrapolating results obtained in vitro to the in vivo situation, that type I IFNs may have the potential to impede trafficking of inflammatory cells into the CNS, although whether enough IFN-β-1b enters the circulation after s.c. injection to actually alter adhesion molecule expression on endothelial cells is doubtful (see later).

Th1 cells in MS

DTH responses involve activation of Th1-type T cells bearing the CD4 marker and, through them, activation of macrophages. Th1 cells have been identified in MS [41–43]. Activated Th1-type T cells release IL-2, IFN-γ, and lymphotoxin (LT). IL-2 provides a proliferative signal to T cells acting in both an autocrine and paracrine manner. Its activating capacity is not confined to Th1-type T cells and the amount recoverable from supernatants of PBMC cultures exposed to IFN-β-1b is increased over the amount recoverable from control PBMC cultures. Increased levels of IL-2 can be detected in the circulation when MS is active [44]. IL-2 stimulates IFN-γ production by Th1-type T cells. Increased levels of circulating IFN-γ can be detected just prior to and during MS attacks [45], as can increased numbers of circulating IFN-γ-secreting cells [46]. IFN-γ can be detected in active MS plaques [47, 48].

T cell lines from HLA-DR2–positive donors produce more LT in response to specific antigen than T cell lines that are not HLA-DR2 positive. The HLA-DR2 haplotype is significantly overrepresented in the MS population compared to the normal population [49]. LT production is increased by IL-2, by IFN-γ, and most markedly by the combination of IL-2 and IFN-γ [50]. LT is known to be toxic to oligodendrocytes in vitro [51]. The cytotoxic potential of LT can be potentiated by IFN-γ [52]. The ability of T cell lines to adoptively transfer experimental allergic encephalomyelitis (EAE) correlates with their capacity to secrete LT [53]. LT can be detected in T cells and microglia in active MS plaques [54] and is thought to have a major role in tissue damage in MS.

IFN-γ released by Th1–type T cells activates macrophages and primes them for release of IL-1 and TNF. IL-1 levels are increased in one-third of chronic progressive MS patients [44]. TNF can be detected in the circulation and the spinal fluid during MS attacks [24, 25]. Elevated TNF levels in the circulation of MS patients correlate with elevated levels of IL-2 [44]. TNF is present in astrocytes and foamy macrophages in active MS plaques [55], and in endothelial cells at the lesion edge. TNF binds to the same receptors as LT and shares with LT the ability to damage oligodendrocytes in vitro [56, 57]. It is probable that it does the same in vivo since antibody to TNF and LT inhibits EAE, the best available animal model for MS [58]. TNF stimulates astrogliosis, a prominent feature within established MS plaques [59]. IFN-γ and IL-1 also stimulate astrogliosis [60, 61]. TNF increases IFN-γ-induced major histocompatibility complex (MHC) class II expression on macrophages in tissue culture as discussed later [62].

When MS plaques are examined by immunohistochemical techniques, IFN-γ, IL-2, LT, IL-1, and TNF can all be detected in those plaques in which an ongoing inflammatory response is present [47, 48, 54, 55]. IFN-β can also be detected. It is found on astrocytes, lymphocytes, and macrophages in active MS lesions. This finding points to a role at some level for endogenous IFN-β in the control of the MS process [47, 48]. Note that TNF, LT, and IL-2 can all induce IFN-α and IFN-β production by macrophages [63–65] and that all are found in active MS plaques. IFN-α is also present in MS plaques being found on MHC class II-positive macrophages [47, 48]. From serial blood samples from MS patients 21% exhibited elevated levels of IFN-α [66] although without any obvious correlation with overt disease activity. It would be of interest to know if "interferonemia α" ties to a lack of activity as determined by MRI scanning.

IFN-β modulates immune responses. It is not an immunosuppressive agent. Rather it augments some immune responses while inhibiting others. MHC class II protein expression is crucial for antigen presentation to CD4 T cells. IFN-γ is the major inducer of MHC class II protein expression [67]. Astrocytes in active MS foci express MHC class II proteins ([47, 48], but for a contrary view see 68). Presumably MHC class II proteins are induced on astrocytes by locally released IFN-γ and this induction is reinforced by locally released TNF [62]. Class II protein-expressing astrocytes could function as antigen-presenting cells (APC). In vitro, IFN-β lessens the expression of MHC class II proteins induced by IFN-γ, i.e., it directly counteracts one of the immune-augmenting effects of released IFN-γ [69]. IFN-β inhibits IFN-γ release from mitogen-stimulated PBMC cultures [70, 71], and ultimately does so in vivo as well, thus pointing to a second potential mechanism, albeit indirect, by which MHC class II protein expression and, hence antigen presentation, might be curtailed.

In vivo, however, the situation is more complex. When IFN-β treatment is begun, flu-like symptoms with fever, chills, and muscle aches are the rule [72]. The magnitude of these systemic effects is directly related to dose. Systemic symptoms begin 4–6 h after drug injection and last for 4–8 h. They abate over the first month or two in most patients. No decline in attack frequency was noted over the 1st month of IFN-β-1b treatment in the pivotal trial. During the 2nd month and at later times attack frequency was reduced. No decline in MRI-determined disease activity was observed in a trial of IFN-β-1A during the 1st month on treatment, although after 2 months of treatment disease activity had fallen dramatically. Efficacy apparently surfaces as systemic side effects abate.

IFN-β-1b-induced fever is attenuated by glucocorticoid or azathioprine treatment, suggesting that fever is mediated indirectly [73]. Fever bespeaks the generation of endogenous pyrogens. The major endogenous pyrogens are IL-1, TNF, and IL-6. Release of IL-1 and TNF from monocytes/macrophages is primed for by IFN-γ [74]. This suggested to us that IFN-β-1b might actually induce IFN-γ release at early time points. This postulate was reinforced by the observation that MHC class II expression on circulating monocytes rises promptly on IFN-β induction [75]. In addition, IFN-β is known to activate NK cells and activated NK cells secrete IFN-γ. Neopterin levels in the blood rise promptly and substantially when IFN-β treatment is begun, only to fall back after 1–2 months, although not to baseline [76]. Neopterin secretion from monocytes and macrophages is potently stimulated by IFN-γ and to a far lesser extent by IFN-α and IFN-β [77]. Neopterin is a pyrazino-pyrimidine compound derived from guanosine triphosphate within the biosynthetic pathway of tetrahydrobiopterin, an essential co-factor for tyrosine and tryptophan hydroxylation. Its role in immune responses is unclear. It augments TNF production by macrophages that have been primed with IFN-γ, suggesting that it may synergize with, or provide a co-signal for, IFN-γ [78]. IL-4 from Th2 cells decreases neopterin secretion and inhibits IFN-γ release from Th1 cells, suggesting a possible role for neopterin in Th1–Th2 cell balance [79].

Based on the considerations outlined above, we counted circulating IFN-γ-secreting cells in MS patients immediately prior to their first dose of IFN-β-1b and at various times thereafter. Mitogen-activated IFN-γ-secreting cell numbers rose within 24 h and remained significantly elevated for 2 months before falling back to baseline values [80]. The IFN-γ-secreting cell surge was prevented by low-dose prednisone treatment during the 1st month on IFN-β-1b. IFN-γ is secreted by Th1-type CD4 cells, by CD8 cells and, as mentioned, by NK cells. IFN-γ-secreting cell numbers rise in all three types when IFN-β-1b treatment is begun [81].

Since IFN-γ provokes MS attacks, the prospect arises that IFN-β-1b induction might, by increasing IFN-γ-secreting cells, pose an increased risk for MS attacks during the first month or two on treatment. This was not seen in the pivotal MS trial, but no decrease in attacks was seen over this time period either. Perhaps some other action of IFN-β-1b counterbalances the potential risk posed by the IFN-γ-secreting cell surge. LT-secreting cell numbers fall significantly in the blood within 1 week of starting IFN-β-1b treatment [81]. Since LT is toxic to oligodendrocytes, it is possible that the LT fall counters the IFN-γ-secreting cell rise.

EAE, like MS, is a DTH disease and is mediated by Th1-type CD4 cells acting in conjunction with macrophages. IFN-γ-secreting Th1 cells increase in the lymph nodes that drain the antigen injection site 6–8 days after immunization, appear in the CNS just before the anticipated onset of disease and, in strains subject to relapses, reappear in the CNS as a herald of relapse [82–84].

What ends MS attacks?

MS attacks run a finite course. Attacks of EAE, the animal model for MS, also run a finite course. What causes an attack of MS or of EAE to end is far from clear. Several possibilities emerge. One is that CD8 suppressor cells become activated as the attack evolves, pass into the CNS, and release inhibitory cytokines such as transforming growth factor-β1 (TGF-β1). CD8 cells are present in MS plaques but their role within the CNS is uncertain. It is known, however, that TGF-β1 administration will attenuate EAE [85–87].

In our view, a role for CD8 suppressor cells within the CNS seems unlikely, at least in the case of EAE, for the following reason. Most strains of mice that develop EAE recover from their initial attack but fail to relapse. If spleen cells from such recovered animals are stimulated in vitro with a nonspecific lectin they will adoptively transfer EAE to another mouse of the same strain. It follows that recovered animals harbor T cells capable of causing EAE but that they are somehow held in check in the milieu of the recovered animal. CD8 knockout mice (i.e., mice that lack CD8 cells) develop EAE normally and also recover at a normal tempo [88]. Control EAE mice do not relapse once their initial EAE attack has ended. CD8 knockout mice relapse frequently, but they recover from their relapses at a tempo similar to that observed in their initial attack. These data suggest that CD8 cells have a policing role in the prevention of subsequent EAE attacks, presumably acting as suppressor cells in the periphery, but little if any role in terminating an attack once underway. Might CD8 cells have a similar role in MS?

Nonspecific CD8 suppressor cell function is deficient during MS attacks and recovers as attacks end [89, 90]. In MS patients with progressive disease, CD8 suppressor cell function is persistently subnormal [91]. Both IFN-α and IFN-β-1b restore suppressor function when added to PBMC from MS patients in vitro [92, 93]. Suppressor function was studied in MS patients entered into the pivotal trial. It was subnormal as they entered the trial and was largely restored in those receiving 8 MU IFN-β-1b s.c. every other day [94], the eventually approved dose. No such restoration was observed in placebo-treated patients. Data are given in Table 1.

Generation of CD8 suppressor cells is a multistep and multicell process (reviewed in [5]). Monocytes provide a priming signal, possibly via prostaglandin E_2 (PGE_2). CD4 cells provide a second inductive factor of unknown type but for which IFN-α or IFN-β can substitute. Released suppressor factors may then require a second signal from monocytes to activate (oxidize?) released latent suppressor factors. Failure of CD8 suppressor cell effector function could lie at any point along the path.

CD8 suppressor cells can secrete IL-4 and IL-10, which inactivate Th1 cells as discussed elsewhere. They can also secrete TGF-β1, a potent immunosuppressive cytokine [95–97]. TGF-β1 inhibits Th1 cell proliferation, DTH responses, IL-1, TNF, and reactive oxygen intermediate secretion from monocytes/macrophages, MHC class II protein expression, T cell adhesion to TNF- or IFN-γ-stimulated endothelial cells, IFN-γ gene expression, and IFN-γ production. TGF-β1 has no effect on IL-2 production, spares CD8 cells, and may even stimulate them. TGF-β1 administration protects against EAE [85–87], and TGF-β1-secreting suppressor cells can be retrieved from animals that have recovered from EAE [98]. TGF-β1 inhibits LT + IFN-γ production by disease effector T cells in EAE, and exposure of T cells to TGF-β1 in vitro prevents adoptive transfer of disease [99].

TGF-β1 may also have a suppressive role in MS. TGF-β1 can be detected in the supernatants of PBMC cultures of 4% of healthy controls, 11% of MS patients in remission, 33% of MS patients during attacks, and 87% of MS patients recovering from an attack [100]. IFN-β-1b increases TGF-β1 release from cultured PBMC [101]. Whether it similarly increases TGF-β1 production in vivo is not known.

An inverse relationship between suppressor cell function and IFN-γ secretion has been proposed [102]. The IFN-γ-secreting cell surge seen when IFN-β-1b treatment is begun fades away after a month or two of treatment. Suppressor function rises

Table 1. Interferon (IFN)-β-1b increases suppressor cell function in multiple sclerosis (MS)[a]

Group	% Suppression[b]
1. Control	37.2 ± 2.5 (94)
2. MS baseline	17.5 ± 4.2 (31)
3. Placebo	20.0 ± 3.3 (41)
4. IFN-β-1b (8.0 MU)	32.1 ± 3.6 (49)

[a]For 1 vs 2,3 $P < 0.01$, for 4 vs 2,3 $P < 0.02$
[b]$\pm$ SEM; number of observations in parentheses

coincident with this decline [34]. When IFN-γ is added to PBMC that are also exposed to IFN-β-1b, the rise in suppressor function caused by IFN-β-1b is unaltered [93]. This finding suggests that if IFN-γ and suppressor function are inversely linked then the flow of information is from the suppressor cell to the IFN-γ-secreting cell rather than in the other direction.

Given the ability of IFN-β-1b to improve suppressor function, one strategy to improve its efficacy in MS might be to find agents that potentiate this action of IFN-β-1b and to then provide them as an adjunctive treatment. *Trans*-retinoic acid is known to potentiate some actions of IFN-β-1b (reviewed in [103]). It also lessens the severity of EAE [104, 105]. Accordingly, we tested it for its ability to potentiate suppressor cell function when combined with IFN-β-1b. It does so. Data are given in Table 2.

Th2-type cells are a second class of CD4 cells. Their major role is to provide help to B cells and to facilitate B cell switching from the production of IgM antibodies to the production of antibodies of other isotypes. IL-4 and IL-10 released by Th2 cells have a major role in facilitating B cell function. There is a yin-yang between Th1- and Th2-type T cells, not surprisingly, since, depending on the insult, the immune system must opt for the type of response that will dominate. IFN-γ produced by Th1 cells inhibits Th2 cells, whereas IL-4 and IL-10 produced by Th2 cells inhibit Th1 cells. Mouse strains that are susceptible to EAE exhibit a Th1 cell bias, resistant strains a Th2 cell bias. IL-4, released by Th2 cells, decreases TNF release from macrophages, decreases cell-associated IL-1α expression on them, and decreases release from them of IL-1b, the major secreted form of IL-1 [106–108]. IL-4 decreases IFN-γ-driven H_2O_2 release from macrophages [109]. IL-4 also increases release of IL-1 receptor antagonist [108], an action that should serve to further potentiate the reduction by IL-4 of IL-1 release.

Recognition of the multiple inhibitory actions of IL-4 on Th1 cells and macrophages has led to the postulate that infiltrating Th2 cells provide the "turn off" signal for EAE attacks and possibly also for MS attacks. IL-4 mRNA can be detected in the CNS of animals with EAE at peak of disease and IL-10 mRNA becomes apparent as EAE is abating [110]. Unfortunately, mRNA methodology suffers from problems with quantitation. Presence of mRNA says little about its amount, nor does mRNA necessarily translate into secreted product.

It is possible to count cytokine-secreting cells by the immunospot technique. When cells are teased from the CNS of mice with acute monophasic EAE, large numbers of IFN-γ-secreting cells can be detected, but IL-4-secreting cells are few at all times [111]. Ratios of IFN-γ to IL-4-secreting cells range from 20- to 100-fold in favor of IFN-γ-secreting cells. The number of IL-4-secreting cells is too few in our estimation to permit one to assign them a global role as terminators of a first acute EAE attack.

Table 2. Effect of retinoic acid (RA) on suppressor cell function

Treatment	Multiple sclerosis[a]	Controls
None	9.0 ± 5.8 (14)	17.1 ± 2.8 (14)
IFN-β-1b (1000 μ)	25.2 ± 3.2 (14)	33.9 ± 3.6 (13)
RA	14.6 ± 2.9 (14)	24.2 ± 3.3 (14)
RA + IFN-β-1b	36.1 ± 3.9 (14)	40.5 ± 4.1 (12)

[a] percent suppression $\pm$ S.E.M. Number of subjects in parentheses.

This is not to say that they have no role in prevention of subsequent attacks. CD4 suppressor cells can be retrieved from rats that have recovered from EAE [112]. When given to naive rats prior to induction of EAE they attenuate the severity of disease. They do not prevent adoptively transferred EAE, i.e., they prevent the generation of effector cells but fail to block pre-activated effector cells. Addition of IFN-β-1b to cultured PBMC has no effect on IL-4 production or on IL-10 production by Th2-type cells.

Macrophages quickly become the dominant cell type in EAE inflammatory foci. They are also the predominant invading cell in MS plaques. IFN-γ provides an activating signal to macrophages. IL-2 synergizes with IFN-γ in this regard [113, 114]. IFN-γ increases TNF and IL-1 secretion by macrophages. IFN-γ also increases secretion by macrophages of nitric oxide (NO) and oxygen intermediates both of which contribute to tissue damage. Macrophages are the final effectors of myelin damage in MS, directly stripping myelin from axons. Cytokine secretion by blood monocytes from MS patients is increased compared to controls after exposure to IL-2 and/or IFN-γ, i.e., monocytes of MS patients are pre-primed. IFN-α and IFN-β both decrease release of O_2^- and H_2O_2 from macrophages, at least in mice [115]. Whether the same pertains in vivo in man is not known.

Macrophages can also exert suppressive effects, particularly as they mature. IL-10 is made by mature macrophages and they may well be the source of the IL-10 mRNA detected late in the course of an EAE attack [110]. IL-10 inhibits IFN-γ production by Th1 cells, while, in accord with the yin-yang between Th1 and Th2 cells, IFN-γ from Th1 cells inhibits IL-10 production by Th2 cells [116]. IFN-γ also inhibits IL-10 production by monocytes. IL-10 augments proliferation of CD8 suppressor cells. IL-10 decreases IL-1 and TNF production by monocytes and blocks their antigen-presenting potential. IL-10 stimulates B cells and increases MHC class II protein expression by them [116]. These actions suggest a possible role for IL-10 in the termination of MS attacks. The number of IL-10-secreting macrophages is significantly increased in vitro by exposure to IFN-β-1b, even though IFN-β-1b has no effect on IL-10 secretion by Th2 type cells [81].

PGE_2 is another suppressive molecule released by mature macrophages. PGE_2 decreases IL-2, IFN-γ, and LT production by Th1 cells [117–121] and decreases both MHC class II protein expression by monocytes/macrophages and TNF release from them. A PGE_1 analog, misoprostol, which induces PGE_2 production [122], exerts a protective effect in EAE [123] and has proven effective as a treatment for trigeminal neuralgia on a background of MS [124]. IFN-γ decreases PGE_2 production in keeping with its pro-inflammatory role [125]. Interestingly, IFN-β likewise inhibits PGE_2 release from macrophages, a potentially pro-inflammatory effect [126]. This finding indicates that the beneficial effect of IFN-β-1b in MS is not mediated via PGE, but that PGE administration might potentiate the beneficial effect of IFN-β-1b. This postulate is being tested. IFN-β-1b also decreases surface expression of CD14, a

Table 3. Selected actions of IFN-β of possible relevance to multiple sclerosis

Step in DTH response	Parameter	Effect[a]
Activation	Viral protein translation	**Impeded**
	MHC class I expression	**Increased**
	MHC class II expression	**Variable**
	B7-1 expression on B cells	**Decreased**
	B7-2 expression on monocytes	**Increased**
	Deficient suppressor cell function	**Restored**
Recruitment	Circulating lymphocytes	**Lowered**
	Cell entry into lymph nodes	**Increased**
Expansion	IL-2 production by Th1 cells	*Increased*
	IL-2 receptor expression	*Decreased*
	T cell proliferation	**Lowered**
Trafficking	Cell exit from lymph nodes	**Lessened**
	Endothelial cell adhesion molecule expression	*Lessened*
	IFN-γ production	**Increased initially; falls later**
Tissue Damage	LT production by Th1 cells	**Decreased**
	TNF production by macrophages	*Decreased*
	Macrophage cytolytic activity	*Variable*
	Release of mitogen and oxygen intermediates from macrophages	*Variable*
	CD14 expression on macrophages	*Decreased*
Recovery/other	IL-10 production by macrophages	*Increased*
	TGF-β1 production by PBMC	*Increased*
	PGE2 release from macrophages	*Decreased*
	Antibody synthesis	**Variable – overall stimulatory**
	Cytotoxic T cell function	**Increased**

[a]Effects documented in vivo in bold face; effects shown only in vitro in italics
DTH, delayed-type hypersensitivity; MHC, major histocompatibility complex; LT lymphotoxin; TNF, tumor necrosis factor; TGF, transforming growth factor; PGE_2, prostaglandin E_2

protein involved in monocyte/macrophage activation and in TNF release [127, 128]. This action of IFN-β-1b should be anti-inflammatory.

Presentation of antigen by MHC class II proteins does not suffice for T cell activation. Co-stimulatory signals are also essential. Major co-stimulatory signals are provided by B7-1 and B7-2, proteins expressed on so-called "professional" APC such as B cells, monocytes, and macrophages [129]. B7-1 and B7-2 bind to CD28 and to CTLA4 on T cells. B7-1 ligation to CD28 activates Th1-type T cells, whereas B7-2 ligation activates Th2-type T cells, at least in the EAE model [130]. The number of B cells expressing B7-1 is significantly increased when MS is active compared to times of inactivity or to normal controls [131]. IFN-β-1b significantly down-regulates the number of B7-1-expressing B cells in patients receiving the agent. This action may serve to attenuate Th1 cell responses. B7-1 expression on monocytes is not altered by IFN-β-1b treatment, but B7-2 expression on monocytes of MS patients is increased by IFN-β-1b treatment. This skewing of the ratio between B7-1 and B7-2 in favor of B7-2 may perhaps lead to preferential activation of Th2 cells with, as an indirect consequence, inhibition of Th1 cells. IL-10 decreases B7 expression [132]. Inasmuch as IL-10 secretion from monocytes in vitro is increased by IFN-β-1b exposure [81], the effect of IFN-β-1b on B7 expression may be mediated indirectly, i.e., via IL-10.

Cultured astrocytes express MHC class II molecules when stimulated by IFN-γ [133, 134]. Astrocytes in animals with EAE and within active MS plaques are also reported to express MHC class II proteins. Presumably this expression is induced by IFN-γ and further potentiated by TNF, both of which are detected in active MS foci. Astrocytes do not express co-stimulatory molecules. Astrocytes thus could function, at least theoretically, as "non-professional" APC. If astrocytes do in fact function as non-professional APC in the CNS, they could induce T cell anergy rather than T cell activation. Anergy is described operationally as a state which follows antigen presentation by non-professional APC that lack the co-stimulatory molecules requisite for T cell activation. In anergy, Th1 cell proliferation stops and production of Th1 cytokines, such as IL-2, ceases (although IFN-γ production may be relatively spared). Anergic cells can be rescued for proliferation by provision of exogenous IL-2. In acute monophasic EAE in the PL/J mouse, cells recovered from the CNS at the peak of disease proliferate poorly, if at all, in response to lectin or antigen stimulation, make no IL-2, release substantial amounts of IFN-γ, and are to some extent rescuable by exogenous IL-2 [111]. Accordingly, they can be viewed as anergic. Whether antigen presentation by astrocytes in a manner that promotes anergy accounts for this finding is not known, although evidence in support of this notion has been brought forward [135, 136]. Another possible inhibitory role for astrocytes can be envisaged. They may be induced by cytokine exposure to secrete inhibitory molecules, including PGs and TGF-β2 [135].

Whether the IFN-β found in MS lesions contributes to anergy is unknown. What effect IFN-β might have on MHC class II protein expression in the CNS is not known either. IFN-β does lessen IFN-γ-driven MHC expression by cultured astrocytes [133], but if given alone it augments rather than lessens MHC class II antigen expression by cultured astrocytes [134]. The fact that IFN-β-1b lowers B7-1 expression on B cells and skews the B7-1/B7-2 ratio on monocytes towards B7-2 might lessen the effectiveness of professional APC and permit a more active role for APC with anergizing potential.

An off signal to Th1 cell responses may also be provided by glucocorticoids. Activated monocytes/macrophages release IL-1, which enters the circulation and acts at the hypothalamic level to cause corticotrophin releasing factor (CRF) to be released. CRF in turn causes release of adrenocorticotropic hormone (ACTH) from the pituitary. Released ACTH triggers glucocorticoid release from the adrenal cortex. A glucocorticoid surge that lessens EAE severity occurs during acute EAE [137], and glucocorticoids are used to treat MS attacks, often successfully. Glucocorticoids selectively inhibit Th1-type T cells and cause their apoptosis. Glucocorticoid levels were measured in MS patients receiving IFN-β-1b but were normal [138].

Finally, Th1 effector cells that have entered the CNS may be destined to die. Apoptotic T cells are readily detected in evolved EAE lesions and in active MS lesions [14, 139]. The signal for apoptosis may be provided in part by glucocorticoids (as above) but could also be intrinsic to the Th1 cells themselves. High concentrations of IFN-γ can, under some circumstances, inhibit Th1 cell proliferation [140] and cause Th1 cell death [141]. High concentrations of IFN-γ can also induce production by macrophages of TGF-β1, an immunosuppressive cytokine. It is known that IFN-γ, administered in large dose either systemically or intrathecally to animals about to develop EAE, attenuates disease [142, 143]. Conversely, anti-IFN-γ antibody administration at disease onset leads to a worsening of disease severity [144]. While various explanations for these findings are possible, an acceleration of possibly pre-programmed cell death by IFN-γ and a deceleration by anti-IFN-γ antibody could account for them. Whether IFN-β can induce or contribute to T cell apoptosis is not known.

Why does MS become progressive?

FK506 is an immunosuppressive molecule widely used in clinical organ transplantation. This agent inhibits IL-2 and IFN-γ production, although probably not that of LT. FK506 also inhibits apoptosis. When FK506 is given to rats with ongoing EAE, disease severity is lessened and few animals relapse. FK506 given after immunization but prior to disease onset prevents acute EAE, but the animals go on to develop a progressive demyelinating disorder that is ultimately fatal. These data suggest that prevention of apoptosis by FK506 switches an acute monophasic and self-limited process to a progressive one, albeit indolent initially (presumably because of the concomitant inhibitory effect of the agent on IL-2 and IFN-γ production) but one that ultimately becomes vicious. The issue may bear on MS since, after a period of time characterized by attacks and remissions, MS, more often than not, becomes a progressive disorder. Might some T cells, as they pass through numerous cycles of replication, change or mutate so that a population is selected that is refractory to apoptotic signals? The question is unresolved but testable. Nothing is known concerning the effect of IFN-β on apoptotic mechanisms.

The role of antibodies

B cells and plasma cells are present within the CNS in MS. They secrete immunoglobulins which can be detected in the spinal fluid. The B cells secret IgG_1 for the most part, indicating that they have undergone Th2 cell-mediated isotype switching. An elevated spinal fluid IgG level (usually with oligoclonal banding which bespeaks clonal expansion of B cells) is a useful adjunct when a diagnosis of MS is suspected, although the finding is by no means specific for MS. Individual patients show oligoclonal bands unique to that patient which persist over time. Even identical twins concordant for MS show different oligoclonal band patterns [146]. The antigenic specificities of the oligoclonal IgG bands are not known, nor is the reason for a continuing secretion of IgG by particular B cell clones in the absence of any obvious antigenic driving force understood. B cells tend to be sparse or absent in early active lesions and their presence is not essential to the progression of disease. Cases are on record in which neither oligoclonal bands nor increased IgG were detected in the spinal fluid during life and no B cells were found in the brain at autopsy, although the MS pathology was otherwise typical [147].

IFN-α treatment is associated with increased antibody responses and the development of antibody-mediated autoimmune processes (reviewed in [5]). Thyroid antibodies, anti-nuclear antibodies, hemolytic anemia, thrombocytopenia, worsening of rheumatoid arthritis, and development of a rheumatoid arthritis/systemic lupus erythematosus overlap syndrome have all been recorded in persons receiving IFN-α. Many of those who developed high titers of anti-thyroid antibodies while being treated, had

had low level anti-thyroid antibodies prior to treatment. The suggestion has been advanced that class I IFNs enhance pre-existing autoimmunity rather than causing it. No antibody-mediated autoimmunity has been observed in patients treated with IFN-β-1b but clinicians should be alert to the possibility that this may occur.

Antibodies that neutralized the ability of IFN-β-1b to impede viral replication in vitro were detected in some 35% of MS patients receiving IFN-β-1b in the pivotal trial that led to approval of the agent. Most who developed antibodies did so during the 1st year on the agent. Low titer anti-IFN-neutralizing antibodies can be detected in 10% of the normal population [148]. Whether those with such antibodies prior to treatment are at increased risk for development of high titers of anti-IFN antibodies during treatment has not been studied. There is some evidence that efficacy of IFN-β-1b treatment was less beyond 2 years on treatment in those who developed antibodies than in those who did not, although why there should be a delay of a year or more between appearance of antibody and lessening of efficacy is not apparent. Antibody titers did not rise inexorably over time; in many patients they declined over time. Concern has been raised as to the accuracy and validity of the antibody assay employed, and the issue is being re-examined with a battery of new antibody assay techniques. Accordingly, the evidence given above must be viewed as tentative. Should the suspicion that development of anti-IFN antibodies lessens efficacy be sustained by ongoing studies, attempts will have to be made to find means to prevent generation of antibodies and/or to lower their titer once they appear. On a happier note, if those with antibodies do less well then, the two-thirds of patients who do not develop antibodies do even better than the overall data amassed to date have indicated.

In an early IFN-α trial in MS, elevated serum IgG levels were detected in patients receiving this agent [149, 150]. Increased intrathecal IgG synthesis was observed in 6 of 12 cases and linked to clinical course. While on IFN-α the 6 patients with increased intrathecal IgG synthesis had two mild attacks, whereas the other 6 patients had six major attacks while on IFN-α. Both groups had seven attacks while on placebo. The data, while fragmentary, suggest that increased intrathecal IgG synthesis is unrelated to disease pathogenesis, and may even be protective.

Additional actions of IFN-β

Intravenous IFN-β treatment lessens the severity of both actively induced and adoptively transferred EAE [151, 152]. When T cells from EAE animals are exposed to IFN-β in vitro prior to adoptive transfer into naive recipients, the severity of adoptively transferred disease is reduced [153, 154]. The effect was ascribed to the antiproliferative action of IFN-β (see below), or possibly to generation of suppressor cells. Intraventricular IFN-β does not attenuate the severities of actively induced or adoptively transferred disease [155]. Lactic dehydrogenase virus (LDHV)-infected mice are protected from EAE [156]. LDHV is a potent inducer of IFN-α and IFN-β and its disease-attenuating action has been ascribed to this induction [157]. Unfortunately, the mechanisms by which IFN-β attenuates EAE are unknown.

IFN-β-1b is hydrophobic. When IFN-β-1b is given s.c. none can be found in the blood. The sites of injection remain erythematous for weeks, pointing to an ongoing local effect. The blood-brain barrier is highly impermeable to IFNs [158]. These findings suggest that relatively little, if any, s.c. injected IFN-β-1b finds its way to the CNS and that the drug is more likely to exert its beneficial effect in the periphery than in the CNS. There are numerous mechanisms that can be involved under these circumstances in addition to those already discussed. IFN-β alters cell trafficking (reviewed in [5]). IFN-β facilitates exit of lymphocytes from the blood and causes lymphadenopathy (entry of lymphocytes into nodes). IFN-β impedes exit of lymphocytes from the lymph nodes (trapping). These actions could lessen access of activated T cells to the CNS

IFN-β-1b also exerts an antiproliferative action, lengthening the duration of the cell cycle. Proliferation of both Th1 and Th2 cells is inhibited, at least in vitro [159, 160]. Since lymphocyte proliferation is a prerequisite for an effective immune response this action of IFN-β-1b could contribute to its beneficial effect in MS. IFN-β-1b downregulates IL-2 receptors, transferrin receptors, and alters CD2 function [161]. These actions may contribute to the anti-proliferative effect since all three molecules are important to the proliferative response. Lymphopenia, seen with IFN-β-1b treatment, depends in part on effects on cell trafficking and in part on an antiproliferative effect.

IFN-β augments expression of MHC class I molecules throughout the body. This is not surprising since CD8 cells, the function of which is increased by IFN-β [162], recognize antigen presented by MHC class I proteins. IFN-β also increases IL-2 production [163]. This again is not unexpected since IL-2 provides a stimulatory signal for CD8 cells in addition to CD4 cells.

From the foregoing, it is evident that IFN-β-1b has many actions that could be responsible for its beneficial effect in MS. A listing of those we consider potentially relevant to MS is provided in Table 3. IFN-β-1b is only partially effective in MS. There is a need to do better. A clearer understanding of those of its actions that matter in MS should improve our understanding of the disease and provide a rationale for new approaches to therapy.

Acknowledgement. Original work cited in this review was supported by PHS-PO1 NS 24575 (B.G.W.A. and A.R.) and gifts from the Brain Research and Butz Foundations (B.G.W.A.). As always we thank Ms. Madeline Murphy for typing the manuscript and for much appreciated editing.

References

1. IFNB Multiple Sclerosis Study Group (1993) Interferon beta-1b is effective in relapsing-remitting multiple sclerosis: I. Clinical results of a multicenter, randomized, double-blind, placebo-controlled trial. Neurology 43:655
2. Arnason BGW (1993) Interferon beta in multiple sclerosis. Neurology 43:641
3. IFNB Multiple Sclerosis Study Group and the University of British Columbia MS/MRI Analysis Group (1995) Interferon beta-1B in the treatment of multiple sclerosis: final outcome of the randomized controlled trial. Neurology 45:1277
4. Paty DW, Li DKB, UBC MS/MRI Study Group, IFNB Multiple Sclerosis Study Group (1993) Interferon beta-1b is effective in relapsing-remitting multiple sclerosis II. MRI analysis results of a multicenter, randomized, double-blind, placebo-controlled trial. Neurology 43:662
5. Arnason BGW, Reder AT (1994) Interferons and multiple sclerosis. Clin Neuropharmacol 17:495
6. Durelli L, Bongioanni MR, Cavallo R, Ferrero B, Ferri R, Ferrio MF, Bradac GB, Riva A, Vai S, Geuna M, Bergamini L, Bergamasco B (1994) Chronic systemic high dose recombinant interferon alfa-2a reduces exacerbation rate, MRI signs of disease activity, and lymphocyte interferon gamma production in relapsing-remitting multiple sclerosis. Neurology 44:406

7. Jacobs L, Cookfair D, Rudick R, Herndon R, Richert J, Salazar A, Fischer J, Granger C, Simon J, Goodkin D, MSCR Group (1994) Results of a phase III trial of intramuscular recombinant beta interferon as treatment for multiple sclerosis (abstract). Ann Neurol 36:259
8. Reference deleted
9. Willoughby EW, Grochowski E, Li DKB, Oger J, Kastrukoff LF, Paty DW (1989) Serial magnetic resonance scanning in multiple sclerosis: a second prospective study in relapsing patients. Ann Neurol 25:43
10. Noronha ABC, Richman DP, Arnason BGW (1980) Detection of in vivo stimulated cerebrospinal fluid lymphocytes in multiple sclerosis by flow cytometry. N Engl J Med 303:713
11. Prineas JW, Barnard RO, Kwon EE, Sharer LR, Cho E-S (1993) Multiple sclerosis: remyelination of nascent lesions. Ann Neurol 33:137
12. Prineas JW, Kwon EE, Goldenberg PZ, Ilyas AA, Quarles RH, Benjamins JA, Sprinkle TJ (1989) Multiple sclerosis: oligodendrocyte proliferation and differentiation in fresh lesions. Lab Invest 61:489
13. Rodriguez M, Scheithauer BW, Forbes G, Kelly PJ (1993) Oligodendrocyte injury is an early event in lesions of multiple sclerosis. Mayo Clinic Proc 68:627
14. Ozawa K, Suchanek G, Breitschopf H, Brück W, Budka H, Jellinger K, Lassmann H (1994) Patterns of oligodendroglia pathology in multiple sclerosis. Brain 117:1311
15. Brück W, Schmied M, Suchanek G, Brück Y, Breitschopf H, Poser S, Piddlesden S, Lassmann H (1994) Oligodendrocytes in the early course of multiple sclerosis. Ann Neurol 35:65
16. Van Walderveen MAA, Barkhof F, Hommes OR, Polman CH, Tobi H, Frequin STFM, VALK J (1995) Correlating MRI and clinical disease activity in multiple sclerosis: relevance of hypointense lesions on short-TR/short-TE (T_1-weighted) spin-echo images. Neurology 45:1684
17. Paty DW, Li DKB, Oger JJ-F, Kastrukoff L, Koopmans R, Tanton E, Zhao GJ (1994) Magnetic resonance imaging in the evaluation of clinical trials in multiple sclerosis. Ann Neurol 36:595
18. Miller DH (1994) Magnetic resonance in monitoring the treatment of multiple sclerosis. Ann Neurol 36:591
19. Filippi M, Horsfield MA, Morissey SP, MacManus DG, Rudge P, McDonald WI, Miller DH (1994) Quantitative brain MRI lesion load predicts the course of clinically isolated syndromes suggestive of multiple sclerosis. Neurology 44:635
20. Weinshenker BG, Bass B, Rice GPA, Noseworthy J, Carriere W, Baskerville J, Ebers GC (1989) The natural history of multiple sclerosis: a geographically based study. I. Clinical course and disability. Brain 112:133
21. Weinshenker BG, Bass B, Rice GPA, Noseworthy J, Carriere W, Baskerville J, Ebers GC (1989) The natural history of multiple sclerosis: a geographically based study. 2. Predictive value of the early clinical course. Brain 112:1419
22. Greenwood BG, cited in: Psychosocial factors in multiple sclerosis. Proceedings of the MS Forum Modern Management Workshop, Rome 1995, Professional Postgraduate Services, Worthing U.K., p 28
23. Cella DF, Dineen K, Arnason B, Reder A, Webster KA, Karabatsos G, Chang C-H, Lloyd S, Mo F, Stewart J, Stefoski D (1996) Validation of the functional assessment of multiple sclerosis (FAMS) quality of life instrument. Neurology (in press)
24. Sharief MK, Hentges R (1991) Association between tumor necrosis factor-α and disease progression in patients with multiple sclerosis. N Engl J Med 325:467
25. Maimone D, Gregory S, Arnason BGW, Reder AT (1991) Cytokine levels in the cerebrospinal fluid and sera of patients with multiple sclerosis. J Neuroimmunol 32:67
26. Rothwell NJ, Relton JK (1993) Involvement of cytokines in acute neurodegeneration in the CNS. Neurosci Biobehav Rev 17:217
27. Pliskin NH, Towle VL, Hamer DP, Reder AT, Noronha A, Pietre S, Arnason BGW (1994) The effects of interferon-beta on cognitive function in multiple sclerosis (abstract). Ann Neurol 36:326
28. Sibley WA, Bamford CR, Clark K (1985) Clinical viral infections and multiple sclerosis. Lancet I:1313
29. Panitch HS, Bever CT, Katz E, Johnson KP (1991) Upper respiratory tract infections trigger attacks of multiple sclerosis in patients treated with interferon-β (abstract). J Neuroimmunol 35[Suppl 1]:125
30. Andersen O, Lygner P-E, Bergström T, Andersson M, Vahlne A (1993) Viral infections trigger multiple sclerosis relapses: a prospective seroepidemiological study. J Neurol 240:417
31. Reich N, Pine R, Levy D, Darnell JE Jr (1988) Transcription of interferon-stimulated genes is induced by adenovirus particles but is suppressed by E1A gene products. J Virol 62:114

32. Panitch HS (1994) Influence of infection on exacerbations of multiple sclerosis. Ann Neurol 36:S25
33. Panitch HS, Hirsch RL, Haley AS, Johnson KP (1987) Exacerbations of multiple sclerosis with gamma interferon. Lancet I:893
34. Reference deleted
35. Yu C-L, Haskard DO, Cavender D, Johnson AR, Ziff M (1985) Human gamma interferon increases the binding of T lymphocytes to endothelial cells. Clin Exp Immunol 62:554
36. May MJ, Ager A (1992) ICAM-1-independent lymphocyte transmigration across high endothelium: differential up-regulation by interferon-γ, tumor necrosis factor-α and interleukin-1β. Eur J Immunol 22:219
37. Pober JS, Gimbrone MA Jr, Lapierre LA, et al (1986) Overlapping patterns of activation of human endothelial cells by interleukin 1, tumor necrosis factor, and immune interferon. J Immunol 137:1893
38. Thorhill MH, Aung UK, Haskard DO (1990) IL-4 increases human endothelial cell adhesiveness for T cells but not for neutrophils. J Immunol 144:3050
39. McCarron RM, Wang L, Racke MK, McFarlin DE, Spatz M (1993) Cytokine-regulated adhesion between encephalitogenic T lymphocytes and cerebrovascular endothelial cells. J Neuroimmunol 43:23
40. Eguchi K, Kawakami A, Nakashima M, Ida H, Sakito S, Matsuoka N, Terada K, Sakai M, Kawabe Y, Fukada T, Ishimaru T, Kurouji K, Fujita N, Aoyaji T, Maeda K, Nagataki S (1992) Interferon-alpha and dexamethasone inhibit adhesion of T cells to endothelial cells and synovial cells. Clin Exp Immunol 88:448
41. Rotteveel FTM, Kuenen B, Kokkelink I, Meager A, Lucas CJ (1990) Relative increase of inflammatory $CD4^+$ T cells in the cerebrospinal fluid of multiple sclerosis patients and control individuals. Clin Exp Immunol 79:15
42. Brod SA, Benjamin D, Hafler DA (1991) Restricted T cell expression of IL-2/IFN-γ mRNA in human inflammatory disease. J Immunol 147:810
43. Benvenuto R, Paroli M, Buttinelli C, Franco A, Barnaba V, Fieschi C, Balsano F (1991) Tumor necrosis factor-alpha synthesis by cerebrospinal-fluid-derived T cell clones from patients with multiple sclerosis. Clin Exp Immunol 84:97
44. Trotter JL, Collins KG, Veen RC van der (1991) Serum cytokine levels in chronic progressive multiple sclerosis: interleukin-2 levels parallel tumor necrosis factor-α levels. J Neuroimmunol 33:29
45. Beck J, Rodot P, Catinot L, Falcoff E, Kirchner H, Wietzerbin J (1988) Increased production of interferon gamma and tumor necrosis factor precedes clinical manifestation in multiple sclerosis: do cytokines trigger off exacerbations? Acta Neurol Scand 78:318
46. Lu C-Z, Jensen MA, Arnason BGW (1993) Interferon gamma- and interleukin-4-secreting cells in multiple sclerosis. J Neuroimmunol 46:123
47. Traugott U, Lebon P (1988) Multiple sclerosis: involvement of interferons in lesion pathogenesis. Ann Neurol 24:243
48. Traugott U, Lebon P (1988) Interferon-γ and Ia antigen are present on astrocytes in active chronic multiple sclerosis lesions. J Neurol Sci 84:257
49. Zipp F, Weber F, Huber S, Sorgiu S, Czlonkowska A, Holler E, Albert E, Weiss EH, Wekerle H, Hohlfeld R (1995) Genetic control of multiple sclerosis: increased production of lymphotoxin and tumor necrosis factor-α by HLA-$DR2^+$ T cells. Ann Neurol 38:723
50. Nedwin GE, Svedersky LP, Bringman TS, Palladino MA Jr, Goeddel DV (1985) Effect of interleukin 2, interferon-γ, and mitogens on the production of tumor necrosis factors α and β. J Immunol 135:2492
51. Selmaj K, Raine CS, Farooq M, Norton WT, Brosnan CF (1991) Cytokine cytotoxicity against oligodendrocytes: apoptosis induced by lymphotoxin. J Immunol 147:1522
52. Stone-Wolff DS, Yip YK, Kelker HC, Le M-M, Henriksen-Destefano D, Rubin BY, Rinderknecht E, Aggarwal BB, Vilček J (1984) Interrelationships of human interferon-gamma with lymphotoxin and monocyte cytotoxin. J Exp Med 159:828
53. Powell MB, Mitchell D, Lederman J, Buchmeier J, Zamvil SS, Graham M, Ruddle NH, Steinman L (1990) Lymphotoxin porduction by myelin basic protein-specific T cell clones correlates with encephalitogenicity. Int Immunol 2:539
54. Selmaj K, Raine CS, Cannella B, Brosnan CF (1991) Identification of lymphotoxin and tumor necrosis factor in multiple sclerosis lesions. J Clin Invest 87:949
55. Hofman FM, Hinton DR, Johnson K, Merrill JE (1989) Tumor necrosis factor identified in multiple sclerosis brain. J Exp Med 170:607

56. Selmaj KW, Raine CS (1988) Tumor necrosis factor mediates myelin and oligodendrocyte damage in vitro. Ann Neurol 23:339
57. Soliven B, Szuchet S, Nelson DJ (1991) Tumor necrosis factor inhibits K^+ current expression in cultured oligodendrocytes. J Membr Biol 124:127
58. Ruddle NH, Bergman CM, McGrath KM, Lingenheld EG, Grunnet ML, Padula SJ, Clark RB (1990) An antibody to lymphotoxin and tumor necrosis factor prevents transfer of experimental allergic encephalomyelitis. J Exp Med 172:1193
59. Selmaj KW, Farooq M, Norton WT, Raine CS, Brosnan CF (1990) Proliferation of astrocytes in vitro in response to cytokines: a primary role for tumor necrosis factor. J Immunol 144:129
60. Yong VW, Moumdijan R, Yong FP, Ruijs TC, Freedman MS, Cashman N, Antel JP (1991) γ-Interferon promotes proliferation of adult human astrocytes in vitro and reactive gliosis in the adult mouse brain in vivo. Proc Natl Acad Sci USA 88:7016
61. Giulian D, Lachman LB (1985) Interleukin-1 stimulation of astroglial proliferation after brain injury. Science 228:497
62. Benveniste EN, Sparacio SM, Bethea JR (1989) Tumor necrosis factor-α on the expression of an Ia antigen on a murine macrophage cell line. J Immunol 137:2853
63. Jacobsen H, Mestan J, Mittnacht S, Dieffenbach CW (1989) Beta interferon subtype 1 induction by tumor necrosis factor. Mol Cell Biol 9:9037
64. Rubin BY, Anderson SL, Lunn RM, Richardson NK, Hellerman GR, Smith LJ, Old LJ (1988) Tumor necrosis factor and IFN induce a common set of proteins. J Immunol 141:1180
65. Reyes VE, Ballas ZK, Singh H, Klimpel GR (1986) Interleukin-2 induces IFN-α/β production in murine bone marrow cells. Cell Immunol 102:374
66. Hertzog PJ, Wright A, Harris G, Linnane AW, MacKay IR (1991) Intermittent interferonemia and interferon responses in multiple sclerosis. Clin Immunol Immunopathol 58:18
67. Kelley VE, Fiers W, Strom TB (1984) Cloned human interferon-γ, but not interferon-β or α, induces expression of HLA-DR determinants by fetal monocytes and myeloid leukemic cell lines. J Immunol 132:240
68. Meinl E, Aloisi F, Ertl B, Weber F, deWall Malefyt R, Wekerle H, Hohlfeld R (1994) Multiple sclerosis: immunomodulatory effects of human astrocytes on T cells. Brain 117:1323
69. Ling PD, Warren MK, Vogel SN (1985) Antagonistic effect of interferon-beta on the interferon-gamma-induced expression of Ia antigen in murine macrophages. J Immunol 135:1857
70. Noronha A, Toscas A, Jensen MA (1993) Interferon β decreases T cell activation and interferon γ production in multiple sclerosis. J Neuroimmunol 46:145
71. Panitch HS, Folus JS, Johnson KP (1987) Recombinant beta interferon inhibits gamma interferon production in multiple sclerosis. Ann Neurol 22:139
72. Billiau A, De Somer P, Edy VG, De Clercq E, Heremans H (1979) Human fibroblast interferon for clinical trials: pharmacokinetics and tolerability in experimental animals and humans. Antimicrob Agents Chemother 16:56
73. DeSomer P, Edy VG, Billau A (1977) Interferon-induced skin reactivity in man. Lancet II:47
74. Collart MA, Belin D, Vassalli J-D, Kossodo S de, Vassalli P (1986) γ Interferon enhances macrophage transcription of the tumor necrosis factor/cachetin, interleukin 1, and urokinase genes, which are controlled by short-lived repressors. J Exp Med 164:2113
75. Spear GT, Paulnock DM, Jordan RL, Meltzer DM, Merritt JA, Borden EC (1987) Enhancement of monocyte class I and II histocompatibility antigen expression in man by in vivo β-interferon. Clin Exp Immunol 69:107
76. Chiang J, Gloff CA, Yoshizawa CN, Williams GJ (1993) Pharmacokinetics of recombinant human interferon-β_{ser} in healthy volunteers and its effect on serum neopterin. Pharm Res 10:567
77. Huber C, Batchelor JR, Fuchs D, Hausen A, Lang A, Niedenwieser D, Reibnegger G, Swetly P, Troppmair J, Wachter H (1984) Immune response-associated production of neopterin: release from macrophages primarily under control of interferon-gamma. J Exp Med 160:310
78. Barak M, Gruener N (1991) Neopterin augmentation of tumor necrosis factor production. Immunol Lett 30:101
79. Moutabarrik A, Takahara S, Namiki M, Kameoka H, Seguchi T, Yokokawa K, Takano Y, Sonada T, Ishibashi M, Zaid D, et al (1992) Contrasting effects of interferon-gamma and interleukin-4 on neopterin generation from human adhesion monocytes. Lymphokine Cytokine Res 11:327
80. Dayal A, Jensen MA, Lledo A, Arnason BGW, (1995) Interferon gamma-secreting cells in multiple sclerosis patients treated with interferon beta-1B. Neurology 45:2173

81. Dayal A, Qu Z-X, Jensen M, Arnason BGW (1996) Transretinoic acid reverses induction of interferon γ-secreting cells by interferon β-1B. Neurology (in press)
82. Olsson T (1992) Cytokines in neuroinflammatory disease: role of myelin autoreactive T cell production of interferon-gamma. J Neuroimmunol 40:211
83. Mustafa MI, Diener P, Höjeberg B, Van der Meide P, Olsson T (1991) T cell immunity and interferon-γ secretion during experimental allergic encephalomyelitis in Lewis rats. J Neuroimmunol 31:165
84. Mustafa MI, Diener P, Sun J-B, Link H, Olsson T (1993) Immunopharmacologic modulation of experimental allergic encephalomyelitis: low-dose cyclosporin-A treatment causes disease relapse and increased system T and B cell-mediated myelin-directed autoimmunity. Scand J Immunol 38:499
85. Johns L, Franders KC, Sriram S (1991) Successful treatment of experimental allergic encephalomyelitis with transforming growth factor-β_1. Neurology 41:318
86. Racke MK, Dhib-Jalbut S, Cannella B, Albert PS, Raine CS, McFarlin DE (1991) Prevention and treatment of chronic relapsing experimental allergic encephalomyelitis by transforming growth factor-β1. J Immunol 146:3012
87. Kuruvilla AP, Shah R, Hochwald GM, Liggitt HD, Palladino MA, Thorbecke GJ (1991) Protective effect of transforming growth factor β_1 on experimental autoimmune diseases in mice. Proc Natl Acad Sci USA 88:2918
88. Koh D-R, Fung-Leung W-P, Ho A, Gray D, Acha-Orbea H, Mak T-W (1992) Less mortality but more relapses in experimental allergic encephalomyelitis in CD8$^{--/--}$ mice. Science 256:1210
89. Arnason BGW, Antel JP (1978) Suppressor cell function in multiple sclerosis. Ann Immunol (Paris) 129C:159
90. Antel JP, Arnason BGW, Medof ME (1979) Suppressor cell function in multiple sclerosis: correlation with clinical disease activity. Ann Neurol 5:338
91. Antel JP, Bania MB, Reder A, Cashman N (1985) Activated suppressor cell dysfunction in progressive multiple sclerosis. J Immunol 137:137
92. Noronha A, Toscas A, Jensen MA (1990) Interferon beta augments suppressor cell function in multiple sclerosis. Ann Neurol 27:207
93. Noronha A, Toscas A, Jensen MA (1992) Contrasting effects of alpha, beta and gamma interferons on nonspecific suppressor function in multiple sclerosis. Ann Neurol 31:103
94. Noronha A, Toscas A, Arnason BGW, Jensen M (1994) IFN-beta augments in vivo suppressor function in MS (abstract). Neurology 44[Suppl 2]:A212
95. Tsunawaki S, Sporn M, Ding A, Nathan C (1988) Deactivation of macrophages by transforming growth factor-β. Nature 334:260
96. Fontana A, Constam DB, Frei K, Malipiero U, Pfister HW (1992) Modulation of the immune response by transforming growth factor beta. Int Arch Allergy Immunol 99:1
97. Gamble JR, Vadas MA (1991) Endothelial cell adhesiveness for human T lymphocytes is inhibited by transforming growth factor 146:1149
98. Karpus WJ, Swanborg RH (1991) $CD4^+$ suppressor cells inhibit the function of effector cells of experimental autoimmune encephalomyelitis through a mechanism involving transforming growth factor-beta. J Immunol 146:1163
99. Stevens DB, Gould KE, Swanborg RH (1994) Transforming growth factor-β_1 inhibits tumor necrosis factor α/lymphotoxin production and adoptive transfer of disease by effector cells of autoimmune encephalomyelitis. J Neuroimmunol 51:77
100. Beck J, Rondot P, Jullien P, Wietzerbin J, Lawrence DA (1991) TGF-β-like activity produced during regression of exacerbations in multiple sclerosis. Acta Neurol Scand 84:452
101. Noronha A, Jensen M, Toscas A (1993) TGF-β activity in MS: effect of IFN-β. Neurology 43[Suppl]:355
102. Panitch HS, Folus JS, Johnson KP (1991) Activated suppressor cells inhibit synthesis of interferon γ in patients with multiple sclerosis and normal subjects (abstract). J Neuroimmunol 36:S186
103. Qu Z-X, Jensen MA, Arnason BGW (1995) Retinoic acid potentiates the ability of interferon beta-1B to augment suppressor cell function. Neuroscience 21 (part 2):1151
104. Massacesi L, Abbamondi AL, Giorgi C, Sarlo F, Lolli F, Amaducci L (1987) Suppression of experimental allergic encephalomyelitis by retinoic acid. J Neurol Sci 80:55
105. Racke MK, Burnette D, Pak S-H, McFarlin DE, Scott DE (1995) Retinoid treatment of experimental allergic encephalomyelitis: IL-4 production correlates with improved disease course. J Immunol 154:450

106. Hart PH, Vitti GF, Burgess DR, Whitty GA, Piccoli DS, Hamilton JA (1989) Potential antiinflammatory effects of interleukin 4: suppression of human monocyte tumor necrosis factor α, interleukin 1, and prostaglandin E_2. Proc Natl Acad Sci USA 86:3803
107. Wong HL, Lotze MT, Wahl LM, Wahl SM (1992) Administration of recombinant IL-4 to humans regulates gene expression, phenotype, and function in circulating monocytes. J Immunol 148:2118
108. Vannier E, Miller LC, Dinarello CA (1992) Coordinated antiinflammatory effects of interleukin 4: interleukin 4 suppresses interleukin 1 production but up-regulates gene expression and synthesis of interleukin 1 receptor antagonist. Proc Natl Acad Sci USA 89:4076
109. Lehn M, Weiser WY, Engelhorn S, Gills S, Remold HG (1989) IL-4 inhibits H_2O_2 production and antileishmanial capacity of human cultured monocytes mediated by IFN-γ. J Immunol 143:3020
110. Kennedy MK, Torrance DS, Picha KS, Mohler KM (1992) Analysis of cytokine mRNA expression in the central nervous system of mice with experimental autoimmune encephalomyelitis reveals that IL-10 mRNA expression correlates with recovery. J Immunol 149:2496
111. Jensen MA, Noronha A, Toscas A, Arnason BGW (1996) Antigenically non-specific global suppression of IL-2 and IFN-γ synthesis and an anergic like state in central nervous system infiltrating mononuclear cells precedes recovery from acute monophasic experimental allergic encephalomyelitis. Autoimmunity (in press)
112. Karpus WJ, Swanborg RH (1989) $CD4^+$ suppressor cells differentially affect the production of IFN-γ by effector cells of experimental autoimmune encephalomyelitis. J Immunol 143:3492
113. Nedwin GE, Svedersky LP, Bringman TS, Palladino MA Jr, Goeddel DV (1985) Effect of interleukin 2, interferon-γ, and mitogens on the production of tumor necrosis factors α and β. J Immunol 135:2492
114. Herrmann F, Cannistra SA, Lindemann A, Blohm D, Rambaldi A, Mertelsmann RH, Griffin JD (1989) Functional consequences of monocyte IL-2 receptor expression: induction of IL-1β secretion by IFN-γ and IL-2. J Immunol 142:139
115. Boraschi D, Pasqualetto E, Ghezzi P, Salmona M, Bartalini M, Barbarulli G, Censini S, Soldateschi D, Tagliabuc A (1982) Dissociation between macrophage tumoricidal capacity and suppressive activity: analysis with macrophage-defective mouse strains. J Immunol 131:1707
116. Go NF, Castle BE, Barrett R, Kastelein A, Dang W, Mosmann TR, Moore KW, Howard M (1990) Interleukin 10, a novel B cell stimulatory factor: unresponsiveness of X chromosome-linked immunodeficiency B cells. J Exp Med 172:1625
117. Chouaib S, Fradelizi D (1982) The mechanism of inhibition of human IL 2 production. J Immunol 129:2463
118. Betz M, Fox BS (1991) Prostaglandin E_2 inhibits production of Th1 lymphokines but not of Th2 lymphokines. J Immunol 146:108
119. Chouaib S, Welte K, Mertelsmann R, Dupont B (1985) Prostaglandin E2 acts at two distinct pathways of T lymphocyte activation: inhibition of interleukin-2 production and down-regulation of transferrin receptor expression. J Immunol 135:1172
120. Rappaport RS, Dodge GR (1982) Prostaglandin E inhibits the production of human interleukin 2. J Exp Med 155:943
121. Ferreri NR, Sarr T, Askenase PW, Ruddle NH (1992) Molecular regulation of tumor necrosis factor-alpha and lymphotoxin production in T cells: inhibition by prostaglandin E_2. J Biol Chem 267: 9443
122. Widomski DL, Walsh RE, Baron DA, Hidvegi MI, Fretland DJ, Collins PW, Gaginella TS (1991) Effects of the prostaglandin analogue misoprostol on inflammatory mediator release by human monocytes. Agents Actions 34:30
123. Reder AT, Thapar M, Sapugay AM, Jensen MA (1995) Eicosenoids modify experimental allergic encephalomyelitis. Am J Ther 2:1
124. Reder AT, Arnason BGW (1995) Trigeminal neuralgia in multiple sclerosis relieved by a prostaglandin E analogue. Neurology 45:1097
125. Dore-Duffy P, Perry W, Kuo H-H (1983) Interferon-mediated inhibition of prostaglandin synthesis in human mononuclear leukocytes. Cell Immunol 79:232
126. Boraschi D, Soldateschi D, Tagliabue A (1982) Macrophage activation by interferon: dissociation between tumoricidal capacity and suppressive activity. Eur J Immunol 12:320
127. Reder AT (1991) IFN-β and IFN-γ modify expression of monocyte surface proteins. J Interferon Res 11[Suppl 1]:121

128. Dentener MA, Bazil V, Von Asmuth EJU, Ceska M, Buurman WA (1993) Involvement of CD14 in lipopolysaccharide-induced tumor necrosis factor-α, IL-6 and IL-8 release by human monocytes and alveolar macrophages. J Immunol 150:2885
129. Gimmi CD, Freeman GJ, Gribben JG, Sugita K, Freedman AS, Morimoto C, Nadler LM (1991) B-cell surface antigen B7 provides a costimulatory signal that induces T cells to proliferate and secrete interleukin 2. Proc Natl Acad Sci USA 88:6575
130. Kuchroo VK, Das MP, Brown JA, Ranger AM, Zamvil SS, Sobel RA, Weiner HL, Nabavi N, Glimcher LH (1995) B7-1 and B7-2 costimulatory molecules activate differentially the Th1/Th2 developmental pathways: application to autoimmune disease therapy. Cell 80:707
131. Genç K, Reder AT (1995) IFN-β-1b reverses elevated lymphocyte B7-1 in MS (abstract). J Neuroimmunol 52 [Suppl 1]:20
132. Ding L, Linsley PS, Huang L-Y, Germain RN, Shevach EM (1993) IL-10 inhibits macrophage (Mo) co-stimulatory activity by selectively inhibiting the upregulation of B7 expression. J Immunol 150:180A
133. Ransohoff RM, Devajyothi C, Estes ML, Babcock G, Rudick RA, Frohman EM, Barna BP (1991) Interferon-β specifically inhibits interferon-γ-induced class II major histocompatibility complex gene transcription in a human astrocytoma cell line. J Neuroimmunol 33:103
134. Joseph J, Knobler RL, D'Imperio C, Lublin FD (1988) Down-regulation of interferon-γ-induced class II expression on human glioma cells by recombinant interferon-β: effects of dosage treatment schedule. J Neuroimmunol 20:39
135. Matsumoto Y, Hanawa H, Tsuchida M, Abo T (1993) In situ inactivation of infiltrating T cells in the central nervous system with autoimmune encephalomyelitis: the role of astrocytes. Immunology 79:381
136. Weber F, Meinl E, Aloisi F, Nevinny-Stickel C, Albert E, Wekerle H (1994) Human astrocytes are only partially competent antigen presenting cells: possible implications for lesion development in multiple sclerosis. Brain 117:59
137. MacPhee IAM, Antoni FA, Mason DW (1989) Spontaneous recovery of rats from experimental allergic encephalomyelitis is dependent on regulation of the immune system by endogenous adrenal corticosteroids. J Exp Med 169:431
138. Reder AT, Lowy MT (1992) Interferon-beta treatment does not elevate cortisol in multiple sclerosis. J Interferon Res 12:195
139. Pender MP, McCombe PA, Yoong G, Nguyen KB (1992) Apoptosis of $\alpha\beta$T lymphocytes in the nervous system in experimental autoimmune encephalomyelitis. J Autoimmun 5:401
140. Hecht TT, Longo DL, Matis LA (1983) The relationship between immune interferon production and proliferation in antigen-specific MHC-restricted T cell lines and clones. J Immunol 131:1049
141. Liu Y, Janeway CA Jr (1990) Interferon γ plays a critical role in induced cell death of effector T cells: a possible third mechanism of self-tolerance. J Exp Med 172:1735
142. Heremans H, Dillen C, Dijkmans R, Grau G, Billiau A (1989) The role of cytokines in various animals models of inflammation. Lymphokine Res 8:329
143. Voorthuis JAC, Uitdehaag BMJ, DeGroot CJA, Goede PH, Van Der Meide PH, Dijkstra CD (1990) Suppression of experimental allergic encephalomyelitis by intraventricular administration of interferon-gamma in Lewis rats. Clin Exp Immunol 81:183
144. Billiau A, Heremans H, Vandekerckhove F, Dijkmans R, Sobis H, Meulepas E, Carton H (1988) Enhancement of experimental allergic encephalomyelitis in mice by antibodies against IFN-gamma. J Immunol 140:1506
145. Reference deleted
146. Xian-hao X, McFarlin DE (1984) Oligoclonal bands in CSF: twins with MS. Neurology 34:769
147. Farrell MA, Kaufmann JCE, Gilbert JJ, Noseworthy JH, Armstrong HA, Ebers GC (1985) Oligoclonal bands in multiple sclerosis: clinical-pathologic correlation. Neurology 35:212
148. Ross C, Hansen MB, Schyberg T, Berg K (1990) Autoantibodies to crude human leucocyte interferon (IFN), native human IFN, recombinant human IFN-alpha 2b and human IFN-gamma in healthy blood donors. Clin Exp Immunol 82:57
149. Panitch HS, Francis GS, Hooper CJ, Merigan TC, Johnson KP (1985) Serial immunological studies in multiple sclerosis patients treated systematically with human alpha interferon. Ann Neurol 18:434
150. Rice GP, Woelfel EL, Talbot PJ, Braheny SL, Sipe JC, Knobler AL, Merigan TC, Oldstone MB (1985) Immunological complications in multiple sclerosis patients receiving interferon. Ann Neurol 18:439

151. Abreu SL, Thampoe I, Kaplan P (1986) Interferon in experimental autoimmune encephalomyelitis: intraventricular administration. J Interferon Res 6:627
152. Abreu SL (1982) Suppression of experimental allergic encephalomyelitis by interferon. Immunol Commun 11:1
153. Abreu SL (1985) Interferon in experimental autoimmune encephalomyelitis (EAE): effects of exogenous interferon on the antigen-enhanced adoptive transfer of EAE. Int Arch Allergy Appl Immunol 76:302
154. Abreu SL, Tondreau J, Levine S, Sowinski R (1983) Inhibition of passive localized experimental allergic encephalomyelitis by interferon. Int Arch Allergy Apply Immunol 72:30
155. Hertz F, Deghenghi R (1985) Effect of rat and β-human interferons on hyperacute experimental allergic encephalomyelitis in rats. Agents Actions 16:347
156. Inada T, Mims CA (1986) Infection of mice with lactic dehydrogenase virus prevents development of experimental allergic encephalomyelitis. J Neuroimmunol 11:53
157. Heremans H, Billiau A, Coutelier JP, De Somer P (1987) The inhibition of endotoxin-induced local inflammation by LDH virus or LDH virus-infected tumors is mediated by itnerferon. Proc Soc Exp Biol Med 185:6
158. Habif DV, Lipton R, Cantell K (1975) Interferon crosses blood-cerebrospinal fluid barrier in monkeys. Proc Soc Exp Biol Med 149:287
159. Noronha A, Toscas A, Jensen MA (1993) Interferon β decreases T cell activation and interferon γ production in multiple sclerosis. J Neuroimmunol 46:145
160. Rudick RA, Carpenter CS, Cookfair DL, Tuohy VK, Ransohoff RM (1993) In vitro and in vivo inhibition of mitogen-driven T-cell activation by recombinant interferon beta. Neurology 43:2080
161. Klimpel GR, Infante AJ, Patterson J, Hess CB, Asuncion M (1990) Virus-induced interferon α/β (IFN-α/β) production by T cells and by Th1 and Th2 helper T cell clones: a study of the immunoregulatory actions of IFN- versus IFN-α/β on functions of different T cell populations. Cell Immunol 128:603
162. Zarling JM, Sosman J, Eskra L, Borden EC, Horoszewicz JS, Carter WA (1978) Enhancement of T cell cytotoxic responses by purified human fibroblast interferon. J Immunol 121:2002
163. Rosztóczy I, Siroki O, Béládi I (1986) Effects of interferons-α, -β, and -γ on human interleukin-2 production. J Interferon Res 6:581

MIX
Papier aus verantwortungsvollen Quellen
Paper from responsible sources
FSC® C105338

If you have any concerns about our products,
you can contact us on
ProductSafety@springernature.com

In case Publisher is established outside the EU,
the EU authorized representative is:
Springer Nature Customer Service Center GmbH
Europaplatz 3, 69115 Heidelberg, Germany

Printed by Libri Plureos GmbH
in Hamburg, Germany